STUDENT STUDY GUIDE AND SOLUTIONS MANUAL

Richard N. Aufmann
Palomar College

Vernon C. Barker
Palomar College

Richard D. Nation
Palomar College

Christine S. Verity

COLLEGE TRIGONOMETRY

SIXTH EDITION

Aufmann/Barker/Nation

HOUGHTON MIFFLIN COMPANY   BOSTON   NEW YORK

Publisher: Richard Stratton
Senior Sponsoring Editor: Molly Taylor
Senior Marketing Manager: Jennifer Jones
Editorial Associate: Andrew Lipsett
Editorial Assistant: Joanna Carter
Marketing Associate: Mary Legere
New Title Project Manager: Susan Brooks-Peltier

Printed in the U.S.A.

ISBN: 978-0-618-82508-0

1 2 3 4 5 6 7 8 9 – VHO – 11 10 09 08 07

Contents

Study Tips

The skills you will learn in any mathematics course will be important in your future career—no matter what career you choose. In your textbook, we have provided you with the tools to master these skills. There's no mystery to success in this course; a little hard work and attention to your instructor will pay off. Here are a few tips to help ensure your success in this class.

Know Your Instructor's Requirements

To do your best in this course, you must know exactly what your instructor requires. If you don't, you probably will not meet his or her expectations and are not likely to earn a good grade in the course.

Instructors ordinarily explain course requirements during the first few days of class. Course requirements may be stated in a *syllabus*, which is a printed outline of the main topics of the course, or they may be presented orally. When they are listed in a syllabus or on other printed pages, keep them in a safe place. When they are presented orally, make sure to take complete notes. In either case, understand them completely and follow them exactly.

Attend Every Class

Attending class is vital if you are to succeed in this course. Your instructor will provide not only information but also practice in the skills you are learning. Be sure to arrive on time. You are responsible for everything that happens in class, even if you are absent. If you *must* be absent from a class session:

1. Deliver due assignments to the instructor as soon as possible.
2. Contact a classmate to learn about assignments or tests announced in your absence.
3. Hand copy or photocopy notes taken by a classmate while you were absent.

Take Careful Notes in Class

You need a notebook in which to keep class notes and records about assignments and tests. Make sure to take complete and well-organized notes. Your instructor will explain text material that may be difficult for you to understand on your own and may supply important information that is not provided in the textbook. Be sure to include in your notes everything that is written on the chalkboard.

Information recorded in your notes about assignments should explain exactly what they are, how they are to be done, and when they are due. Information about tests should include exactly what text material and topics will be covered on each test and the dates on which the tests will be given.

Survey the Chapter

Before you begin reading a chapter, take a few minutes to survey it. Glancing through the chapter will give you an overview of its contents and help you see how the pieces fit together as you read.

Begin by reading the chapter title. The title summarizes what the chapter is about. Next, read the section headings. The section headings summarize the major topics presented in the chapter. Then read the topic headings that appear in green within each section. The topic headings describe the concepts for that section. Keep these headings in mind as you work through the material. They provide direction as you study.

Use the Textbook to learn the Material

For each concept studied, read very carefully all of the material from the topic heading to the examples provided for that concept. As you read, note carefully the formulas and words printed in **boldface** type. It is important for you to know these formulas and the definitions of these words.

You will note that each example references an exercise. The example is worked out for you; the exercise, which is highlighted in red in the section's exercise set, is left for you to do. After studying the example, do the exercise. Immediately look up the answer to this exercise in the Solutions section at the back of the text. If your answer is correct, continue. If your answer is incorrect, check your solution against the one given in the Solutions section. It may be helpful to review the worked-out example also. Determine where you made your mistakes.

Next, do the other problems in the exercise set that correspond to the concept just studied. The answers to all the odd-numbered exercises appear in the answer section in the back of the text, and the solutions appear in this Study Guide. Check your answers to the exercises against these.

If you have difficulty solving problems in the exercise set, review the material in the text. Many examples are solved within the text material. Review the solutions to these problems. Reread the examples provided for the concept. If, after checking these sources and trying to find your mistakes, you are still unable to solve a problem correctly, make a note of the exercise number so that you can ask someone for help with that problem.

Review Material

Reviewing material is the repetition that is essential for learning. Much of what we learn is soon forgotten unless we review it. If you find that you do not remember information that you studied previously, you probably have not reviewed it sufficiently. *You will remember best what you review most.*

One method of reviewing material is to begin a study session by reviewing a concept you have studied previously. For example, before trying to solve a new type of problem, spend a few minutes solving a kind of problem you already know how to solve. Not only will you provide yourself with the review practice you need, but you are also likely to put yourself in the right frame of mind for learning how to solve the new type of problem.

Use the End-of-Chapter Material

To help you review the material presented within a chapter, a Chapter Review appears at the end of each chapter. In the Chapter Review, the main concepts of each section are summarized. Included are important definitions and formulas. After completing a chapter, be sure to read the Chapter Review. Use it to check your understanding of the material presented and to determine what concepts you need to review. Return to any section that contains a concept you need to study again.

Each chapter ends with Chapter Review Exercises and a Chapter Test. The problems these contain summarize what you should have learned when you have finished the chapter. Do these exercises as you prepare for an examination. Check your answers against those in the back of the text. Answers to the odd-numbered Chapter Review Exercises and all the Chapter Test exercises are provided there. For any problem you answer incorrectly, review the material corresponding to that concept in the textbook. Determine *why* your answer was wrong.

Finding Good Study Areas

Find a place to study where you are comfortable and can concentrate well. Many students find the campus library to be a good place. You might select two or three places at the college library where you like to study. Or there may be a small, quiet lounge on the third floor of a building where you find you can study well. Take the time to find places that promote good study habits.

Determining When to Study

Spaced practice is generally superior to massed practice. For example, four half-hour study periods will produce more learning than one two-hour study session. The following suggestions may help you decide when to study.

1. A free period immediately before class is the best time to study about the lecture topic for the class.
2. A free period immediately after class is the best time to review notes taken during the class.
3. A brief period of time is good for reciting or reviewing information.

4. A long period of an hour or more is good for doing challenging activities such as learning to solve a new type of problem.

5. Free periods just before you go to sleep are good times for learning information. (There is evidence that information learned just before sleep is remembered longer than information learned at other times.)

Determining How Much to Study

Instructors often advise students to spend twice as much time outside of class studying as they spend in the classroom. For example, if a course meets for three hours each week, instructors customarily advise students to study for six hours each week outside of class.

Any mathematics course requires the learning of skills, which are abilities acquired through practice. It is often necessary to practice a skill more than a teacher requires. For example, this textbook may provide 50 practice problems on a specific concept, and the instructor may assign only 25 of them. However, some students may need to do 30, 40, or all 50 problems.

If you are an accomplished athlete, musician, or dancer, you know that long hours of practice are necessary to acquire a skill. Do not cheat yourself of the practice you need to develop the abilities taught in this course.

Study followed by reward is usually productive. Schedule something enjoyable to do following study sessions. If you know that you only have two hours to study because you have scheduled an enjoyable activity for yourself, you may be inspired to make the best use of the two hours that you have set aside for studying.

Keep Up to Date with Course Work

College terms start out slowly. Then they gradually get busier and busier, reaching a peak of activity at final examination time. If you fall behind in the work for a course, you will find yourself trying to catch up at a time when you are very busy with all of your other courses. Don't fall behind—keep up to date with course work.

Keeping up with course work is doubly important for a course in which information and skills learned early in the course are needed to learn information and skills later in the course. Any mathematics course falls into this category. Skills must be learned immediately and reviewed often.

Your instructor gives assignments to help you acquire a skill or understand a concept. Do each assignment as it is assigned, or you may well fall behind and have great difficulty catching up. Keeping up with course work also makes it easier to prepare for each exam.

Be Prepared for Tests

The Chapter Test at the end of a chapter should be used to prepare for an examination. Additional Chapter Tests are also provided in this Study Guide. We suggest that you try a Chapter Test a few days before your actual exam. Do these exercises in a quiet place, and try to complete the exercises in the same amount of time as you will be allowed for your exam. When completing the exercises, practice the strategies of successful test takers: 1) look over the entire test before you begin to solve any problem; 2) write down any rules or formulas you may need so they are readily available; 3) read the directions carefully; 4) work the problems that are easiest for you first; 5) check your work, looking particularly for careless errors.

When you have completed the exercises in the Chapter Test, check your answers. If you missed a question, review the material in the appropriate section and then rework some of the exercises from that concept. This will strengthen your ability to perform the skills in that concept.

Get Help for Academic Difficulties

If you do have trouble in this course, teachers, counselors, and advisers can help. They usually know of study groups, tutors, or other sources of help that are available. They may suggest visiting an office of academic skills, a learning center, a tutorial service or some other department or service on campus.

Students who have already taken the course and who have done well in it may be a source of assistance. If they have a good understanding of the material, they may be able to help by explaining it to you.

Solutions Manual

Chapter 1
Functions and Graphs

1.
$$2x + 10 = 40$$
$$2x = 40 - 10$$
$$2x = 30$$
$$x = 15$$

3.
$$5x + 2 = 2x - 10$$
$$5x - 2x = -10 - 2$$
$$3x = -12$$
$$x = -4$$

5.
$$2(x - 3) - 5 = 4(x - 5)$$
$$2x - 6 - 5 = 4x - 20$$
$$2x - 11 = 4x - 20$$
$$2x - 4x = -20 + 11$$
$$-2x = -9$$
$$x = \frac{9}{2}$$

7.
$$\frac{3}{4}x + \frac{1}{2} = \frac{2}{3}$$
$$12\left(\frac{3}{4}x + \frac{1}{2}\right) = 12\left(\frac{2}{3}\right)$$
$$9x + 6 = 8$$
$$9x = 8 - 6$$
$$9x = 2$$
$$x = \frac{2}{9}$$

9.
$$\frac{2}{3}x - 5 = \frac{1}{2}x - 3$$
$$6\left(\frac{2}{3}x - 5\right) = 6\left(\frac{1}{2}x - 3\right)$$
$$4x - 30 = 3x - 18$$
$$4x - 3x = -18 + 30$$
$$x = 12$$

11.
$$0.2x + 0.4 = 3.6$$
$$0.2x = 3.2$$
$$x = 16$$

13.
$$\frac{3}{5}(n + 5) - \frac{3}{4}(n - 11) = 0$$
$$20\left[\frac{3}{5}(n + 5) - \frac{3}{4}(n - 11)\right] = 20 - 0$$
$$12(n + 5) - 15(n - 11) = 0$$
$$12n + 60 - 15n + 165 = 0$$
$$-3n = -225$$
$$n = 75$$

15.
$$3(x + 5)(x - 1) = (3x + 4)(x - 2)$$
$$3x^2 + 12x - 15 = 3x^2 - 2x - 8$$
$$14x = 7$$
$$x = \frac{1}{2}$$

17.
$$0.08x + 0.12(4000 - x) = 432$$
$$0.08x + 480 - 0.12x = 432$$
$$-0.04x = -48$$
$$x = 1200$$

19.
$$x + 2y = 8$$
$$2y = -x + 8$$
$$y = -\frac{1}{2}x + 4$$

21.
$$2x + 5y = 10$$
$$2x = -5y + 10$$
$$x = -\frac{5}{2}y + 5$$

23.
$$ay - by = c$$
$$y(a - b) = c$$
$$y = \frac{c}{a - b}$$

25.
$$x = \frac{y}{1 - y}$$
$$x(1 - y) = y$$
$$x - xy = y$$
$$x = xy + y$$
$$x = y(x + 1)$$
$$\frac{x}{x + 1} = y$$

27.
$$x^2 - 2x - 15 = 0$$
$$a = 1 \quad b = -2 \quad c = -15$$
$$x = \frac{-(-2) \pm \sqrt{(-2)^2 - 4(1)(-15)}}{2(1)}$$
$$x = \frac{2 \pm \sqrt{4 + 60}}{2}$$
$$x = \frac{2 \pm \sqrt{64}}{2} = \frac{2 \pm 8}{2}$$
$$x = \frac{2 + 8}{2} = \frac{10}{2} = 5 \quad \text{or}$$
$$x = \frac{2 - 8}{2} = \frac{-6}{2} = -3$$

29. $x^2 + x - 1 = 0$

$a = 1 \quad b = 1 \quad c = -1$

$$x = \frac{-1 \pm \sqrt{1^2 - 4(1)(-1)}}{2}$$

$$x = \frac{-1 \pm \sqrt{1 + 4}}{2}$$

$$x = \frac{-1 \pm \sqrt{5}}{2}$$

31. $2x^2 + 4x + 1 = 0$

$a = 2 \quad b = 4 \quad c = 1$

$$x = \frac{-4 \pm \sqrt{4^2 - 4(2)(1)}}{2(2)}$$

$$x = \frac{-4 \pm \sqrt{16 - 8}}{4}$$

$$x = \frac{-4 \pm \sqrt{8}}{4} = \frac{-4 \pm 2\sqrt{2}}{4}$$

$$x = \frac{-2 \pm \sqrt{2}}{2}$$

33. $3x^2 - 5x - 3 = 0$

$a = 3 \quad b = -5 \quad c = -3$

$$x = \frac{-(-5) \pm \sqrt{(-5)^2 - 4(3)(-3)}}{2(3)}$$

$$x = \frac{5 \pm \sqrt{25 + 36}}{6}$$

$$x = \frac{5 \pm \sqrt{61}}{6}$$

35. $\frac{1}{2}x^2 + \frac{3}{4}x - 1 = 0$

$a = \frac{1}{2} \quad b = \frac{3}{4} \quad c = -1$

$$x = \frac{-\frac{3}{4} \pm \sqrt{\left(\frac{3}{4}\right)^2 - 4\left(\frac{1}{2}\right)(-1)}}{2\left(\frac{1}{2}\right)}$$

$$x = \frac{-\frac{3}{4} \pm \sqrt{\frac{9}{16} + 2}}{1}$$

$$x = -\frac{3}{4} \pm \sqrt{\frac{41}{16}}$$

$$x = -\frac{3}{4} \pm \frac{\sqrt{41}}{4}$$

$$x = \frac{-3 \pm \sqrt{41}}{4}$$

37. $\sqrt{2}x^2 + 3x + \sqrt{2} = 0$

$a = \sqrt{2} \quad b = 3 \quad c = \sqrt{2}$

$$x = \frac{-3 \pm \sqrt{3^2 - 4 \cdot \sqrt{2} \cdot \sqrt{2}}}{2\sqrt{2}}$$

$$x = \frac{-3 \pm \sqrt{9 - 8}}{2\sqrt{2}}$$

$$x = \frac{-3 + 1}{2\sqrt{2}} = \frac{-2}{2\sqrt{2}} = -\frac{\sqrt{2}}{2} \quad \text{or}$$

$$x = \frac{-3 - 1}{2\sqrt{2}} = \frac{-4}{2\sqrt{2}} = -\sqrt{2}$$

39. $x^2 = 3x + 5$

$x^2 - 3x - 5 = 0$

$a = 1 \quad b = -3 \quad c = -5$

$$x = \frac{-(-3) \pm \sqrt{(-3)^2 - 4(1)(5)}}{2(1)}$$

$$x = \frac{3 \pm \sqrt{9 + 20}}{2}$$

$$x = \frac{3 \pm \sqrt{29}}{2}$$

41. $x^2 - 2x - 15 = 0$

$(x + 3)(x - 5) = 0$

$x + 3 = 0 \quad$ or $\quad x - 5 = 0$

$\quad x = -3 \qquad\qquad x = 5$

43. $8y^2 + 189y - 72 = 0$

$(8y - 3)(y + 24) = 0$

$8y - 3 = 0 \quad$ or $\quad y + 24 = 0$

$\quad y = \frac{3}{8} \qquad\qquad y = -24$

45. $3x^2 - 7x = 0$

$x(3x - 7) = 0$

$x = 0 \quad$ or $\quad 3x - 7 = 0$

$\qquad\qquad\qquad x = \frac{7}{3}$

47. $(x - 5)^2 - 9 = 0$

$[(x - 5) - 3][(x - 5) + 3] = 0$

$x - 8 = 0 \quad$ or $\quad x - 2 = 0$

$\quad x = 8 \qquad\qquad x = 2$

49. $2x + 3 < 11$

$\quad 2x < 8$

$\quad\ \ x < 4$

51. $x + 4 > 3x + 16$

$\quad -2x > 12$

$\qquad\ x < -6$

53. $-6x + 1 \geq 19$

$\quad -6x \geq 18$

$\qquad\ x \leq -3$

55. $-3(x + 2) \leq 5x + 7$

$\quad -3x - 6 \leq 5x + 7$

$\qquad\quad -8 \leq 13$

$\qquad\qquad x \geq -\frac{13}{8}$

57. $-4(3x - 5) > 2(x - 4)$

$\quad -12x + 20 > 2x - 8$

$\qquad\quad -14x > -28$

$\qquad\qquad\ x < 2$

59. $x^2 + 7x > 0$

$x(x+7) > 0$

The product is positive.
The critical values are 0 and –7.

$x(x-7)$ + + + + | – – – – – – – | + + + +

-7   0

$(-\infty, -7) \cup (0, \infty)$

61. $x^2 + 7x + 10 < 0$

$(x+5)(x+2) < 0$

The product is negative.
The critical values are –5 and –2.

$(x+5)(x+2)$ + + + + | – – – | + + + + +

-5   -2   0

$(-5, -2)$

63. $x^2 - 3x \ge 28$

$x^2 - 3x - 28 \ge 0$

$(x+4)(x-7) \ge 0$

The product is positive or zero.
The critical values are –4 and 7.

$(x+4)(x-7)$ + + | – – – – – – – – – – | + +

-4   0   7

$(-\infty, -4] \cup [7, \infty)$

65. $6x^2 - 4 \le 5x$

$6x^2 - 5x - 4 \le 0$

$(3x-4)(2x+1) \le 0$

The product is negative or zero.
The critical values are $\frac{4}{3}$ and $-\frac{1}{2}$.

$(3x-4)(2x+1)$ + + + + | – – – | + + + + +

$-\frac{1}{2}$ 0 $\frac{4}{3}$

$$\left[-\frac{1}{2}, \frac{4}{3}\right]$$

67. $|x| < 4$

$-4 < x < 4$

$(-4, 4)$

69. $|x-1| < 9$

$-9 < x-1 < 9$

$-8 < x < 10$

$(-8, 10)$

71. $|x+3| > 30$

$x+3 < -30$ or $x+3 > 30$

$x < -33$ $x > 27$

$(-\infty, -33) \cup (27, \infty)$

73. $|2x-1| > 4$

$2x-1 < -4$ or $2x-1 > 4$

$2x < -3$ $2x > 5$

$x < -\dfrac{3}{2}$ $x > \dfrac{5}{2}$

$$\left(-\infty, -\frac{3}{2}\right) \cup \left(\frac{5}{2}, \infty\right)$$

75. $|x+3| \ge 5$

$x+3 \le -5$ or $x+3 \ge 5$

$x \le -8$ $x \ge 2$

$(-\infty, -8] \cup [2, \infty)$

77. $|3x-10| \le 14$

$-14 \le 3x-10 \le 14$

$-4 \le 3x \le 24$

$-\dfrac{4}{3} \le x \le 8$

$$\left[-\frac{4}{3}, 8\right]$$

79. $|4-5x| \ge 24$

$4-5x \le 24$ or $4-5x \ge 24$

$-5x \le -28$ $-5x \ge 20$

$x \ge \dfrac{28}{5}$ $x \le -4$

$$(-\infty, -4] \cup \left[\frac{28}{5}, \infty\right)$$

81. $|x-5| \ge 0$

Because an absolute value is always nonnegative, the inequality is always true. The solution set consists of all real numbers.

$(-\infty, \infty)$

83. $|x-4| \le 0$

Because an absolute value is always nonnegative, the inequality $|x-4| < 0$ has no solution. Thus the only solution of the inequality $|x-4| \le 0$ is the solution of the equation $x - 4 = 0$.

$x = 4$

85.
$$A = 35$$
$$A = LW$$
$$LW = 35$$
$$L = \frac{35}{W}$$

$$P = 27$$
$$P = 2L + 2W$$
$$2L + 2W = 27$$
$$2\left(\frac{35}{W}\right) + 2W = 27$$
$$70 + 2W^2 = 27W$$
$$2W^2 - 27W + 70 = 0$$
$$(2W - 7)(W - 10) = 0$$

$$W = \frac{7}{2} \quad \text{or} \quad W = 10$$

$$35 = \frac{7}{2}L \qquad 35 = LW$$
$$70 = 7L \qquad 35 = 10L$$
$$10 = L \qquad 3.5 = L$$

The rectangle measures 3.5 cm by 10 cm.

87.
$$A = 1500 = lw$$
$$P = 600 - 2l + 3w$$
$$l = \frac{15000}{w}$$

$$2l + 3w = 600$$
$$2\left(\frac{15000}{w}\right) + 3w = 600$$
$$30{,}000 + 3w^2 = 600w$$
$$3w^2 - 600w + 30{,}000 = 0$$
$$3(w^2 - 200w + 10{,}000) = 0$$
$$3(w - 100)(w - 100) = 0$$

$$w = 100 \text{ ft}$$
$$l = \frac{15000}{100} = 150 \text{ ft}$$

The dimensions are 100 feet by 150 feet.

89. Plan A: $5 + 0.01x$
Plan B: $1 + 0.08x$
$$5 + 0.01x < 1 + 0.08x$$
$$4 < .07x$$
$$57.1 < x$$

Plan A is less expensive if you use at least 58 checks.

91. Plan A: $100 + 8x$
Plan B: $250 + 3.5x$
$$100 + 8x > 250 + 3.5x$$
$$4.5x > 150$$
$$x > 33.3$$

Plan A pays better if at least 34 sales are made.

93.
$$68 \le F \le 104$$
$$68 \le \frac{9}{5}C + 32 \le 104$$
$$36 \le \frac{9}{5}C \le 72$$
$$20 \le C \le 40$$

•••

Connecting Concepts

95.
$$253 = \frac{1}{2}n(n+1)$$
$$n^2 + n - 506 = 0$$
$$(n + 23)(n - 22) = 0$$
$$n = 22$$

So $1 + 2 + 3 + \cdots + 21 + 22 = 253$.

97.
$$R = 420x - 2x^2$$
$$420x - 2x^2 > 0$$
$$2x(210 - x) > 0$$

The product is positive.
The critical values are 0 and 210.

$$2x(210 - x)$$

$- - - -|+ + + + + + + + +|- - -$

0 210

$(0, 210)$

99. $s = -16t^2 + v_0t + s_0$ $s > 48$, $v_0 = 64$, $s_0 = 0$

$-16t + 64t > 48$

$-16t^2 + 64t - 48 > 0$

$-16(t^2 - 4t + 3) > 0$

$-16(t - 3)(t - 1) > 0$

The product is positive.
The critical values are $t = 3$ and $t = 1$.

$-16(t - 3)(t - 1)$ |-|+ +|- - - -

                         0 1   3

$1 \text{ second} < t < 3 \text{ seconds}$

The ball is higher than 48 ft between 1 and 3 seconds.

101. **a.** $|s - 4.25| \le 0.01$

b. $s - 4.25 = 0.01$,   or $s - 4.25 = -0.01$

$s = 4.26$ $s = 4.24$ critical values

$4.24 \le s \le 4.26$

• •

PS1. $\dfrac{4 + (-7)}{2} = \dfrac{-3}{2}$

PS2. $\sqrt{50} = \sqrt{25}\sqrt{2} = 5\sqrt{2}$

PS3. $y = 3x - 2$

$\overset{?}{5 = 3(-1) - 2}$

$5 \ne -5$ No, the equation is not true.

PS4. $y = (-3)^2 - 3(-3) - 2$

$y = 9 + 9 - 2$

$y = 16$

PS5. $|-3 - (-1)| = |-3 + 1| = |-2| = 2$

PS6. $\sqrt{(-3)^2 - 4(-2)(2)} = \sqrt{9 + 16} = \sqrt{25} = 5$

Section 1.2

1.

3. **a.** $31,500
b. Increase from 2004 to 2005
$33.5 - 32.9 = 0.6$
Increase from 2005 to 2006
$33.50 + 0.6 = 34.1$
The per capita income for 2006 would be $34,100.
c. Percent increase from 2004 to 2005
$\dfrac{33.5 - 32.9}{32.9} = 1.824\%$
Percent increase from 2005 to 2006
$33.50(1.01824) = 34.111$
The per capita income for 2006 would be $34,111.

5. $d = \sqrt{(-8 - 6)^2 + (11 - 4)^2}$

$= \sqrt{(-14)^2 + (7)^2}$

$= \sqrt{196 + 49}$

$= \sqrt{245}$

$= 7\sqrt{5}$

7. $d = \sqrt{(-10 - (-4))^2 + (15 - (-20))^2}$

$= \sqrt{(-6)^2 + (35)^2}$

$= \sqrt{36 + 1225}$

$= \sqrt{1261}$

9. $d = \sqrt{(0 - 5)^2 + (0 - (-8))^2}$

$= \sqrt{(-5)^2 + (8)^2}$

$= \sqrt{25 + 64}$

$= \sqrt{89}$

11.
$$d = \sqrt{(\sqrt{12}-\sqrt{3})^2 + (\sqrt{27}-\sqrt{8})^2}$$
$$= \sqrt{(2\sqrt{3}-\sqrt{3})^2 + (3\sqrt{3}-2\sqrt{2})^2}$$
$$= \sqrt{(\sqrt{3})^2 + (3\sqrt{3}-2\sqrt{2})^2}$$
$$= \sqrt{3 + (27 - 12\sqrt{6} + 8)}$$
$$= \sqrt{3 + 27 - 12\sqrt{6} + 8}$$
$$= \sqrt{38 - 12\sqrt{6}}$$

13.
$$d = \sqrt{(-a-a)^2 + (-b-b)^2}$$
$$= \sqrt{(-2a)^2 + (-2b)^2}$$
$$= \sqrt{4a^2 + 4b^2}$$
$$= \sqrt{4(a^2 + b^2)}$$
$$= 2\sqrt{a^2 + b^2}$$

15.
$$d = \sqrt{(-2x-x)^2 + (3x-4x)^2} \text{ with } x < 0$$
$$= \sqrt{(-3x)^2 + (-x)^2}$$
$$= \sqrt{9x^2 + x^2}$$
$$= \sqrt{10x^2}$$
$$= -x\sqrt{10} \quad \text{(Note: since } x < 0, \sqrt{x^2} = -x)$$

17.
$$\sqrt{(4-x)^2 + (6-0)^2} = 10$$
$$\left(\sqrt{(4-x)^2 + (6-0)^2}\right)^2 = 10^2$$
$$16 - 8x + x^2 + 36 = 100$$
$$x^2 - 8x - 48 = 0$$
$$(x-12)(x+4) = 0$$
$$x = 12 \quad \text{or} \quad x = -4$$
The points are $(12, 0)$, $(-4, 0)$.

19.
$$M = \left(\frac{x_1 + x_2}{2}, \frac{y_1 + y_2}{2}\right)$$
$$= \left(\frac{1+5}{2}, \frac{-1+5}{2}\right)$$
$$= \left(\frac{6}{2}, \frac{4}{2}\right)$$
$$= (3, 2)$$

21.
$$M = \left(\frac{6+6}{2}, \frac{-3+11}{2}\right)$$
$$= \left(\frac{12}{2}, \frac{8}{2}\right)$$
$$= (6, 4)$$

23.
$$M = \left(\frac{1.75 + (-3.5)}{2}, \frac{2.25 + 5.57}{2}\right)$$
$$= \left(-\frac{1.75}{2}, \frac{7.82}{2}\right)$$
$$= (-0.875, 3.91)$$

25.

27.

29.

31.

33.

35.

37.

39. Intercepts: $\left(0, \frac{12}{5}\right), (6, 0)$

41. $\left(0, \sqrt{5}\right), \left(0, -\sqrt{5}\right), (5, 0)$

$$x = -y^2 + 5$$

43. $(0, 4), (0, -4), (-4, 0)$

$$x = |y| - 4$$

45. $(0, \pm 2), (\pm 2, 0)$

$x^2 + y^2 = 4$

47. $(0, \pm 4), (\pm 4, 0)$

$|x| + |y| = 4$

49. center $(0, 0)$, radius 6

51. center $(1, 3)$, radius 7

53. center $(-2, -5)$, radius 5

55. center $(8, 0)$, radius $\frac{1}{2}$

57. $(x-4)^2 + (y-1)^2 = 2^2$

59. $\left(x - \frac{1}{2}\right)^2 + \left(y - \frac{1}{4}\right)^2 = \left(\sqrt{5}\right)^2$

61.
$$(x-0)^2 + (y-0)^2 = r^2$$
$$(-3-0)^2 + (4-0)^2 = r^2$$
$$(-3)^2 + 4^2 = r^2$$
$$9 + 16 = r^2$$
$$25 = 5^2 = r^2$$
$$(x-0)^2 + (y-0)^2 = 5^2$$

63.
$$(x+2)^2 + (y-5)^2 = r^2$$
$$(x-1)^2 + (y-3)^2 = r^2$$
$$(4-1)^2 + (-1-3)^2 = r^2$$
$$3^2 + (-4)^2 = r^2$$
$$9 + 16 = r^2$$
$$25 = 5^2 = r^2$$
$$(x-1)^2 + (y-3)^2 = 5^2$$

65.
$$x^2 - 6x \quad + y^2 = -5$$
$$x^2 - 6x + 9 + y^2 = -5 + 9$$
$$(x-3)^2 + y^2 = 2^2$$
center $(3, 0)$, radius 2

67.
$$x^2 - 14x \quad + y^2 + 8y \quad = -56$$
$$x^2 - 14x + 49 + y^2 + 8y + 16 = -56 + 49 + 16$$
$$(x-7)^2 + (y+4)^2 = 3^2$$
center $(7, -4)$, radius 3

69.
$$4x^2 + 4x \quad + 4y^2 = 63$$
$$x^2 + x \quad + y^2 = \frac{63}{4}$$
$$x^2 + x + \frac{1}{4} \quad + y^2 = \frac{63}{4} + \frac{1}{4}$$
$$\left(x + \frac{1}{2}\right)^2 + y^2 = 16$$
$$\left(x + \frac{1}{2}\right)^2 + (y-0)^2 = 4^2$$
center $\left(-\frac{1}{2}, 0\right)$, radius 4

71.
$$x^2 - x \quad + y^2 + \frac{3}{2}y \quad = \frac{15}{4}$$
$$x^2 - x + \frac{1}{4} + y^2 + \frac{3}{2}y + \frac{9}{4} = \frac{15}{4} + \frac{1}{4} + \frac{9}{4}$$
$$\left(x - \frac{1}{2}\right)^2 + \left(y + \frac{3}{2}\right)^2 = \left(\frac{5}{2}\right)^2$$
center $\left(\frac{1}{2}, -\frac{3}{2}\right)$, radius $\frac{5}{2}$

73.
$$d = \sqrt{(-4-2)^2 + (11-3)^2}$$
$$= \sqrt{36 + 64}$$
$$= \sqrt{100}$$
$$= 10$$
Since the diameter is 10, the radius is 5.
The center is the midpoint of the line segment from $(2,3)$ to $(-4,11)$.
$$\left(\frac{2+(-4)}{2}, \frac{3+11}{2}\right) = (-1, 7) \text{ center}$$
$$(x+1)^2 + (y-7)^2 = 5^2$$

75. Since it is tangent to the x-axis, its radius is 11.
$$(x-7)^2 + (y-11)^2 = 11^2$$

77.

79.

81.

83.

85.

87. $\left(\dfrac{x+5}{2}, \dfrac{y+1}{2}\right) = (9, 3)$

therefore $\dfrac{x+5}{2} = 9$ and $\dfrac{y+1}{2} = 3$

$\qquad\qquad x+5 = 18 \qquad\quad y+1 = 6$

$\qquad\qquad\quad x = 13 \qquad\qquad\quad y = 5$

Thus (13, 5) is the other endpoint.

89. $\left(\dfrac{x+(-3)}{2}, \dfrac{y+(-8)}{2}\right) = (2, -7)$

therefore $\dfrac{x-3}{2} = 2$ and $\dfrac{y-8}{2} = -7$

$\qquad\qquad x-3 = 4 \qquad\quad y-8 = -14$

$\qquad\qquad\quad x = 7 \qquad\qquad\quad y = -6$

Thus (7, –6) is the other endpoint.

91. $\sqrt{(3-x)^2 + (4-y)^2} = 5$

$(3-x)^2 + (4-y)^2 = 5^2$

$9 - 6x + x^2 + 16 - 18y + y^2 = 25$

$x^2 - 6x + y^2 - 8y = 0$

93. $\sqrt{(4-x)^2 + (0-y)^2} + \sqrt{(-4-x)^2 + (0-y)^2} = 10$

$(4-x)^2 + (0-y)^2 = 100 - 20\sqrt{(-4-x)^2 + (0-y)^2} + (-4-x)^2 + (-y)^2$

$16 - 8x + x^2 + y^2 = 100 - 20\sqrt{(-4-x)^2 + (-y)^2} + 16 + 8x + x^2 + y^2$

$-16x - 100 = -20\sqrt{(-4-x)^2 + (-y)^2}$

$4x + 25 = 5\sqrt{(-4-x)^2 + (-y)^2}$

$16x^2 + 200x + 625 = 25\left[(-4-x)^2 + (-y)^2\right]$

$16x^2 + 200x + 625 = 25\left[16 + 8x + x^2 + y^2\right]$

$16x^2 + 200x + 625 = 400 + 200x + 25x^2 + 25y^2$

Simplifying yields $9x^2 + 25y^2 = 225$.

95. The center is (-3,3). The radius is 3.

$(x+3)^2 + (y-3)^2 = 3^2$

PS1.
$$y = 3x + 12$$
$$0 = 3x + 12$$
$$-12 = 3x$$
$$-4 = x$$

PS2.
$$y = x^2 - 4x + 3$$
$$0 = x^2 - 4x + 3$$
$$0 = (x-1)(x-3)$$
$$x - 1 = 0 \quad x - 3 = 0$$
$$x = 1 \qquad x = 3$$
1 and 3

PS3.
$$y^2 = x \qquad\qquad y^2 = x$$
$$a^2 = 9 \qquad\qquad b^2 = 9$$
$$a = -3 \text{ or } 3 \qquad b = -3 \text{ or } 3$$

PS4. $d = \sqrt{(3-(-4))^2 + (-2-1)^2} = \sqrt{49+9} = \sqrt{58}$

PS5. −4

PS6. Decrease

Section 1.3

1. Given $f(x) = 3x - 1$,

  a. $f(2) = 3(2) - 1$
  $$= 6 - 1$$
  $$= 5$$

  b. $f(-1) = 3(-1) - 1$
  $$= -3 - 1$$
  $$= -4$$

  c. $f(0) = 3(0) - 1$
  $$= 0 - 1$$
  $$= -1$$

  d. $f\left(\dfrac{2}{3}\right) = 3\left(\dfrac{2}{3}\right) - 1$
  $$= 2 - 1$$
  $$= 1$$

  e. $f(k) = 3(k) - 1$
  $$= 3k - 1$$

  f. $f(k+2) = 3(k+2) - 1$
  $$= 3k + 6 - 1$$
  $$= 3k + 5$$

3. Given $A(w) = \sqrt{w^2 + 5}$,

  a. $A(0) = \sqrt{(0)^2 + 5}$
  $$= \sqrt{5}$$

  b. $A(2) = \sqrt{(2)^2 + 5}$
  $$= \sqrt{9}$$
  $$= 3$$

  c. $A(-2) = \sqrt{(-2)^2 + 5}$
  $$= \sqrt{9}$$
  $$= 3$$

  d. $A(4) = \sqrt{4^2 + 5}$
  $$= \sqrt{21}$$

  e. $A(r+1) = \sqrt{(r+1)^2 + 5}$
  $$= \sqrt{r^2 + 2r + 1 + 5}$$
  $$= \sqrt{r^2 + 2r + 6}$$

  f. $A(-c) = \sqrt{(-c)^2 + 5}$
  $$= \sqrt{c^2 + 5}$$

5. Given $f(x) = \dfrac{1}{|x|}$,

  a. $f(2) = \dfrac{1}{|2|} = \dfrac{1}{2}$

  b. $f(-2) = \dfrac{1}{|-2|} = \dfrac{1}{2}$

  c. $f\left(-\dfrac{3}{5}\right) = \dfrac{1}{\left|-\dfrac{3}{5}\right|}$
  $$= \dfrac{1}{3/5}$$
  $$= 1 + \dfrac{3}{5} = 1 \cdot \dfrac{5}{3} = \dfrac{5}{3}$$

  d. $f(2) + f(-2) = \dfrac{1}{2} + \dfrac{1}{2} = 1$

  e. $f(c^2 + 4) = \dfrac{1}{|c^2 + 4|} = \dfrac{1}{c^2 + 4}$

  f. $f(2 + h) = \dfrac{1}{|2 + h|}$

7. Given $s(x) = \dfrac{x}{|x|}$,

    a. $s(4) = \dfrac{4}{|4|} = \dfrac{4}{4} = 1$

    b. $s(5) = \dfrac{5}{|5|} = \dfrac{5}{5} = 1$

    c. $s(-2) = \dfrac{-2}{|-2|} = \dfrac{-2}{2} = -1$

    d. $s(-3) = \dfrac{-3}{|-3|} = \dfrac{-3}{3} = -1$

    e. Since $t > 0, |t| = t.$

       $s(t) = \dfrac{t}{|t|} = \dfrac{t}{t} = 1$

    f. Since $t < 0, |t| = -t.$

       $s(t) = \dfrac{t}{|t|} = \dfrac{t}{-t} = -1$

9.

    a. Since $x = -4 < 2$, use $P(x) = 3x + 1$.
       $P(-4) = 3(-4) + 1 = -12 + 1 = -11$

    b. Since $x = \sqrt{5} \geq 2$, use $P(x) = -x^2 + 11$.
       $P(\sqrt{5}) = -(\sqrt{5})^2 + 11 = -5 + 11 = 6$

    c. Since $x = c < 2$, use $P(x) = 3x + 1$.
       $P(c) = 3c + 1$

    d. Since $k \geq 1$, then $x = k + 1 \geq 2$,
       so use $P(x) = -x^2 + 11$.
       $P(k+1) = -(k+1)^2 + 11 = -(k^2 + 2k + 1) + 11$
                     $= -k^2 - 2k - 1 + 11$
                     $= -k^2 - 2k + 10$

11. $2x + 3y = 7$
      $3y = -2x + 7$
      $y = -\dfrac{2}{3}x + \dfrac{7}{3}, \; y$ is a function of x.

13. $-x + y^2 = 2$
      $y^2 = x + 2$
      $y = \pm\sqrt{x+2}, \; y$ is a not function of x.

15. $y = 4 \pm \sqrt{x}, \; y$ is not a function of x since for each $x > 0$ there are two values of x.

17. $y = \sqrt[3]{x}, \; y$ is a function of x.

19. $y^2 = x^2$
      $y = \pm\sqrt{x^2}, \; y$ is a not function of x.

21. Function; each x is paired with exactly one y.

23. Function; each x is paired with exactly one y.

25. Function; each x is paired with exactly one y.

27. $f(x) = 3x - 4$ Domain is the set of all real numbers.

29. $f(x) = x^2 + 2$ Domain is the set of all real numbers.

31. $f(x) = \dfrac{4}{x+2}$ Domain is $\{x \mid x \neq -2\}$

33. $f(x) = \sqrt{7+x}$ Domain is $\{x \mid x \geq -7\}$

35. $f(x) = \sqrt{4-x^2}$ Domain is $\{x \mid -2 \leq x \leq 2\}$

37. $f(x) = \dfrac{1}{\sqrt{x+4}}$ Domain is $\{x \mid x > -4\}$

39.

Domain: the set of all real numbers

41.

Domain: the set of all real numbers

43.

Domain: $\{x \mid -6 \le x \le 6\}$

45.

Domain: $\{x \mid -3 \le x \le 3\}$

47. $\dfrac{\text{int}\left[10^2(2.3458)+0.5\right]}{10^2} = \dfrac{\text{int}[235.08]}{100} = \dfrac{235}{100} = 2.35$

49. $\dfrac{\text{int}\left[10^3(34.05622)+0.5\right]}{10^3} = \dfrac{\text{int}[34,056.72]}{1000} = \dfrac{34,056}{1000} = 34.056$

51. $\dfrac{\text{int}\left[10^4(0.08951)+0.5\right]}{10^4} = \dfrac{\text{int}[895.6]}{10,000} = \dfrac{895}{10,000} = 0.0895$

53. **a.**
$$\begin{aligned}
C(2.8) &= 0.39 - 0.34\,\text{int}(1-2.8) \\
&= 0.39 - 0.34\,\text{int}(-1.8) \\
&= 0.39 - 0.34(-2) \\
&= 0.39 + 0.68 \\
&= \$1.07
\end{aligned}$$

b.

$c(w)$

```
2 ┤                    o———
  │              o———•
1 ┤        o———•
  │  o———•
  └──┬──┬──┬──┬──┬──
     1  2  3  4  5
```

55. **a.** Yes; every vertical line intersects the graph in one point.
b. Yes; every vertical line intersects the graph in one point.
c. No; some vertical lines intersect the graph at more than one point.
d. Yes; every vertical line intersects the graph in at most one point.

57. Decreasing on $(-\infty, 0]$; increasing on $[0, \infty)$

59. Increasing on $(-\infty, \infty)$

61. Decreasing on $(-\infty, -3]$; increasing on $[-3, 0]$; decreasing on $[0, 3]$; increasing on $[3, \infty)$

63. Constant on $(-\infty, 0]$; increasing on $[0, \infty)$

65. Decreasing on $(-\infty, 0]$; constant on $[0, 1]$; increasing on $[1, \infty)$

67. g and F are one-to-one since every horizontal line intersects the graph at one point.
f, V, and p are not one-to-one since some horizontal lines intersect the graph at more than one point.

69. **a.**
$$\begin{aligned}
2l + 2w &= 50 \\
2w &= 50 - 2l \\
w &= 25 - l
\end{aligned}$$

b.
$$\begin{aligned}
A &= lw \\
A &= l(25-l) \\
A &= 25l - l^2
\end{aligned}$$

71. $v(t) = 80,000 - 6500t, \quad 0 \le t \le 10$

73. **a.** $C(x) = 5(400) + 22.80x$

$= 2000 + 22.80x$

b. $R(x) = 37.00x$

c. $P(x) = 37.00x - C(x)$

$= 37.00 - [2000 + 22.80x]$

$= 37.00x - 2000 - 22.80x$

$= 14.20x - 2000$

Note x is a natural number.

75. $\dfrac{15}{3} = \dfrac{15-h}{r}$

$5 = \dfrac{15-h}{r}$

$5r = 15 - h$

$h = 15 - 5r$

$h(r) = 15 - 5r$

77. $d = \sqrt{(3t)^2 + (50)^2}$

$d = \sqrt{9t^2 + 2500}$ meters, $0 \le t \le 60$

79. $d = \sqrt{(45 - 8t)^2 + (6t)^2}$ miles

where t is the number of hours after 12:00 noon

81. **a.**

Circle	Square
$C = 2\pi r$	$C = 4s$
$x = 2\pi r$	$20 - x = 4s$
$r = \dfrac{x}{2\pi}$	$s = 5 - \dfrac{x}{4}$
Area $= \pi r^2 = \pi\left(\dfrac{x}{2\pi}\right)^2$	Area $= s^2 = \left(5 - \dfrac{x}{4}\right)^2$
$= \dfrac{x^2}{4\pi}$	$= 25 - \dfrac{5}{2}x + \dfrac{x^2}{16}$

Total Area $= \dfrac{x^2}{4\pi} + 25 - \dfrac{5}{2}x + \dfrac{x^2}{16}$

$= \left(\dfrac{1}{4\pi} + \dfrac{1}{16}\right)x^2 - \dfrac{5}{2}x + 25$

b.

x	0	4	8	12	16	20
Total Area	25	17.27	14.09	15.46	21.37	31.83

c. Domain: [0, 20].

83. **a.**

Left side triangle	Right side triangle
$c^2 = 20^2 + (40 - x)^2$	$c^2 = 30^2 + x^2$
$c = \sqrt{400 + (40 - x)^2}$	$c = \sqrt{900 + x^2}$

Total length $= \sqrt{900 + x^2} + \sqrt{400 + (40 - x)^2}$

b.

x	0	10	20	30	40
Total Length	74.72	67.68	64.34	64.79	70

c. Domain: [0, 40].

85.

x	5	10	12.5	15	20
$Y(x)$	275	375	385	390	394

answers accurate to nearest apple

87. $f(c) = c^2 - c - 5 = 1$

$c^2 - c - 6 = 0$

$(c - 3)(c + 2) = 0$

$c - 3 = 0$ or $c + 2 = 0$

$c = 3$ $c = -2$

89. 1 is not in the range of $f(x)$, since

$1 = \dfrac{x-1}{x+1}$ only if $x + 1 = x - 1$ or $1 = -1$.

91. Set the graphing utility to "dot" mode.

```
Y1☰int X/abs X
Y2 =
Y3 =
Y4 =
Y5 =
Y6 =
Y7 =
Y8 =
```

WINDOW FORMAT
```
Xmin=-4.7
Xmax=4.7
Xscl=1
Ymin=-5
Ymax=2
Yscl=1
```

93.
```
Y1☰X²-2abs X-3
Y2 =
Y3 =
Y4 =
Y5 =
Y6 =
Y7 =
Y8 =
```

WINDOW FORMAT
```
Xmin=-4.7
Xmax=4.7
Xscl=1
Ymin=-5
Ymax=1
Yscl=1
```

95.
```
Y1☰abs (X²-1)-
   abs (X-2)
Y2 =
Y3 =
Y4 =
Y5 =
Y6 =
Y7 =
```

WINDOW FORMAT
```
Xmin=-4.7
Xmax=4.7
Xscl=1
Ymin=-4.7
Ymax=4.7
Yscl=1
```

Connecting Concepts

97. $f(x)\Big|_2^3 = (9-3)-(4-2) = 6-2 = 4$

99. $f(x)\Big|_0^2 = (16-12-2)-0 = 2$

101.
  a. $f(1,7) = 3(1)+5(7)-2 = 3+35-2 = 36$

  b. $f(0,3) = 3(0)+5(3)-2 = 13$

  c. $f(-2,4) = 3(-2)+5(4)-2 = 12$

  d. $f(4,4) = 3(4)+5(4)-2 = 30$

  e. $f(k,2k) = 3(k)+5(2k)-2 = 13k-2$

  f. $f(k+2,k-3) = 3(k+2)+5(k-3)-2 = 3k+6+5k-15-2 = 8k-11$

103. $s = \dfrac{5+8+11}{2} = 12$

$A(5,8,11) = \sqrt{12(12-5)(12-8)(12-11)}$

$\qquad\qquad = \sqrt{12(7)(4)(1)} = \sqrt{336} = 4\sqrt{21}$

105.
$$a^2+3a-3 = a$$
$$a^2+2a-3 = 0$$
$$(a-1)(a+3) = 0$$
$$a=1 \quad \text{or} \quad a=-3$$

107.

PS1.

The graph of g is one unit above the graph of f.

PS2. $f(3) = \dfrac{3(3)^4}{(3)^2 + 1} = \dfrac{243}{10} = 24.3$

 $f(-3) = \dfrac{3(-3)^4}{(-3)^2 + 1} = \dfrac{243}{10} = 24.3$

 $f(3) = f(-3)$

PS3. $f(-2) = 2(-2)^3 - 5(-2) = -16 + 10 = -6$

 $-f(2) = -[2(2)^3 - 5(2)] = -[16 - 10] = -6$

 $f(-2) = -f(2)$

PS4. $f(-2) - g(-2) = (-2)^2 - [-2 + 3] = 4 - 1 = 3$

 $f(-1) - g(-1) = (-1)^2 - [-1 + 3] = 1 - 2 = -1$

 $f(0) - g(0) = (0)^2 - [0 + 3] = 0 - 3 = -3$

 $f(1) - g(1) = (1)^2 - [1 + 3] = 1 - 4 = -3$

 $f(2) - g(2) = (2)^2 - [2 + 3] = 4 - 5 = -1$

PS5. $\dfrac{-a+a}{2} = 0$, $\dfrac{b+b}{2} = b$

 midpoint is $(0, b)$

PS6. $\dfrac{-a+a}{2} = 0$, $\dfrac{-b+b}{2} = 0$

 midpoint is $(0, 0)$

Section 1.4

1.

3.

5.

7.

9.

11.

13. **a.** No   **b.** Yes

15. **a.** No   **b.** No

17. **a.** Yes   **b.** Yes

19. **a.** Yes   **b.** Yes

21. **a.** Yes   **b.** Yes

23. No, since $(-y) = 3(-x) - 2$ simplifies to $(-y) = -3x - 2$, which is not equivalent to the original equation $y = 3x - 2$.

25. Yes, since $(-y) = -(-x)^3$ implies $-y = x^3$ or $y = -x^3$, which is the original equation.

27. Yes, since $(-x)^2 + (-y)^2 = 10$ simplifies to the original equation.

29. Yes, since $-y = \dfrac{-x}{|-x|}$ simplifies to the original equation.

31.

symmetric with
respect to the y-axis

33.

symmetric with
respect to the origin

35.

symmetric with
respect to the origin

37.

symmetric with
respect to the line $x = 4$

39.

symmetric with
respect to the line $x = 2$

41.

no symmetry

43. Even since $g(-x) = (-x)^2 - 7 = x^2 - 7 = g(x).$

45. Odd, since $F(-x) = (-x)^5 + (-x)^3$
$$= -x^5 - x^3$$
$$= -F(x).$$

47. Even

49. Even

51. Even

53. Even

55. Neither

57.

59. a. $f(x+2)$

b. $f(x)+2$

61. a. $f(x+3)$
$(-2-3, 5) = (-5, 5)$
$(0-3, -2) = (-3, -2)$
$(1-3, 0) = (-2, 0)$

b. $f(x)+1$
$(-2, 5+1) = (-2, 6)$
$(0, -2+1) = (0, -1)$
$(1, 0+1) = (1, 1)$

63. **a.** $f(-x)$

b. $-f(x)$

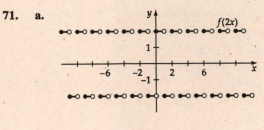

65. **a.** $f(-x)$
$(--1, 3) = (1, 3)$
$(-2, -4)$

b. $-f(x)$
$(-1, -3)$
$(2, --4) = (2, 4)$

67.

69.

$$y = -\frac{1}{2}m(x) + 3$$

71. **a.**

b.

$f\left(\frac{1}{3}x\right)$

73. **a.**

$y = h(2x)$

b.

$y = h\left(\frac{1}{2}x\right)$

75.

$y = \sqrt[3]{x+3}$
$y = \sqrt[3]{x}$
$y = \sqrt[3]{x-1}$

77.

$J(x)$ for
$c = 2$
$c = 0$
$c = -1$

79.

81.

83. a. **b.** **c.**

Connecting **C**oncepts

85. a.
$$f(x) = \frac{2}{(x+1)^2 + 1} + 1$$

b.
$$f(x) = -\frac{2}{(x-2)^2 + 1}$$

Prepare for **S**ection **1.5**

PS1. $f(3) - g(3) = \left(3^2 + 3(3) + 1\right) - \left(4(3) + 5\right)$
$$= 19 - 17$$
$$= 2$$

PS2. $f(-2) \cdot g(-2) = \left(3(-2)^2 - (-2) - 4\right) \cdot \left(2(-2) - 5\right)$
$$= (12 + 2 - 4) \cdot (-9)$$
$$= 10 \cdot (-9)$$
$$= -90$$

PS3. $f(3a) = 2(3a)^2 - 5(3a) + 2$
$$= 18a^2 - 15a + 2$$

PS4. $f(2+h) = 2(2+h)^2 - 5(2+h) + 2$
$$= 2h^2 + 8h + 8 - 5h - 10 + 2$$
$$= 2h^2 + 3h$$

PS5. Domain: all real numbers except $x = 1$

PS6. $2x - 8 = 0$
$$x = 4$$
Domain: $x \geq 4$ or $[4, \infty)$

Section 1.5

1. $f(x)+g(x)=(x^2-2x-15)+(x+3)$

$\quad\quad\quad =x^2-x-12$ Domain all real numbers

$f(x)-g(x)=(x^2-2x-15)-(x+3)$

$\quad\quad\quad =x^2-3x-18$ Domain all real numbers

$f(x)g(x)=(x^2-2x-15)(x+3)$

$\quad\quad\quad =x^3+x^2-21x-45$ Domain all real numbers

$f(x)/g(x)=(x^2-2x-15)/(x+3)$

$\quad\quad\quad =x-5$ Domain $\{x|x\neq-3\}$

3. $f(x)+g(x)=(2x+8)+(x+4)$

$\quad\quad\quad =3x+12$ Domain all real numbers

$f(x)-g(x)=(2x+8)-(x+4)$

$\quad\quad\quad =x+4$ Domain all real numbers

$f(x)g(x)=(2x+8)(x+4)$

$\quad\quad\quad =2x^2+16x+32$ Domain all real numbers

$f(x)/g(x)=(2x+8)/(x+4)$

$\quad\quad\quad =[2(x+4)]/(x+4)$

$\quad\quad\quad =2$ Domain $\{x|x\neq-4\}$

5. $f(x)+g(x)=(x^3-2x^2+7x)+x$

$\quad\quad\quad =x^3-2x^2+8x$ Domain all real numbers

$f(x)-g(x)=(x^3-2x^2+7x)-x$

$\quad\quad\quad =x^3-2x^2+6x$ Domain all real numbers

$f(x)g(x)=(x^3-2x^2+7x)x$

$\quad\quad\quad =x^4-2x^3+7x^2$ Domain all real numbers

$f(x)/g(x)=(x^3-2x^2+7x)/x$

$\quad\quad\quad =x^2-2x+7$ Domain $\{x\,|\,x\neq0\}$

7. $f(x)+g(x)=(4x-7)+(2x^2+3x-5)$

$\quad\quad\quad =2x^2+7x-12$ Domain all real numbers

$f(x)-g(x)=(4x-7)-(2x^2+3x-5)$

$\quad\quad\quad =-2x^2+x-2$ Domain all real numbers

$f(x)g(x)=(4x-7)(2x^2+3x-5)$

$\quad\quad\quad =6x^3-10x^2+12x^2-20x-21x+35$

$\quad\quad\quad =6x^3+2x^2-41x+35$ Domain all real numbers

$f(x)/g(x)=(4x-7)/(2x^2+3x-5)$

$\quad\quad\quad =\dfrac{4x-7}{2x^2+3x-5}$ Domain $\left\{x|x\neq1,x\neq-\dfrac{5}{2}\right\}$

9. $f(x)+g(x)=\sqrt{x-3}+x$ $\quad$ Domain $\{x\,|\,x\geq3\}$

$f(x)-g(x)=\sqrt{x-3}-x$ $\quad$ Domain $\{x\,|\,x\geq3\}$

$f(x)g(x)=x\sqrt{x-3}$ $\quad\quad$ Domain $\{x\,|\,x\geq3\}$

$f(x)/g(x)=\dfrac{\sqrt{x-3}}{x}+x$ $\quad$ Domain $\{x\,|\,x\geq3\}$

11. $f(x)+g(x)=\sqrt{4-x^2}+2+x$ Domain $\{x\,|-2\leq x\leq2\}$

$f(x)-g(x)=\sqrt{4-x^2}-2-x$ Domain $\{x\,|-2\leq x\leq2\}$

$f(x)g(x)=\left(\sqrt{4-x^2}\right)(2+x)$ Domain $\{x\,|-2\leq x\leq2\}$

$f(x)/g(x)=\dfrac{\sqrt{4-x^2}}{2+x}$ $\quad$ Domain $\{x\,|-2\leq x\leq2\}$

13. $(f+g)(x)=x^2-x-2$

$(f+g)(5)=(5)^2-(5)-2$

$\quad\quad\quad =25-5-2$

$\quad\quad\quad =18$

15. $(f+g)(x)=x^2-x-2$

$(f+g)\left(\dfrac{1}{2}\right)=\left(\dfrac{1}{2}\right)^2-\left(\dfrac{1}{2}\right)-2$

$\quad\quad\quad =\dfrac{1}{4}-\dfrac{1}{2}-2$

$\quad\quad\quad =-\dfrac{9}{4}$

17. $(f-g)(x)=x^2-5x+6$

$(f-g)(-3)=(-3)^2-5(-3)+6$

$\quad\quad\quad =9+15+6$

$\quad\quad\quad =30$

19. $(f-g)(x)=x^2-5x+6$

$(f-g)(-1)=(-1)^2-5(-1)+6$

$\quad\quad\quad =1+5+6$

$\quad\quad\quad =12$

21. $(fg)(x)=\left(x^2-3x+2\right)(2x-4)$

$\quad\quad\quad =2x^3-6x^2+4x-4x^2+12x-8$

$\quad\quad\quad =2x^3-10x^2+16x-8$

$(fg)(7)=2(7)^3-10(7)^2+16(7)-8$

$\quad\quad\quad =686-490+112-8$

$\quad\quad\quad =300$

23. $(fg)(x) = 2x^3 - 10x^2 + 16x - 8$

$(fg)\left(\dfrac{2}{5}\right) = 2\left(\dfrac{2}{5}\right)^3 - 10\left(\dfrac{2}{5}\right)^2 + 16\left(\dfrac{2}{5}\right) - 8$

$\qquad = \dfrac{16}{125} - \dfrac{40}{25} + \dfrac{32}{5} - 8$

$\qquad = \dfrac{-384}{125} = -3.072$

25. $\left(\dfrac{f}{g}\right)(x) = \dfrac{x^2 - 3x + 2}{2x - 4}$

$\left(\dfrac{f}{g}\right)(x) = \dfrac{1}{2}x - \dfrac{1}{2}$

$\left(\dfrac{f}{g}\right)(-4) = \dfrac{1}{2}(-4) - \dfrac{1}{2}$

$\qquad = -2 - \dfrac{1}{2}$

$\qquad = -2\dfrac{1}{2} \ \text{or} \ -\dfrac{5}{2}$

27. $\left(\dfrac{f}{g}\right)(x) = \dfrac{1}{2}x - \dfrac{1}{2}$

$\left(\dfrac{f}{g}\right)\left(\dfrac{1}{2}\right) = \dfrac{1}{2}\left(\dfrac{1}{2}\right) - \dfrac{1}{2}$

$\qquad = \dfrac{1}{4} - \dfrac{1}{2}$

$\qquad = -\dfrac{1}{4}$

29. $\dfrac{f(x+h) - f(x)}{h} = \dfrac{[2(x+h)+4] - (2x+4)}{h}$

$\qquad = \dfrac{2x + 2(h) + 4 - 2x - 4}{h}$

$\qquad = \dfrac{2h}{h}$

$\qquad = 2$

31. $\dfrac{f(x+h) - f(x)}{h} = \dfrac{[(x+h) - 6] - (x^2 - 6)}{h}$

$\qquad = \dfrac{x^2 + 2x(h) + (h)^2 - 6 - x^2 + 6}{h}$

$\qquad = \dfrac{2x(h) + h^2}{h}$

$\qquad = 2x + h$

33. $\dfrac{f(x+h) - f(x)}{h} = \dfrac{2(x+h)^2 + 4(x+h) - 3 - (2x^2 + 4x - 3)}{h}$

$\qquad = \dfrac{2x^2 + 4xh + 2h^2 + 4x + 4h - 3 - 2x^2 - 4x + 3}{h}$

$\qquad = \dfrac{4xh + 2h^2 + 4h}{h}$

$\qquad = 4x + 2h + 4$

35. $\dfrac{f(x+h) - f(x)}{h} = \dfrac{-4(x+h)^2 + 6 - (-4x^2 + 6)}{h}$

$\qquad = \dfrac{-4x^2 - 8xh - 4h^2 + 6 + 4x^2 - 6}{h}$

$\qquad = \dfrac{-8xh - 4h^2}{h}$

$\qquad = -8x - 4h$

37.

$(g \circ f)(x) = g[f(x)]$

$\qquad = g[3x + 5]$

$\qquad = 2[3x + 5]$

$\qquad = 6x + 10 - 7$

$\qquad = 6x + 3$

$(f \circ g)(x) = f[g(x)]$

$\qquad = f[2x - 7]$

$\qquad = 3[2x - 7] + 5$

$\qquad = 6x - 21 + 5$

$\qquad = 6x - 16$

39.

$(g \circ f)(x) = g\left[x^2 + 4x - 1\right]$

$\qquad = \left[x^2 + 4x - 1\right] + 2$

$\qquad = x^2 + 4x + 1$

$(f \circ g)(x) = f[x + 2]$

$\qquad = [x + 2]^2 + 4[x + 2] - 1$

$\qquad = x^2 + 4x + 4 + 4x + 8 - 1$

$\qquad = x^2 + 8x + 11$

41.
$$(g \circ f)(x) = g[f(x)]$$
$$= g\left[x^3 + 2x\right]$$
$$= -5\left[x^3 + 2x\right]$$
$$= -5x^3 - 10x$$

$$(f \circ g)(x) = f[g(x)]$$
$$= f[-5x]$$
$$= [-5x]^3 + 2[-5x]$$
$$= -125x^3 - 10x$$

43.
$$(g \circ f)(x) = g[f(x)]$$
$$= g\left[\frac{2}{x+1}\right]$$
$$= 3\left[\frac{2}{x+1}\right] - 5$$
$$= \frac{6}{x+1} - \frac{5(x+1)}{x+1}$$
$$= \frac{6 - 5x - 5}{x+1}$$
$$= \frac{1 - 5x}{x+1}$$

$$(f \circ g)(x) = f[g(x)]$$
$$= f[3x - 5]$$
$$= \frac{2}{[3x-5]+1}$$
$$= \frac{2}{3x-4}$$

45.

$$(g \circ f)(x) = g[f(x)]$$
$$= g\left[\frac{1}{x^2}\right]$$
$$= \sqrt{\left[\frac{1}{x^2}\right] - 1}$$
$$= \sqrt{\frac{1 - x^2}{x^2}}$$
$$= \frac{\sqrt{1 - x^2}}{|x|}$$

$$(f \circ g)(x) = f\left[\sqrt{x-1}\right]$$
$$= \frac{1}{\left[\sqrt{x-1}\right]^2}$$
$$= \frac{1}{x - 1}$$

47.

$$(g \circ f)(x) = g\left[\frac{3}{|5-x|}\right]$$
$$= -\frac{2}{\left[\dfrac{3}{|5-x|}\right]}$$
$$= \frac{-2|5-x|}{3}$$

$$(f \circ g)(x) = f\left[-\frac{2}{x}\right]$$
$$= \frac{3}{\left|5 - \left[-\dfrac{2}{x}\right]\right|}$$
$$= \frac{3}{\left|5 + \dfrac{2}{x}\right|}$$
$$= \frac{3}{\dfrac{|5x+2|}{|x|}}$$
$$= \frac{3|x|}{|5x+2|}$$

Use the results to work Exercises 49 to 64.

49.
$$(g \circ f)(x) = 4x^2 + 2x - 6$$
$$(g \circ f)(4) = 4(4)^2 + 2(4) - 6$$
$$= 64 + 8 - 6$$
$$= 66$$

51.
$$(f \circ g)(x) = 2x^2 - 10x + 3$$
$$(f \circ g)(-3) = 2(-3)^2 - 10(-3) + 3$$
$$= 18 + 30 + 3$$
$$= 51$$

53. $(g \circ h)(x) = 9x^4 - 9x^2 - 4$

$(g \circ h)(0) = 9(0)^4 - 9(0)^2 - 4$

$\qquad = -4$

55. $(f \circ f)(x) = 4x + 9$

$(f \circ f)(8) = 4(8) + 9$

$\qquad = 41$

57. $(h \circ g)(x) = -3x^4 + 30x^3 - 75x^2 + 4$

$(h \circ g)\left(\dfrac{2}{5}\right) = -3\left(\dfrac{2}{5}\right)^4 + 30\left(\dfrac{2}{5}\right)^3 - 75\left(\dfrac{2}{5}\right)^2 + 4$

$\qquad = -\dfrac{48}{625} + \dfrac{240}{125} - \dfrac{300}{25} + 4$

$\qquad = \dfrac{-48 + 1200 - 7500 + 2500}{625}$

$\qquad = -\dfrac{3848}{625}$

59. $(g \circ f)(x) = 4x^2 + 2x - 6$

$(g \circ f)(\sqrt{3}) = 4(\sqrt{3})^2 + 2(\sqrt{3}) - 6$

$\qquad = 12 + 2\sqrt{3} - 6$

$\qquad = 6 + 2\sqrt{3}$

61. $(g \circ f)(x) = 4x^2 + 2x - 6$

$(g \circ f)(2c) = 4(2c)^2 + 2(2c) - 6$

$\qquad = 16c^2 + 4c - 6$

63. $(g \circ h)(x) = 9x^4 - 9x^2 - 4$

$(g \circ h)(k+1) = 9(k+1)^4 - 9(k+1)^2 - 4$

$\qquad = 9(k^4 + 4k^3 + 6k^2 + 4k + 1) - 9k^2 - 18k - 9 - 4$

$\qquad = 9k^4 + 36k^3 + 54k^2 + 36k + 9 - 9k^2 - 18k - 13$

$\qquad = 9k^4 + 36k^3 + 45k^2 + 18k - 4$

65. **a.** $r = 1.5t$ and $A = \pi r^2$

$\qquad$ so $A(t) = \pi \left[r(t) \right]^2$

$\qquad\qquad = \pi(1.5t)^2$

$\qquad A(2) = 2.25\pi(2)^2$

$\qquad\qquad = 9\pi$ square feet

$\qquad\qquad \approx 28.27$ square feet

b. $r = 1.5t$

$\qquad h = 2r = 2(1.5t) = 3t$ and

$\qquad V = \dfrac{1}{3}\pi r^2 h$ so

$\qquad V(t) = \dfrac{1}{3}\pi(1.5t)^2[3t]$

$\qquad\qquad = 2.25\pi t^3$

Note: $V = \dfrac{1}{3}\pi r^2 h = \dfrac{1}{3}(\pi r^2) = \dfrac{1}{3}hA$

$\qquad\qquad = \dfrac{1}{3}(3t)(2.25\pi t^2) = 2.25\pi t^3$

$\qquad V(3) = 2.25\pi(3)^3$

$\qquad\qquad = 60.75\pi$ cubic feet

$\qquad\qquad \approx 190.85$ cubic feet

67. **a.**   Since $d^2 + 4^2 = s^2$,

$\qquad\qquad d^2 = s^2 - 16$

$\qquad\qquad d = \sqrt{s^2 - 16}$

$\qquad\qquad d = \sqrt{(48-t)^2 - 16}$ $\cdot$ $s = 48 - t$

$\qquad\qquad = \sqrt{2304 - 96t + t^2 - 16}$

$\qquad\qquad = \sqrt{t^2 - 96t + 2288}$

b. $s(35) = 48 - 35 = 13$ ft

$\qquad d(35) = \sqrt{35^2 - 96(35) + 2288}$

$\qquad\qquad = \sqrt{153} \approx 12.37$ ft

69. $(Y \circ F)(x) = Y(F(x))$ converts x inches to yards.

F takes x inches to feet, and then Y takes feet to yards.

71. a. On $[0, 1]$, $a = 0$

$\Delta t = 1 - 0 = 1$

$C(a + \Delta t) = C(1) = 99.8$

$C(a) = C(0) = 0$

Average rate of change $= \dfrac{C(1) - C(0)}{1} = 99.8 - 0 = 99.8$

This is identical to the slope of the line through

$(0, C(0))$ and $(1, C(1))$ since $m = \dfrac{C(1) - C(0)}{1 - 0} = C(1) - C(0)$

b. On $[0, 0.5]$, $a = 0$, $\Delta t = 0.5$

Average rate of change $= \dfrac{C(0.5) - C(0)}{0.5} = \dfrac{78.1 - 0}{0.5} = 156.2$

c. On $[1, 2]$, $a = 1$, $\Delta t = 2 - 1 = 1$

Average rate of change $= \dfrac{C(2) - C(1)}{1} = \dfrac{50.1 - 99.8}{1} = -49.7$

d. On $[1, 1.5]$, $a = 1$, $\Delta t = 1.5 - 1 = 0.5$

Average rate of change $= \dfrac{C(1.5) - C(1)}{0.5} = \dfrac{84.4 - 99.8}{0.5} = \dfrac{-15.4}{0.5} = -30.8$

e. On $[1, 1.25]$, $a = 1$, $\Delta t = 1.25 - 1 = 0.25$

Average rate of change $= \dfrac{C(1.25) - C(1)}{0.25} = \dfrac{95.7 - 99.8}{0.25} = \dfrac{-4.1}{0.25} = -16.4$

f. On $[1, 1 + \Delta t]$, $Con(1 + \Delta t) = 25(1 + \Delta t)^3 - 150(1 + \Delta t)^2 + 225(1 + \Delta t)$

$\qquad = 25(1 + 3(\Delta t) + 3(\Delta t)^3) - 150(1 + 2(\Delta t) + (\Delta t)^2) + 225(1 + \Delta t)$

$\qquad = 25 + 75(\Delta t) + 75(\Delta t)^2 + 25(\Delta t)^3 - 150 - 300(\Delta t) - 150(\Delta t)^2 + 225 + 225(\Delta t)$

$\qquad = 100 - 75(\Delta t)^2 + 25(\Delta t)^3$

$Con(1) = 100$

Average rate of change $= \dfrac{Con(1 + \Delta t) - Con(1)}{\Delta t} = \dfrac{100 - 75(\Delta t)^2 + 25(\Delta t)^3 - 100}{\Delta t}$

$\qquad = \dfrac{-75(\Delta t)^2 + 25(\Delta t)^3}{\Delta t}$

$\qquad = -75(\Delta t) + 25(\Delta t)^2$

As Δt approaches 0, the average rate of change over $[1, 1 + \Delta t]$ seems to approach 0.

Connecting Concepts

73.
$\begin{aligned}(g \circ f)(x) &= g[f(x)] \\ &= g[2x + 3] \\ &= 5(2x + 3) + 12 \\ &= 10x + 15 + 12 \\ &= 10x + 27\end{aligned}$
$\qquad$
$\begin{aligned}(f \circ g)(x) &= f[g(x)] \\ &= f[5x + 12] \\ &= 2(5x + 12) + 3 \\ &= 10x + 24 + 3 \\ &= 10x + 27\end{aligned}$

$(g \circ f)(x) = (f \circ g)(x)$

75.

$(g \circ f)(x) = g[f(x)]$

$\quad = g\left[\dfrac{6x}{x-1}\right]$

$\quad = \dfrac{5\left(\dfrac{6x}{x-1}\right)}{\dfrac{6x}{x-1}-2}$

$\quad = \dfrac{\dfrac{30x}{x-1}}{\dfrac{6x-2x+2}{x-1}} = \dfrac{\dfrac{30x}{x-1}}{\dfrac{4x+2}{x-1}}$

$\quad = \dfrac{30x}{x-1} \cdot \dfrac{x-1}{2(2x+1)}$

$\quad = \dfrac{15x}{2x+1}$

$(g \circ f)(x) = (f \circ g)(x)$

$(f \circ g)(x) = f[g(x)]$

$\quad = f\left[\dfrac{5x}{x-2}\right]$

$\quad = \dfrac{6\left(\dfrac{5x}{x-2}\right)}{\dfrac{5x}{x-2}-1}$

$\quad = \dfrac{\dfrac{30x}{x-2}}{\dfrac{5x-x+2}{x-2}} = \dfrac{\dfrac{30x}{x-2}}{\dfrac{4x+2}{x-2}}$

$\quad = \dfrac{30x}{x-2} \cdot \dfrac{x-2}{2(2x+1)}$

$\quad = \dfrac{15x}{2x+1}$

77.

$(g \circ f)(x) = g[f(x)]$

$\quad = g[2x+3]$

$\quad = \dfrac{[2x+3]-3}{2}$

$\quad = \dfrac{2x}{2}$

$\quad = x$

$(f \circ g)(x) = f[g(x)]$

$\quad = f\left[\dfrac{x-3}{2}\right]$

$\quad = 2\left[\dfrac{x-3}{2}\right]+3$

$\quad = x-3+3$

$\quad = x$

79.

$(g \circ f)(x) = g[f(x)]$

$\quad = g\left[\dfrac{4}{x+1}\right]$

$\quad = \dfrac{4-\left[\dfrac{4}{x+1}\right]}{\left[\dfrac{4}{x+1}\right]}$

$\quad = \dfrac{\dfrac{4x+4-4}{x+1}}{\dfrac{4}{x+1}}$

$\quad = \dfrac{4x}{x+1} \cdot \dfrac{x+1}{4}$

$\quad = x$

$(f \circ g)(x) = f[g(x)]$

$\quad = f\left[\dfrac{4-x}{x}\right]$

$\quad = \dfrac{4}{\left[\dfrac{4-x}{x}\right]+1}$

$\quad = \dfrac{4}{\dfrac{4-x+x}{x}}$

$\quad = \dfrac{4}{\dfrac{4}{x}}$

$\quad = 4 \cdot \dfrac{x}{4}$

$\quad = x$

81.

$(g \circ f)(x) = g[f(x)]$

$\quad = g\left[x^3-1\right]$

$\quad = \sqrt[3]{\left[x^3-1\right]+1}$

$\quad = \sqrt[3]{x^3}$

$\quad = x$

$(f \circ g)(x) = f[g(x)]$

$\quad = f\left[\sqrt[3]{x+1}\right]$

$\quad = \left[\sqrt[3]{x+1}\right]^3-1$

$\quad = x+1-1$

$\quad = x$

PS1. $2x + 5y = 15$

$5y = -2x + 15$

$y = -\dfrac{2}{5}x + 3$

PS2. $x = \dfrac{y+1}{y}$

$xy = y + 1$

$xy - y = 1$

$y(x - 1) = 1$

$y = \dfrac{1}{x-1}$

PS3. $f(-1) = \dfrac{2(-1)^2}{(-1)-1} = \dfrac{2}{-2} = -1$

PS4. $(3, 7)$

PS5. All real numbers.

PS6. $x + 2 \geq 0$

$x \geq -2$

$\{x \mid x \geq -2\}$

Section 1.6

1. If $f(3) = 7$, then $f^{-1}(7) = 3$.

3. If $h^{-1}(-3) = -4$, then $h(-4) = -3$.

5. If 3 is in the domain of f^{-1}, then $f[f^{-1}(3)] = 3$.

7. The domain of the inverse function f^{-1} is the range of f.

9.

Yes, the inverse is a function.

11.

Yes, the inverse is a function.

13.

Yes, the inverse is a function.

15.

No, the inverse relation is not a function.

17. $f(x) = 4x; \ g(x) = \dfrac{x}{4}$

$f[g(x)] = f\left(\dfrac{x}{4}\right) = 4\left(\dfrac{x}{4}\right) = x$

$g[f(x)] = g(4x) = \dfrac{4x}{4} = x$

Yes, f and g are inverses of each other.

19. $f(x) = 4x - 1; \ g(x) = \dfrac{1}{4}x + \dfrac{1}{4}$

$f[g(x)] = f\left(\dfrac{1}{4}x + \dfrac{1}{4}\right)$

$= 4\left(\dfrac{1}{4}x + \dfrac{1}{4}\right) - 1 = x + 1 - 1$

$= x$

$g[f(x)] = g(4x - 1)$

$= \dfrac{1}{4}(4x - 1) + \dfrac{1}{4} = x - \dfrac{1}{4} + \dfrac{1}{4}$

$= x$

Yes, f and g are inverses of each other.

21. $f(x) = -\frac{1}{2}x - \frac{1}{2}; \ g(x) = -2x + 1$

$f[g(x)] = f(-2x + 1)$

$\qquad = -\frac{1}{2}(-2x + 1) - \frac{1}{2} = x - \frac{1}{2} - \frac{1}{2}$

$\qquad = x - 1$

$\qquad \neq x$

No, f and g are not inverses of each other.

23. $f(x) = \dfrac{5}{x - 3}; \ g(x) = \dfrac{5}{x} + 3$

$f[g(x)] = f\left(\dfrac{5}{x} + 3\right)$

$\qquad = \dfrac{5}{\dfrac{5}{x} + 3 - 3} = \dfrac{5}{\dfrac{5}{x}} = 5 \cdot \dfrac{x}{5}$

$\qquad = x$

$g[f(x)] = g\left(\dfrac{5}{x - 3}\right)$

$\qquad = \dfrac{5}{\dfrac{5}{x - 3}} + 3 = x - 3 + 3$

$\qquad = x$

Yes, f and g are inverses of each other.

25. $f(x) = x^3 + 2; \ g(x) = \sqrt[3]{x - 2}$

$f[g(x)] = f\left(\sqrt[3]{x - 2}\right)$

$\qquad = \left(\sqrt[3]{x - 2}\right)^3 + 2 = x - 2 + 2$

$\qquad = x$

$g[f(x)] = g\left(x^3 + 2\right)$

$\qquad = \sqrt[3]{x^3 + 2 - 2} = \sqrt[3]{x^3}$

$\qquad = x$

Yes, f and g are inverses of each other.

27. The inverse of $\{(-3, 1), (-2, 2), (1, 5), (4, -7)\}$ is $\{(1, -3), (2, -2), (5, 1), (-7, 4)\}$.

29. The inverse of $\{(0, 1), (1, 2), (2, 4), (3, 8), (4, 16)\}$ is $\{(1, 0), (2, 1), (4, 2), (8, 3), (16, 4)\}$.

31. $f(x) = 2x + 4$

$\qquad x = 2y + 4$

$\qquad x - 4 = 2y$

$\qquad \frac{1}{2}x - 2 = y$

$\qquad f^{-1}(x) = \frac{1}{2}x - 2$

33. $f(x) = 3x - 7$

$\qquad x = 3y - 7$

$\qquad x + 7 = 3y$

$\qquad \frac{1}{3}x + \frac{7}{3} = y$

$\qquad f^{-1}(x) = \frac{1}{3}x + \frac{7}{3}$

35. $f(x) = -2x + 5$

$\qquad x = -2y + 5$

$\qquad x - 5 = -2y$

$\qquad -\frac{1}{2}x + \frac{5}{2} = y$

$\qquad f^{-1}(x) = -\frac{1}{2}x + \frac{5}{2}$

37. $f(x) = \dfrac{2x}{x - 1}, \ x \neq 1$

$\qquad x = \dfrac{2y}{y - 1}$

$\qquad x(y - 1) = xy - x = 2y$

$\qquad xy - 2y = y(x - 2) = x$

$\qquad y = \dfrac{x}{x - 2}$

$\qquad f^{-1}(x) = \dfrac{x}{x - 2}, \ x \neq 2$

39. $f(x) = \dfrac{x - 1}{x + 1}, \ x \neq -1$

$\qquad x = \dfrac{y - 1}{y + 1}$

$\qquad x(y + 1) = xy + x = y - 1$

$\qquad xy - y = -x - 1$

$\qquad y - xy = y(1 - x) = x + 1$

$\qquad y = \dfrac{x + 1}{1 - x}$

$\qquad f^{-1}(x) = \dfrac{x + 1}{1 - x}, \ x \neq 1$

41. $f(x) = x^2 + 1, \ x \geq 0$

$\qquad x = y^2 + 1$

$\qquad x - 1 = y^2$

$\qquad \sqrt{x - 1} = y$

$\qquad f^{-1}(x) = \sqrt{x - 1}, \ x \geq 1$

Note: Do not use $\pm$ with the radical because the domain of f, and thus the range of f^{-1}, is nonnegative.

43.
$$f(x) = \sqrt{x-2}, \; x \geq 2$$
$$x = \sqrt{y-2}$$
$$x^2 = y-2$$
$$x^2 + 2 = y$$
$$f^{-1}(x) = x^2 + 2, \; x \geq 0$$
Note: The range of f, is nonnegative, therefore the domain of f^{-1} is also nonnegative.

45.
$$f(x) = x^2 + 4x, \; x \geq -2$$
$$x = y^2 + 4y$$
$$x + 4 = y^2 + 4y + 4$$
$$x + 4 = (y+2)^2$$
$$\sqrt{x+4} = y + 2$$
$$y = \sqrt{x+4} - 2$$
$$f^{-1}(x) = \sqrt{x+4} - 2, \; x \geq -4$$
Note: The range of f, is non-negative, therefore the domain of f^{-1} is also non-negative.

47.
$$f(x) = x^2 + 4x - 1, \; x \leq -2$$
$$x = y^2 + 4y - 1$$
$$x + 1 = y^2 + 4y$$
$$x + 1 + 4 = y^2 + 4y + 4$$
$$x + 5 = (y+2)^2$$
$$-\sqrt{x+5} = y + 2$$
$$-\sqrt{x+5} - 2 = y$$
$$f^{-1}(x) = -\sqrt{x+5} - 2, \; x \geq -5$$
Note: Because the range of f, is non-positive, the range of f^{-1} must also be non-positive.

49.
$$V(x) = x^3$$
$$x = y^3$$
$$\sqrt[3]{x} = y$$
$$V^{-1}(x) = \sqrt[3]{x}$$
$V^{-1}(x)$ finds the length of a side of a cube given the volume.

51.
$$f(x) = \frac{5}{9}(x - 32)$$
$$x = \frac{5}{9}(y - 32)$$
$$\frac{9}{5}x = y - 32$$
$$\frac{9}{5}x + 32 = y$$
$$f^{-1}(x) = \frac{9}{5}x + 32$$

$f^{-1}(x)$ is used to convert x degrees Celsius to an equivalent Fahrenheit temperature.

53.
$$s(x) = 2x + 24$$
$$x = 2y + 24$$
$$x - 24 = 2y$$
$$\frac{1}{2}x - 12 = y$$
$$s^{-1}(x) = \frac{1}{2}x - 12$$

55.
$$E(s) = 0.05s + 2500$$
$$s = 0.05y + 2500$$
$$s - 2500 = 0.05y$$
$$\frac{1}{0.05}s - \frac{2500}{0.05} = y$$
$$20s - 50{,}000 = y$$
$$E^{-1}(s) = 20s - 50{,}000$$

The executive can use the inverse function to determine the value of the software that must be sold in order to achieve a given monthly income.

57. **a.** $p(10) \approx 0.12 = 12\%$; $p(30) \approx 0.71 = 71\%$

b. The graph of p, for $1 \leq n \leq 60$, is an increasing function. Thus p has an inverse that is a function.

c. $p^{-1}(0.223)$ represents the number of people required to be in the group for a 22.3% probability that at least two of the people will share a birthday.

59. **a.**   D $f(13) = 2(13) - 1 = 25$

O $f(24) = 2(24) - 1 = 47$

(space) $f(36) = 2(36) - 1 = 71$

Y $f(34) = 2(34) - 1 = 67$

O $f(24) = 47$

U $f(30) = 2(30) - 1 = 59$

R $f(27) = 2(27) - 1 = 53$

(space) $f(36) = 71$

H $f(17) = 2(17) - 1 = 33$

O $f(24) = 47$

M $f(22) = 2(22) - 1 = 43$

E $f(14) = 2(14) - 1 = 27$

W $f(32) = 2(32) - 1 = 63$

O $f(24) = 47$

R $f(27) = 53$

K $f(20) = 2(20) - 1 = 39$

The code is
25 47 71 67 47 59 53 71 33 47 43 27 63 47 53 39.

b. $f^{-1}(49) = \dfrac{49+1}{2} = 25$ P

$f^{-1}(33) = \dfrac{33+1}{2} = 17$ H

$f^{-1}(47) = \dfrac{47+1}{2} = 24$ O

$f^{-1}(45) = \dfrac{45+1}{2} = 23$ N

$f^{-1}(27) = \dfrac{27+1}{2} = 14$ E

$f^{-1}(71) = \dfrac{71+1}{2} = 36$ (space)

$f^{-1}(33) = 17$ H

$f^{-1}(47) = 24$ O

$f^{-1}(43) = \dfrac{43+1}{2} = 22$ M

$f^{-1}(27) = 14$ E

The message is PHONE HOME.

c.   Answers will vary.

61. $f(2) = 7$, $f(5) = 12$, and $f(4) = c$. Because f is an increasing linear function, and 4 is between 2 and 5, then $f(4)$ is between $f(2)$ and $f(5)$. Thus, c is between 7 and 12.

63. f is a linear function, therefore f^{-1} is a linear function.

$f(2) = 3 \Rightarrow f^{-1}(3) = 2$

$f(5) = 9 \Rightarrow f^{-1}(9) = 5$

Since 6 is between 3 and 9, $f^{-1}(6)$ is between 2 and 5.

65. g is a linear function, therefore g^{-1} is a linear function.

$g^{-1}(3) = 4 \Rightarrow g(4) = 3$

$g^{-1}(7) = 8 \Rightarrow g(8) = 7$

Since 5 is between 4 and 8, $g(5)$ is between 3 and 7.

•••

Connecting Concepts

67. $f(x) = mx + b$, $m \neq 0$

$y = mx + b$

$x = my + b$

$x - b = my$

$\dfrac{x - b}{m} = y$

$f^{-1}(x) = \dfrac{1}{m}x - \dfrac{b}{m}$

The slope is $\dfrac{1}{m}$ and the y-intercept is $\left(0, -\dfrac{b}{m}\right)$.

69.   The reflection of f across the line given by $y=x$ yields f. Thus f is its own inverse.

71.   There is at most one point where each horizontal line intersects the graph of the function. The function is a one-to-one function.

73.   A horizontal line intersects the graph of the function at more than one point. Thus, the function is not a one-to-one function.

PS1. Slope: $-\dfrac{1}{3}$; y-intercept: $(0, 4)$

PS2. $3x - 4y = 12$

$$y = \frac{3}{4}x - 3$$

Slope: $\dfrac{3}{4}$; y-intercept: $(0, -3)$

PS3. $y = -0.45x + 2.3$

PS4. $y + 4 = -\dfrac{2}{3}(x - 3)$

$$y = -\frac{2}{3}x - 2$$

PS5. $f(2) = 3(2)^2 + 4(2) - 1 = 12 + 8 - 1 = 19$

PS6. $\left|f(x_1) - y_1\right| + \left|f(x_2) - y_2\right| = \left|(2)^2 - 3 - (-1)\right| + \left|(4)^2 - 3 - 14\right|$

$$= \left|4 - 3 + 1\right| + \left|16 - 3 - 14\right|$$
$$= 2 + 1$$
$$= 3$$

Section 1.7

1. The scatter diagram suggests no relationship between x and y.

3. The scatter diagram suggests a linear relationship between x and y.

5. Figure A better approximates a graph that can be modeled by an equation than does Figure B. Thus Figure A has a coefficient of determination closer to 1.

7. Enter the data on your calculator. The technique for a TI-83 calculator is illustrated here.

$y = 2.00862069x + 0.5603448276$

9. Enter the data on your calculator. The technique for a TI-83 calculator is illustrated here.

$y = -0.7231182796x + 9.233870968$

11. Enter the data on your calculator. The technique for a TI-83 calculator is illustrated here.

$y = 2.222641509x - 7.364150943$

13. Enter the data on your calculator. The technique for a TI-83 calculator is illustrated here.

$y = 1.095779221x^2 - 2.69642857x + 1.136363636$

15. Enter the data on your calculator. The technique for a TI-83 calculator is illustrated here.

$y = -0.2987274717x^2 - 3.20998141x + 3.416463667$

17. Enter the data on your calculator. The technique for a TI-83 calculator is illustrated here.

a. $y = 23.55706665x - 24.4271215$

b. $y = 23.55706665(54) - 24.4271215 \approx 1248$ cm

19. Enter the data on your calculator. The technique for a TI-83 calculator is illustrated here.

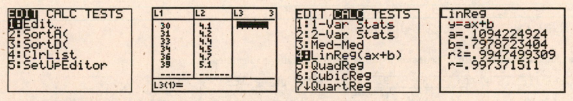

a. $y = 0.194224924x + 0.7978723404$

b. $y = 0.194224924(32) + 0.7978723404 \approx 4.3$ m/s

21. Enter the data on your calculator. The technique for a TI-83 calculator is illustrated here.

a. $y = 0.1628623408x - 0.6875682232$

b. $y = 0.1628623408(158) - 0.6875682232 \approx 25$

23.   Enter the data on your calculator. The technique for a TI-83 calculator is illustrated here.

The value of *r* is close to 0. Therefore, no, there is not a strong linear relationship between the current and the torque.

25.   Enter the data on your calculator. The technique for a TI-83 calculator is illustrated here.

   a.   The value of *r* is close to –1, so, yes, there is a strong linear correlation.

   b. $y = -0.9033088235x + 78.62573529$

   c. $y = -0.9033088235(25) + 78.62573529 \approx 56$ years

27.   Enter the data on your calculator. The technique for a TI-83 calculator is illustrated here.

   $y = 113.3111246x + 21.83605895$

   a.   Positively

   b. $y = 113.3111246(9.5) + 21.83605895 \approx 1098$ calories

29.   Enter the data on your calculator. The technique for a TI-83 calculator is illustrated here.

   $y = -0.6328671329x^2 + 33.6160839x - 379.4405594$

31.   Enter the data on your calculator. The technique for a TI-83 calculator is illustrated here.

   a. $y = -0.0165034965x^2 + 1.366713287x + 5.685314685$

   b. $y = -0.0165034965(50)^2 + 1.366713287(50) + 5.685314685 \approx 32.8$ mpg

33.   **a.**   Enter the data on your calculator. The technique for a TI-83 calculator is illustrated here.

5-lb ball

$y = 0.6130952381t^2 - 0.0714285714t + 0.1071428571$

10-lb ball

$y = 0.6091269841t^2 - 0.0011904762t - 0.3$

15-lb ball

$y = 0.5922619048t^2 + 0.3571428571t - 1.520833333$

   b.   All the regression equations are approximately the same. Therefore, there is one equation of motion.

1.   a, c, d, e

2.
$$f[g(x)] = 3(2x+4)+8$$
$$= 6x+12+8$$
$$= 6x+20$$

$$g[f(x)] = 2(3x+8)+4$$
$$= 6x+16+4$$
$$= 6x+20$$

Thus $f[g(x)] = 6x+20 = g[f(x)]$.

To be inverse functions,
$$f[g(x)] = x = g[f(x)].$$
No. They are not inverse functions.

3.
$$f(2) = 3$$
$$f(x) = f(x+4)$$
$$f(2) = f(2+4) = f(6)$$
$$f(6) = f(6+4) = f(10)$$
$$f(10) = f(10+4) = f(14)$$
$$f(14) = f(14+4) = f(18)$$
Thus, $f(18) = f(2) = 3$.

4. $|x+2| < 3$

5.   It is the slope of the line between $(a, f(a))$ and $(b, f(b))$.

6. $(3, -2)$

7. $(7, 3)$

8. $(-3, 6)$

9. $(3, 4)$

10.   Yes. The slope of the regression line is negative.

1.
$$3 - 4z = 12$$
$$-4z = 9$$
$$z = -\frac{9}{4}$$

2.
$$4y - 3 = 6y + 5$$
$$-2y = 8$$
$$y = -4$$

3.
$$2x - 3(2 - 3x) = 14x$$
$$2x - 6 + 9x = 14x$$
$$-6 = 3x$$
$$-2 = x$$

4.
$$5 - 2(3m + 2) = 3(1 - m)$$
$$5 - 6m - 4 = 3 - 3m$$
$$1 - 6m = 3 - 3m$$
$$-3m = 2$$
$$m = -\frac{2}{3}$$

5.
$$y^2 - 3y - 18 = 0$$
$$(y - 6)(y + 3) = 0$$

$$y - 6 = 0 \quad \text{or} \quad y + 3 = 0$$

$$y = 6 \quad \text{or} \quad y = -3$$

6.
$$2z^2 - 9z + 4 = 0$$
$$(2z - 1)(z - 4) = 0$$

$$2z - 1 = 0 \quad \text{or} \quad z - 4 = 0$$
$$2z = 1 \qquad\qquad z = 4$$
$$z = \frac{1}{2}$$

7.
$$3v^2 + v = 1$$
$$3v^2 + v - 1 = 0$$

$$v = \frac{-1 \pm \sqrt{1^2 - 4(3)(-1)}}{2(3)}$$

$$v = \frac{-1 \pm \sqrt{13}}{6}$$

8.
$$3s = 4 - 2s^2$$
$$2s^2 + 3s - 4 = 0$$

$$s = \frac{-3 \pm \sqrt{3^2 - 4(2)(-4)}}{2(2)}$$

$$s = \frac{-3 \pm \sqrt{9 + 32}}{4}$$

$$s = \frac{-3 \pm \sqrt{41}}{4}$$

9.
$$3c - 5 \leq 5c + 7$$
$$-2c \leq 12$$
$$c \geq -6$$

10.
$$7a > 5 - 2(3a - 4)$$
$$7a > 5 - 6a + 8$$
$$13a > 13$$
$$a > 1$$

11.
$$x^2 - x - 12 \geq 0$$
$$(x - 4)(x + 3) \geq 0$$

Critical values are 4 and -3.

$$(-\infty, -3] \cup [4, \infty)$$

12.
$$2x^2 - x < 1$$
$$2x^2 - x - 1 < 0$$
$$(2x + 1)(x - 1) < 0$$

Critical values are $-\frac{1}{2}$ and 1.

$$-\frac{1}{2} < x < 1$$

13.
$$|2x - 5| > 3$$
$$2x - 5 > 3 \quad \text{or} \quad 2x - 5 < -3$$
$$2x > 8 \qquad\qquad 2x < 2$$
$$x > 4 \qquad\qquad x < 1$$
$$(-\infty, 1) \cup (4, \infty)$$

14.
$$|1 - 3x| \leq 4$$
$$-4 \leq 1 - 3x \leq 4$$
$$-5 \leq -3x \leq 3$$
$$\frac{5}{3} \geq x \geq -1$$

15.
$$c^2 = a^2 + b^2$$
$$c^2 = 6^2 + 8^2$$
$$c = \sqrt{36 + 64}$$
$$c = \sqrt{100} = 10$$

16.
$$c^2 = a^2 + b^2$$
$$20^2 = 11^2 + b^2$$
$$b = \sqrt{400 - 121}$$
$$b = \sqrt{279}$$
$$b \approx 16.7$$

17.
$$c^2 = a^2 + b^2$$
$$13^2 = a^2 + 12^2$$
$$a = \sqrt{169 - 144}$$
$$a = \sqrt{25}$$
$$a = 5$$

18.
$$c^2 = a^2 + b^2$$
$$c^2 = 7^2 + 14^2$$
$$c = \sqrt{49 + 196}$$
$$c = \sqrt{245}$$
$$c \approx 15.7$$

19.
$$d = \sqrt{(7 - (-3))^2 + (11 - 2)^2}$$
$$= \sqrt{100 + 81}$$
$$= \sqrt{181}$$

20.
$$d = \sqrt{(-3 - 5)^2 + (-8 - (-4))^2}$$
$$= \sqrt{64 + 16}$$
$$= \sqrt{80} = 4\sqrt{5}$$

21.
$$\left(\frac{2 + (-3)}{2}, \frac{8 + 12}{2}\right) = \left(-\frac{1}{2}, 10\right)$$

22. $\left(\dfrac{-4+8}{2}, \dfrac{7+(-11)}{2}\right) = (2,-2)$

23.

24.

25. $y = x^2 - 7$

Replace y with $-y$.

$$-y = x^2 - 7$$

$$y = -x^2 + 7$$

Thus, y is not symmetric with respect to the x-axis.

Replace x with $-x$.

$$y = (-x)^2 - 7$$

$$y = x^2 - 7$$

Thus, y is symmetric with respect to the y-axis.

Replace x with $-x$ and replace y with $-y$.

$$-y = (-x)^2 - 7$$

$$y = -x^2 + 7$$

Thus, y is not symmetric with respect to the origin.

Therefore, the graph of $y = x^2 - 7$ is symmetric with respect to the y-axis.

26. $x = y^2 + 3$

$$y^2 = x - 3$$

$$y = \pm\sqrt{x-3}$$

Replace y with $-y$.

$$x = (-y)^2 + 3$$

$$x = y^2 + 3$$

$$y^2 = x - 3$$

$$y = \pm\sqrt{x-3}$$

Thus, y is symmetric with respect to the x-axis.

Replace x with $-x$.

$$-x = y^2 + 3$$

$$y^2 = -x - 3$$

$$y = \pm\sqrt{-x-3}$$

Thus, y is not symmetric with respect to the y-axis.

Replace x with $-x$ and replace y with $-y$.

$$-x = (-y)^2 + 3$$

$$-x = y^2 + 3$$

$$y^2 = -x - 3$$

$$y = \pm\sqrt{-x-3}$$

Thus, y is not symmetric with respect to the origin.

Therefore, the graph of $x = y^2 + 3$ is symmetric with respect to the x-axis.

27. $y = x^3 - 4x$

Replace y with $-y$.

$$-y = x^3 - 4x$$
$$y = -x^3 + 4x$$

Thus, y is not symmetric with respect to the x-axis.
Replace x with $-x$.

$$y = (-x)^3 - 4(-x)$$
$$y = -x^3 + 4x$$

Thus, y is not symmetric with respect to the y-axis.
Replace x with $-x$ and replace y with $-y$.

$$-y = (-x)^3 - 4(-x)$$
$$-y = -x^3 + 4x$$
$$y = x^3 - 4x$$

Thus, y is symmetric with respect to the origin.

Therefore, the graph of $y = x^3 - 4x$ is symmetric with respect to the origin.

28. $y^2 = x^2 + 4$

$$y = \pm\sqrt{x^2 + 4}$$

Replace y with $-y$.

$$(-y)^2 = x^2 + 4$$
$$y^2 = x^2 + 4$$
$$y = \pm\sqrt{x^2 + 4}$$

y is symmetric with respect to the x-axis.
Replace x with $-x$.

$$y^2 = (-x)^2 + 4$$
$$y^2 = x^2 + 4$$
$$y = \pm\sqrt{x^2 + 4}$$

Thus, y is symmetric with respect to the y-axis.
Replace x with $-x$ and replace y with $-y$.

$$(-y)^2 = (-x)^2 + 4$$
$$y^2 = x^2 + 4$$
$$y = \pm\sqrt{x^2 + 4}$$

Thus, y is symmetric with respect to the origin.

Therefore, the graph of $y^2 = x^2 + 4$ is symmetric with respect to the x-axis, the y-axis, and the origin.

29. $\dfrac{x^2}{3^2} + \dfrac{y^2}{4^2} = 1$

Replace y with $-y$.

$$\frac{x^2}{3^2} + \frac{(-y)^2}{4^2} = 1 \Rightarrow \frac{x^2}{3^2} + \frac{y^2}{4^2} = 1$$

Thus, y is symmetric with respect to the x-axis.
Replace x with $-x$.

$$\frac{(-x)^2}{3^2} + \frac{y^2}{4^2} = 1 \Rightarrow \frac{x^2}{3^2} + \frac{y^2}{4^2} = 1$$

Thus, y is symmetric with respect to the y-axis.
Replace x with $-x$ and replace y with $-y$.

$$\frac{(-x)^2}{3^2} + \frac{(-y)^2}{4^2} = 1 \Rightarrow \frac{x^2}{3^2} + \frac{y^2}{4^2} = 1$$

Thus, y is symmetric with respect to the origin.

Therefore, the graph of $\dfrac{x^2}{3^2} + \dfrac{y^2}{4^2} = 1$ is symmetric with respect to the x-axis, the y-axis, and the origin.

30. $xy = 8$

$$y = \frac{8}{x}$$

Replace y with $-y$.

$$x(-y) = 8$$
$$y = -\frac{8}{x}$$

Thus, y is not symmetric with respect to the x-axis.
Replace x with $-x$.

$$-xy = 8$$
$$y = -\frac{8}{x}$$

Thus, y is not symmetric with respect to the y-axis.
Replace x with $-x$ and replace y with $-y$.

$$-x(-y) = 8$$
$$y = \frac{8}{x}$$

Thus, y is symmetric with respect to the origin.

Therefore, the graph of $xy = 8$ is symmetric with respect to the origin.

31. $|y| = |x|$

Replace y with $-y$.

$$|-y| = |x| \Rightarrow |y| = |x|$$

Thus, y is symmetric with respect to the x-axis.

Replace x with $-x$.

$$|y| = |-x| \Rightarrow |y| = |x|$$

Thus, y is symmetric with respect to the y-axis.

Replace x with $-x$ and replace y with $-y$.

$$|-y| = |-x| \Rightarrow |y| = |x|$$

Thus, y is symmetric with respect to the origin.

Therefore, the graph of $|y| = |x|$ is symmetric with respect to the x-axis, the y-axis, and the origin.

32. $|x + y| = 4$

Replace y with $-y$.

$$|x + (-y)| = 4 \Rightarrow |x - y| = 4$$

Thus, y is not symmetric with respect to the x-axis.

Replace x with $-x$.

$$|(-x) + y| = 4 \Rightarrow |y - x| = 4$$

Thus, y is not symmetric with respect to the y-axis.

Replace x with $-x$ and replace y with $-y$.

$$|(-x) + (-y)| = 4 \Rightarrow |-(x + y)| = 4 \Rightarrow |x + y| = 4$$

Thus, y is symmetric with respect to the origin.

Therefore, the graph of $|x + y| = 4$ is symmetric with respect to the origin.

33. center $(3, -4)$, radius 9

34.
$$x^2 + 10x \quad\quad + y^2 + 4y \quad\quad = -20$$
$$x^2 + 10x + 25 + y^2 + 4y + 4 = 25 + 4 - 20$$
$$(x + 5)^2 \quad + (y + 2)^2 \quad = 9$$

center $(-5, \ -2)$, radius 3

35. $(x - 2)^2 + (y + 3)^2 = 5^2$

36. $(x + 5)^2 + (y - 1)^2 = 8^2$, radius $= |-5 - (3)| = 8$

37.

a. $f(1) = 3(1)^2 + 4(1) - 5$
$\quad = 2$

b. $f(-3) = 27 - 12 - 5$
$\quad = 10$

c. $f(t) = 3t^2 + 4t - 5$

d. $f(x + h) = 3(x + h)^2 + 4(x + h) - 5$
$\quad\quad = 3x^2 + 6xh + 3h^2 + 4x + 4h - 5$

e. $3f(t) = 9t^2 + 12t - 15$

f. $f(3t) = 3(3t)^2 + 4(3t) - 5$
$\quad\quad = 27t^2 + 12t - 5$

38.

a. $g(3) = \sqrt{64 - 3^2}$
$\quad = \sqrt{55}$

b. $g(-5) = \sqrt{64 - (-5)^2}$
$\quad\quad = \sqrt{64 - 25}$
$\quad\quad = \sqrt{39}$

c. $g(8) = \sqrt{64 - 8^2}$
$\quad = 0$

d. $g(-x) = \sqrt{64 - (-x)^2}$
$\quad\quad = \sqrt{64 - x^2}$

e. $2g(t) = 2\sqrt{64 - t^2}$

f. $g(2t) = \sqrt{64 - (2t)^2}$
$\quad\quad = \sqrt{64 - 4t^2}$
$\quad\quad = 2\sqrt{16 - t^2}$

39. **a.** $(f \circ g)(3) = f[g(3)]$
$= f[3-8]$
$= f[-5]$
$= (-5)^2 + 4(-5)$
$= 5$

    b. $(g \circ f)(-3) = g[f(-3)]$
$= g[(-3)^2 + 4(-3)]$
$= g[-3]$
$= [-3 - 8]$
$= -11$

    c. $(f \circ g)(x) = f[g(x)]$
$= f[x-8]$
$= (x-8)^2 + 4(x-8)$
$= x^2 - 16x + 64 + 4x - 32$
$= x^2 - 12x + 32$

    d. $(g \circ f)(x) = g[f(x)]$
$= g[x^2 + 4x]$
$= [x^2 + 4x] - 8$
$= x^2 + 4x - 8$

40. **a.** $(f \circ g)(-5) = f[g(-5)]$
$= f\left[\,|(-5)-1|\,\right]$
$= f[6]$
$= 2(6)^2 + 7$
$= 72 + 7$
$= 79$

    b. $(g \circ f)(-5) = g[f(-5)]$
$= g[2(-5)^2 + 7]$
$= g[57]$
$= |57 - 1|$
$= 56$

    c. $(f \circ g)(x) = f[g(x)]$
$= f\left[\,|x-1|\,\right]$
$= 2\left(|x-1|\right)^2 + 7$
$= 2(x-1)^2 + 7$
$= 2(x^2 - 2x + 1) + 7$
$= 2x^2 - 4x + 2 + 7$
$= 2x^2 - 4x + 9$

    d. $(g \circ f)(x) = g[f(x)]$
$= g[2x^2 + 7]$
$= |2x^2 + 7 - 1|$
$= |2x^2 + 6|$
$= 2x^2 + 6$

41. $\dfrac{f(x+h) - f(x)}{h} = \dfrac{4(x+h)^2 - 3(x+h) - 1 - (4x^2 - 3x - 1)}{h}$

$= \dfrac{4x^2 + 8xh + 4h^2 - 3x - 3h - 1 - 4x^2 + 3x + 1}{h}$

$= \dfrac{8xh + 4h^2 - 3h}{h}$

$= 8x + 4h - 3$

42. $\dfrac{g(x+h) - g(x)}{h} = \dfrac{(x+h)^3 - (x+h) - (x^3 - x)}{h}$

$= \dfrac{x^3 + 3x^2h + 3xh^2 + h^3 - x - h - x^3 + x}{h}$

$= \dfrac{3x^2h + 3xh^2 + h^3 - h}{h}$

$= 3x^2 + 3xh + h^2 - 1$

43. $f(x) = -2x^2 + 3$

Domain: All real numbers

44. $f(x) = \sqrt{6-x}$

$6 - x \geq 0$

$-x \geq -6$

$x \leq 6$

Domain: $\{x | x \leq 6\}$

45. $f(x) = \sqrt{25 - x^2}$

$25 - x^2 \geq 0$

$(5-x)(5+x) \geq 0$

Critical values −5 and 5.

Domain: $\{x | -5 \leq x \leq 5\}$

46. $f(x) = \dfrac{3}{x^2 - 2x - 15}$

$x^2 - 2x - 15 = 0$

$(x+3)(x-5) = 0$

$x = -3 \qquad x = 5$

Domain: $\{x | x \neq -3 \text{ and } x \neq 5\}$

47.

f is increasing on $[3, \infty)$

f is decreasing on $(-\infty, 3]$

48.

f is increasing on $[0, \infty)$

f is decreasing on $(-\infty, 0]$

49.

f is increasing on $[-2, 2]$

f is constant on $(-\infty, -2] \cup [2, \infty)$

50.

f is constant on $...,[-6, -5), [-5, -4), [-4, -3),$
$[-3, -2), [-2, -1), [-1, 0), [0, 1),...$

51.

f is increasing on $(-\infty, \infty)$

52.

f is increasing on $(-\infty, \infty)$

53.

a. Domain $\{x | x \text{ is a real number}\}$
   Range $\{y | y \leq 4\}$

b. g is an even function

54.

a. Domain all real numbers
   Range all real numbers

b. g is neither even nor odd

55.

a. Domain all real numbers
   Range $\{y | y \geq 4\}$

b. g is an even function

56.

a. Domain $\{x | -4 \leq x \leq 4\}$
   Range $\{y | 0 \leq y \leq 4\}$

b. g is an even function

57.

a. Domain $\{x \mid x$ is a real number$\}$

Range $\{y \mid y$ is a real number$\}$

b. g is an odd function

58.

a. Domain $\{x \mid x$ is a real number$\}$

Range $\{y \mid y$ is an even integer$\}$

b. g is neither even or odd.

59. $(f+g)(x) = x^2 - 9 + x + 3$

$= x^2 + x - 6$

Domain of $(f+g)(x)$ is $\{x \mid x$ is a real number$\}$.

$(f-g)(x) = x^2 - 9 - (x+3)$

$= x^2 - x - 12$

Domain of $(f-g)(x)$ is $\{x \mid x$ is a real number$\}$.

$(fg)(x) = (x^2 - 9)(x+3)$

$= x^3 + 3x^2 - 9x - 27$

Domain of $(fg)(x)$ is $\{x \mid x$ is a real number$\}$.

$\left(\dfrac{f}{g}\right)(x) = \dfrac{x^2 - 9}{x+3}$

$= x - 3$

Domain of $\left(\dfrac{f}{g}\right)(x)$ is $\{x \mid x \neq -3\}$.

60. $(f+g)(x) = x^3 + 8 + x^2 - 2x + 4$

$= x^3 + x^2 - 2x + 12$

Domain of $(f+g)(x)$ is the set of all real numbers.

$(f-g)(x) = x^3 + 8 - (x^2 - 2x + 4)$

$= x^3 + 8 - x^2 + 2x - 4$

$= x^3 - x^2 + 2x + 4$

Domain of $(f-g)(x)$ is the set of all real numbers.

$(fg)(x) = (x^3 + 8)(x^2 - 2x + 4)$

$= x^5 - 2x^4 + 4x^3 + 8x^2 - 16x + 32$

Domain of $(fg)(x)$ is the set of all real numbers.

$\left(\dfrac{f}{g}\right)(x) = \dfrac{x^3 + 8}{x^2 - 2x + 4}$

$= \dfrac{(x+2)(x^2 - 2x + 4)}{x^2 - 2x + 4}$

$= x + 2$

Domain of $\left(\dfrac{f}{g}\right)(x)$ is the set of all real numbers.

61. $F[G(x)] = 2\left(\dfrac{x+5}{2}\right) - 5$ and $G[F(x)] = \dfrac{(2x-5)+5}{2}$

$= x + 5 - 5$ $\qquad\qquad\qquad\qquad = \dfrac{2x}{2}$

$= x$ $\qquad\qquad\qquad\qquad\qquad\quad = x$

Because $F[G(x)] = x$ and $G[F(x)] = x$ for all real numbers x, F and G are inverses.

62. $h[k(x)] = \sqrt{x^2} = x$ Since $x \geq 0$

$k[h(x)] = \left(\sqrt{x}\right)^2 = x$

Because $h[k(x)] = x$ for all x in the domain of k and $k[h(x)] = x$ for all x in the domain of h, we have shown that h and k are inverses.

63. $l[m(x)] = \dfrac{\dfrac{3}{x-1} + 3}{\dfrac{3}{x-1}}$ $m[l(x)] = m\left[\dfrac{x+3}{x}\right]$

$= \dfrac{3 + 3x - 3}{x-1} \cdot \dfrac{x-1}{3}$ $= \dfrac{3}{\dfrac{x+3}{x} - 1}$

$= \dfrac{3x}{3}$ $= \dfrac{3}{\dfrac{x+3-x}{x}}$

$= x$ $= 3 \cdot \dfrac{x}{3} = x$

Thus, l and m are inverse functions.

64. $p[q(x)] = \dfrac{\dfrac{2x}{x-5} - 5}{2\left(\dfrac{2x}{x-5}\right)}$

$= \dfrac{\dfrac{2x-5x+25}{x-5}}{\dfrac{4x}{x-5}}$

$= \dfrac{-3x+25}{x-5} \cdot \dfrac{x-5}{4x}$

$= \dfrac{-3x+25}{4x}$

Thus, p and q are not inverse functions.

65. $f(x) = 3x - 4$

$y = 3x - 4$

$x = 3y - 4$

$x + 4 = 3y$

$\dfrac{x+4}{3} = y$

Thus $f^{-1}(x) = \dfrac{x+4}{3} = \dfrac{1}{3}x + \dfrac{4}{3}$.

66. $g(x) = -2x + 3$

$y = -2x + 3$

$x = -2y + 3$

$x - 3 = -2y$

$\dfrac{x-3}{-2} = y$

$\dfrac{3-x}{2} = y$

Thus, $g^{-1}(x) = \dfrac{3-x}{2} = -\dfrac{1}{2}x + \dfrac{3}{2}$.

67. $h(x) = -\dfrac{1}{2}x - 2$

$y = -\dfrac{1}{2}x - 2$

$x = -\dfrac{1}{2}y - 2$

$x + 2 = -\dfrac{1}{2}y$

$-2x - 4 = y$

Thus, $h^{-1}(x) = -2x - 4$.

68. $k(x) = \dfrac{1}{x}$

$y = \dfrac{1}{x}$

$x = \dfrac{1}{y}$

$y = \dfrac{1}{x}$

Thus $k^{-1}(x) = k(x) = \dfrac{1}{x}$.

both k
and k^{-1}

69. Let $x =$ one of the numbers and $50 - x =$ the other number. Their product y is given by

$$y = x(50 - x) = 50x - x^2 = x^2 + 50x$$

Now y takes on its maximum value when

$$x = \frac{-b}{2a} = \frac{-50}{2(-1)} = 25$$

Thus, the two numbers are 25 and $(50 - 25) = 25$. That is, both numbers are 25.

70. Let $x =$ the smaller number. Let $x + 10 =$ the larger number. The sum of their squares y is given by

$$y = x^2 + (x + 10)^2$$
$$= x^2 + x^2 + 20x + 100$$
$$= 2x^2 + 20x + 100$$

Now y takes on its minimum value when

$$x = \frac{-b}{2a} = \frac{-20}{2(2)} = -5$$

Thus, the numbers are -5 and $(-5 + 10) = 5$.

71. $h(t) = -16t^2 + 220$ and $h(t) = 0$ when

$$-16t^2 + 220 = 0$$
$$220 = 16t^2$$
$$\frac{220}{16} = t^2$$
$$\sqrt{\frac{220}{16}} = t$$
$$\frac{2\sqrt{55}}{4} = t$$

Thus, $t \approx 3.7$ seconds.

72. **a.** $h(1050) = \dfrac{1}{8820}(1050)^2 + 25$

$= \dfrac{1,102,500}{8,820} + 25$

$= 125 + 25$

$= 150$ feet

b. $h(2100) = \dfrac{1}{8820}(2100)^2 + 25$

$= \dfrac{4,410,000}{8,820} + 25$

$= 500 + 25$

$= 525$ feet

73. **a.** Enter the data on your calculator. The technique for a TI-83 is illustrated here.

$y = 0.018024687x + 0.00050045744$

b. Yes. $r \approx 0.999$, which is very close to 1.

c. $y = 0.018024687(100) + 0.00050045744$

$= 1.8024687 + 0.00050045744$

≈ 1.8 seconds

74. **a.** Enter the data on your calculator. The technique for a TI-83 is illustrated here.

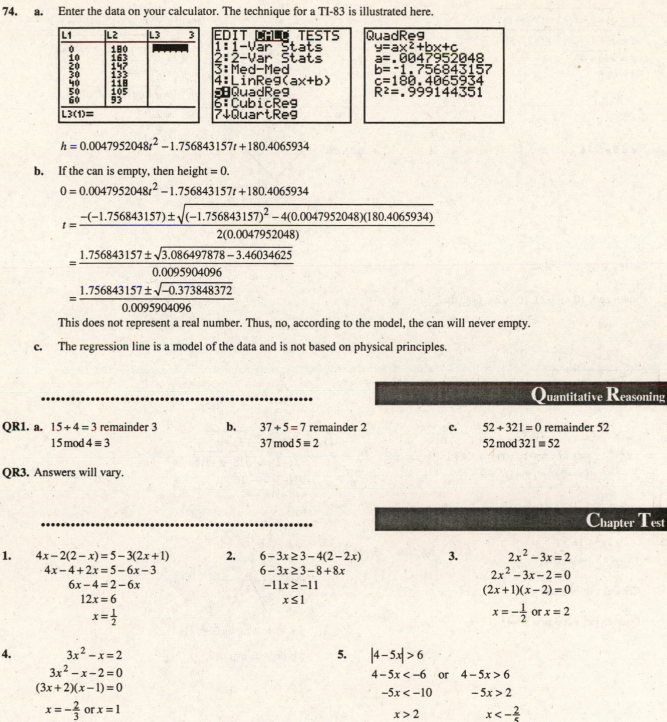

$$h = 0.0047952048t^2 - 1.756843157t + 180.4065934$$

b. If the can is empty, then height = 0.

$$0 = 0.0047952048t^2 - 1.756843157t + 180.4065934$$

$$t = \frac{-(-1.756843157) \pm \sqrt{(-1.756843157)^2 - 4(0.0047952048)(180.4065934)}}{2(0.0047952048)}$$

$$= \frac{1.756843157 \pm \sqrt{3.086497878 - 3.46034625}}{0.0095904096}$$

$$= \frac{1.756843157 \pm \sqrt{-0.373848372}}{0.0095904096}$$

This does not represent a real number. Thus, no, according to the model, the can will never empty.

c. The regression line is a model of the data and is not based on physical principles.

●   **Q**uantitative **R**easoning

QR1. a. $15 \div 4 = 3$ remainder 3   **b.** $37 \div 5 = 7$ remainder 2   **c.** $52 \div 321 = 0$ remainder 52
$15 \bmod 4 \equiv 3$ $37 \bmod 5 \equiv 2$ $52 \bmod 321 \equiv 52$

QR3. Answers will vary.

●   **C**hapter **T**est

1. $4x - 2(2 - x) = 5 - 3(2x + 1)$ **2.** $6 - 3x \geq 3 - 4(2 - 2x)$ **3.** $2x^2 - 3x = 2$
$4x - 4 + 2x = 5 - 6x - 3$ $6 - 3x \geq 3 - 8 + 8x$ $2x^2 - 3x - 2 = 0$
$6x - 4 = 2 - 6x$ $-11x \geq -11$ $(2x + 1)(x - 2) = 0$
$12x = 6$ $x \leq 1$ $x = -\frac{1}{2}$ or $x = 2$
$x = \frac{1}{2}$

4. $3x^2 - x = 2$ **5.** $|4 - 5x| > 6$
$3x^2 - x - 2 = 0$ $4 - 5x < -6$ or $4 - 5x > 6$
$(3x + 2)(x - 1) = 0$ $-5x < -10$ $-5x > 2$
$x = -\frac{2}{3}$ or $x = 1$ $x > 2$ $x < -\frac{2}{5}$

$$\left(-\infty, -\frac{2}{5}\right) \cup (2, \infty)$$

6.
$$d = \sqrt{(x_2 - x_1)^2 + (y_2 - y_1)^2}$$
$$d = \sqrt{[4-(-2)]^2 + (-2-5)^2}$$
$$d = \sqrt{36+49}$$
$$d = \sqrt{85}$$

7.
$$x_m = \frac{4-2}{2} = 1 \qquad \text{length} = \sqrt{(x_2 - x_1)^2 + (y_2 - y_1)^2}$$
$$y_m = \frac{-1+3}{2} = 1 \qquad\qquad = \sqrt{[4-(-2)]^2 + (-1-3)^2}$$
$$\qquad\qquad\qquad = \sqrt{6^2 + (-4)^2}$$
$$\text{midpoint} = (1,\ 1) \qquad\quad = \sqrt{36+16} = \sqrt{52}$$
$$\qquad\qquad\qquad = 2\sqrt{13}$$

8.
$$x = 2y^2 - 4$$

$$x = 0 \qquad 2y^2 - 4 = 0$$
$$2y^2 = 4$$
$$y^2 = 2$$
$$y = \pm\sqrt{2}$$
If $y = 0$, $\qquad x = -4$

intercepts $(0,\ -\sqrt{2})$, $(0,\ \sqrt{2})$, $(-4,\ 0)$

9.

10.
$$x^2 - 4x + y^2 + 2y - 4 = 0$$
$$(x^2 - 4x + 4) + (y^2 + 2y + 1) = 4 + 4 + 1$$
$$(x-2)^2 + (y+1)^2 = 9$$

center: $(2, -1)$ radius: 3

11.
$$f(x) = -\sqrt{25 - x^2}$$
$$f(-3) = -\sqrt{25 - (-3)^2}$$
$$f(-3) = -\sqrt{16}$$
$$f(-3) = -4$$

12.
$$x^2 - 16 \ge 0$$
$$(x-4)(x+4) \ge 0$$

Critical values 4 and −4.

Domain $\{x \mid x \ge 4 \text{ or } x \le -4\}$

13.

f is increasing on $(-\infty, 2]$

f is decreasing on $[2, \infty)$

14. First shift the graph of $f(x)$ horizontally 2 units to the left. Next, reflect the graph across the x-axis. Finally, shift the graph vertically down 1 unit.

15.

a.
$$f(-x) = (-x)^4 - (-x)^2$$
$$= x^4 - x^2$$
$$= f(x)$$

The function of $f(x) = x^4 - x^2$ is an even function.

b.
$$f(-x) = (-x)^3 - (-x)$$
$$= -x^3 + x$$
$$= -(x^3 - x)$$
$$= -f(x)$$

The function $f(x) = x^3 - x$ is an odd function.

c. $f(-x) = -x - 1$

The function $f(x) = x - 1$ is neither an even nor an odd function.

Thus, only **b** defines an odd function.

16.
$$(f+g)(x) = (x^2 - 1) + (x - 2)$$
$$= x^2 + x - 3$$
$$\left(\frac{f}{g}\right)(x) = \frac{x^2 - 1}{x - 2}, \quad x \neq 2$$

17.
$$\frac{f(x+h) - f(x)}{h} = \frac{\left[(x+h)^2 + 1\right] - (x^2 + 1)}{h}$$
$$= \frac{x^2 + 2xh + h^2 + 1 - x^2 - 1}{h}$$
$$= \frac{2xh + h^2}{h}$$
$$= 2x + h$$

18.
$$(f \circ g)(x) = (\sqrt{x-2})^2 - 2\sqrt{x-2} + 1$$
$$= x - 2 - 2\sqrt{x-2} + 1$$
$$= x - 2\sqrt{x-2} - 1$$

19. $y = \dfrac{x}{x+1}$

Interchange x and y. Then solve for y.
$$x = \frac{y}{y+1}$$
$$x(y+1) = y$$
$$xy + x = y$$
$$xy - y = -x$$
$$y(x-1) = -x$$
$$y = \frac{-x}{x-1} = \frac{x}{1-x}$$
$$f^{-1}(x) = \frac{x}{1-x}$$

20.

a. Enter the data on your calculator. The technique for a TI-83 is illustrated here.

$$y = -7.98245614x + 767.122807$$

b. $y = -7.98245614(89) + 767.122807$

≈ 57 Calories

Chapter 2
Trigonometric Functions

1. $90° - 15° = 75°$

$180° - 15° = 165°$

3.
$$\begin{array}{cc} 90° & 89°60' \\ \underline{-70°15'} = \underline{-70°15'} \\ & 19°45' \end{array} \qquad \begin{array}{cc} 180° & 179°60' \\ \underline{-70°15'} = \underline{-70°15'} \\ & 109°45' \end{array}$$

5.
$$\begin{array}{cc} 90° & 89°59'60'' \\ \underline{-56°33'15''} = \underline{-56°33'15''} \\ & 33°26'45'' \end{array} \qquad \begin{array}{cc} 180° & 179°59'60'' \\ \underline{-56°33'15''} = \underline{-56°33'15''} \\ & 123°26'45'' \end{array}$$

7. $\dfrac{\pi}{2} - 1$

$\pi - 1$

9. $\dfrac{\pi}{2} - \dfrac{\pi}{4} = \dfrac{\pi}{4}$

$\pi - \dfrac{\pi}{4} = \dfrac{3\pi}{4}$

11. $\dfrac{\pi}{2} - \dfrac{2\pi}{5} = \dfrac{\pi}{10}$

$\pi - \dfrac{2\pi}{5} = \dfrac{3\pi}{5}$

13. $610° = 250° + 360°$

α is a quadrant III angle coterminal with an angle of measure 250°.

15. $-975° = 105° - 3 \cdot 360°$

α is a quadrant II angle coterminal with an angle of measure 105°.

17. $2456° = 296° + 6 \cdot 360°$

α is a quadrant IV angle coterminal with an angle of measure 296°.

19. On a TI-83 graphing calculator, the degree symbol, °, and the DMS function are located in the ANGLE menu.

$24.56° = 24°33'36''$

21. On a TI-83 graphing calculator, the degree symbol, °, and the DMS function are located in the ANGLE menu.

$64.158° = 64°9'28.8''$

23. On a TI-83 graphing calculator, the degree symbol, °, and the DMS function are located in the ANGLE menu.

$3.402° = 3°24'7.2''$

25. A TI-83 calculator needs to be in degree mode to convert a DMS measure to its equivalent degree measure. On a TI-83 both the degree symbol, °, and the minute symbol, ', are located in the ANGLE menu. The second symbol, '', is entered by pressing ALPHA followed by ['] which is located on the plus sign, [+], key.

$25°25'12'' = 25.42°$

27. A TI-83 calculator needs to be in degree mode to convert a DMS measure to its equivalent degree measure. On a TI-83 both the degree symbol, °, and the minute symbol, ', are located in the ANGLE menu. The second symbol, ", is entered by pressing ALPHA followed by ["] which is located on the plus sign, [+], key.

```
183°33'36"
              183.56
```

$183°33'36'' = 183.56°$

29. A TI-83 calculator needs to be in degree mode to convert a DMS measure to its equivalent degree measure. On a TI-83 both the degree symbol, °, and the minute symbol, ', are located in the ANGLE menu. The second symbol, ", is entered by pressing ALPHA followed by ["] which is located on the plus sign, [+], key.

```
211°46'48"
              211.78
```

$211°46'48'' = 211.78°$

31. $30° = 30°\left(\dfrac{\pi}{180°}\right) = \dfrac{\pi}{6}$

33. $90° = 90°\left(\dfrac{\pi}{180°}\right) = \dfrac{\pi}{2}$

35. $165° = 165°\left(\dfrac{\pi}{180°}\right) = \dfrac{11\pi}{12}$

37. $420° = 420°\left(\dfrac{\pi}{180°}\right) = \dfrac{7\pi}{3}$

39. $585° = 585°\left(\dfrac{\pi}{180°}\right) = \dfrac{13\pi}{4}$

41. $-9° = -9°\left(\dfrac{\pi}{180°}\right) = -\dfrac{\pi}{20}$

43. $\dfrac{7\pi}{3} = \dfrac{7\pi}{3}\left(\dfrac{180°}{\pi}\right) = 420°$

45. $\dfrac{\pi}{5} = \dfrac{\pi}{5}\left(\dfrac{180°}{\pi}\right) = 36°$

47. $\dfrac{\pi}{6} = \dfrac{\pi}{6}\left(\dfrac{180°}{\pi}\right) = 30°$

49. $\dfrac{3\pi}{8} = \dfrac{3\pi}{8}\left(\dfrac{180°}{\pi}\right) = 67.5°$

51. $\dfrac{11\pi}{3} = \dfrac{11\pi}{3}\left(\dfrac{180°}{\pi}\right) = 660°$

53. $-\dfrac{5\pi}{12} = -\dfrac{5\pi}{12}\left(\dfrac{180°}{\pi}\right) = -75°$

55. $1.5 = 1.5\left(\dfrac{180°}{\pi}\right) \approx 85.94°$

57. $133° = 133°\left(\dfrac{\pi}{180°}\right) \approx 2.32$

59. $8.25 = 8.25\left(\dfrac{180°}{\pi}\right) \approx 472.69°$

61. $\theta = \dfrac{s}{r}$

$= \dfrac{8}{2} = 4$

$= 4\left(\dfrac{180°}{\pi}\right) \approx 229.18°$

63. $\theta = \dfrac{s}{r}$

$= \dfrac{12.4}{5.2} \approx 2.38$

$= 2.38\left(\dfrac{180°}{\pi}\right) \approx 136.63°$

65. $s = r\theta$

$= (8)\dfrac{\pi}{4}$

≈ 6.28 in.

67. $s = r\theta$

$= 25 \cdot (42°)\left(\dfrac{\pi}{180°}\right)$

≈ 18.33 cm

69. $\theta = \dfrac{3}{2}(2\pi)$

$= 3\pi$

71. $\theta_2 = \dfrac{r_1}{r_2}\theta_1$

$= \dfrac{14}{28}(150°)\left(\dfrac{\pi}{180°}\right)$

$= \dfrac{5\pi}{12}$ radians or $75°$

73. $\omega = \dfrac{\theta}{t}$

$= \dfrac{2\pi}{60}$

$= \dfrac{\pi}{30}$ radian/sec

75. $\omega = \dfrac{\theta}{t}$

$= \dfrac{50(2\pi)}{60}$

$= \dfrac{5\pi}{3}$ radians/sec

77. $\omega = \dfrac{\theta}{t}$

$= \dfrac{2\pi\left(33\frac{1}{3}\right)}{60}$

$= \dfrac{10\pi}{9}$ radians/sec

≈ 3.49 radians per second

79. $v = \omega r$

$$= \frac{450 \cdot 2\pi \cdot 60 \cdot 15}{12 \cdot 5280}$$

$$\approx 40 \text{ mph}$$

81.
$$r_1\theta_1 = r_2\theta_2$$
$$(3.5)(150 \cdot 2\pi) = (1.75)\theta_2$$
$$600\pi = \theta_2$$
$$300(2\pi) = \theta_2$$

The rear gear is making 300 revolutions.
The rear gear and tire are making same
number of revolutions
Tire is 12 inches = 1 ft
$s = 1 \text{ ft}(300)(2\pi) = 1885 \text{ ft}$

83.   Speed for outer ring: Speed for inner ring:

$v = \dfrac{s}{t}$ $v = \dfrac{s}{t}$

$= \dfrac{2\pi(32)}{3.75} \cdot \dfrac{60^2}{5280}$ $= \dfrac{2\pi(38)}{3.75} \cdot \dfrac{60^2}{5280}$

$\approx 36.5567 \text{ mph}$ $\approx 43.4111 \text{ mph}$

The outer swing has a greater speed of
$43.4111 - 36.5567 \approx 6.9 \text{ mph}$.

85. $s = r\theta$

$$= 93,000,000(31')\left(\frac{1°}{60'}\right)\left(\frac{\pi}{180°}\right)$$

$$\approx 840,000 \text{ mi}$$

87.   **a.** $\omega = \dfrac{\theta}{t}$

$$= \frac{2\pi}{1.61 \text{ hours}}$$

$$\approx 3.9 \text{ radians per hour}$$

     b. $v = \dfrac{s}{t}$

$$= \frac{2\pi r}{1.61 \text{ hours}}$$

$$= \frac{2\pi(625 \text{ km} + 6370 \text{ km})}{1.61 \text{ hours}}$$

$$= \frac{2\pi(6995 \text{ km})}{1.61 \text{ hours}}$$

$$\approx 27,300 \text{ km per hour}$$

89.   **a.**   When the rear tire makes one revolution, the bicycle travels $s = r\theta = 2\pi r = 2\pi(30 \text{ inches}) = 60\pi$ inches.

The angular velocity of point A is

$$\omega = \frac{\theta}{t} = \frac{2\pi}{t} = 2\left(\frac{\pi}{t}\right).$$

When the bicycle travels 60π inches, point B on the front tire travels though an angle of

$$\theta = \frac{s}{r} = \frac{60\pi \text{ inches}}{20 \text{ inches}} = 3\pi.$$

The angular velocity of point B is

$$\omega = \frac{\theta}{t} = \frac{3\pi}{t} = 3\left(\frac{\pi}{t}\right).$$

Thus, point B has the greater angular velocity.

     b.   Point A and point B travel a linear distance of 60π inches in the same amount of time. Therefore, both points have the same linear velocity.

91.   **a.** $1' = 1' \cdot \left(\dfrac{1°}{60'}\right) \cdot \left(\dfrac{\pi}{180°}\right) = \dfrac{\pi}{10,800}$ radians

1 nautical mile $= s = r\theta = (3960 \text{ statute miles})\left(\dfrac{\pi}{10,800}\right) \approx 1.15$ statute miles

     b.   Earth's circumference $= 2\pi r \approx 2\pi(3960 \text{ statute miles})\left(\dfrac{1 \text{ nautical mile}}{1.15 \text{ statute miles}}\right) \approx \dfrac{7920\pi}{1.15}$ nautical miles

The question, then, is what percent of $\dfrac{7920\pi}{1.15}$ is 2217.

$$\frac{2217}{7920\pi/1.15} \approx 0.10 = 10\%$$

93.
$$A = \frac{1}{2}r^2\theta$$
$$= \frac{1}{2}(5^2)\left(\frac{\pi}{3}\right)$$
$$\approx 13 \text{ in}^2$$

95.
$$A = \frac{1}{2}r^2\theta$$
$$= \frac{1}{2}(120)^2\, 0.65$$
$$= 4680 \text{ cm}^2$$

97.
$$25°47' = 25° + 47'\left(\frac{1°}{60'}\right)$$
$$= 25\frac{47}{60}°$$

Convert to radians.

$$25\frac{47}{60}° \cdot \frac{\pi}{180°} = \frac{1547\pi}{10,800} \text{ radians}$$
$$s = r\theta$$
$$s = 3960\left(\frac{1547\pi}{10,800}\right)$$
$$\approx 1780$$

To the nearest 10 miles, Miami is 1780 miles north of the equator.

PS1. $\dfrac{1}{\sqrt{3}} \cdot \dfrac{\sqrt{3}}{\sqrt{3}} = \dfrac{\sqrt{3}}{3}$

PS2. $\dfrac{2}{\sqrt{2}} \cdot \dfrac{\sqrt{2}}{\sqrt{2}} = \dfrac{2\sqrt{2}}{2} = \sqrt{2}$

PS3. $a \div \left(\dfrac{a}{2}\right) = a \cdot \left(\dfrac{2}{a}\right) = 2$

PS4.
$$\left(\frac{a}{2}\right) \div \left(\frac{\sqrt{3}}{2}a\right) = \left(\frac{a}{2}\right) \cdot \left(\frac{2}{a\sqrt{3}}\right)$$
$$= \frac{1}{\sqrt{3}} \cdot \frac{\sqrt{3}}{\sqrt{3}}$$
$$= \frac{\sqrt{3}}{3}$$

PS5.
$$\frac{\sqrt{2}}{2} = \frac{x}{5}$$
$$\frac{5\sqrt{2}}{2} = x$$
$$x \approx 3.54$$

PS6.
$$\frac{\sqrt{3}}{3} = \frac{x}{18}$$
$$\frac{18\sqrt{3}}{3} = x$$
$$x = 6\sqrt{3}$$
$$x \approx 10.39$$

Section 2.2

1.
$$r = \sqrt{5^2 + 12^2}$$
$$r = \sqrt{25 + 144} + \sqrt{169}$$
$$r = 13$$

$\sin\theta = \dfrac{y}{r} = \dfrac{12}{13}$ $\csc\theta = \dfrac{r}{y} = \dfrac{13}{12}$

$\cos\theta = \dfrac{x}{r} = \dfrac{5}{13}$ $\sec\theta = \dfrac{r}{x} = \dfrac{13}{5}$

$\tan\theta = \dfrac{y}{x} = \dfrac{12}{5}$ $\cot\theta = \dfrac{x}{y} = \dfrac{5}{12}$

3.
$$x = \sqrt{7^2 - 4^2}$$
$$x = \sqrt{49 - 16} = \sqrt{33}$$

$\sin\theta = \dfrac{y}{r} = \dfrac{4}{7}$ $\csc\theta = \dfrac{r}{y} = \dfrac{7}{4}$

$\cos\theta = \dfrac{x}{r} = \dfrac{\sqrt{33}}{7}$ $\sec\theta = \dfrac{r}{x} = \dfrac{7}{\sqrt{33}} = \dfrac{7\sqrt{33}}{33}$

$\tan\theta = \dfrac{y}{x} = \dfrac{4}{\sqrt{33}} = \dfrac{4\sqrt{33}}{33}$ $\cot\theta = \dfrac{x}{y} = \dfrac{\sqrt{33}}{4}$

5. $r = \sqrt{2^2 + 5^2}$

$r = \sqrt{4 + 25} = \sqrt{29}$

$\sin\theta = \dfrac{y}{r} = \dfrac{5}{\sqrt{29}} = \dfrac{5\sqrt{29}}{29}$

$\cos\theta = \dfrac{x}{r} = \dfrac{2}{\sqrt{29}} = \dfrac{2\sqrt{29}}{29}$

$\tan\theta = \dfrac{y}{x} = \dfrac{5}{2}$

$\csc\theta = \dfrac{r}{y} = \dfrac{\sqrt{29}}{5}$

$\sec\theta = \dfrac{r}{x} = \dfrac{\sqrt{29}}{2}$

$\cot\theta = \dfrac{x}{y} = \dfrac{2}{5}$

7. $x = \sqrt{2 + \left(\sqrt{3}\right)^2}$

$x = \sqrt{4 + 3} = \sqrt{7}$

$\sin\theta = \dfrac{y}{r} = \dfrac{\sqrt{3}}{\sqrt{7}} = \dfrac{\sqrt{21}}{7}$

$\cos\theta = \dfrac{x}{r} = \dfrac{2}{\sqrt{7}} = \dfrac{2\sqrt{7}}{7}$

$\tan\theta = \dfrac{y}{x} = \dfrac{\sqrt{3}}{2}$

$\csc\theta = \dfrac{r}{y} = \dfrac{\sqrt{7}}{\sqrt{3}} = \dfrac{\sqrt{21}}{3}$

$\sec\theta = \dfrac{r}{x} = \dfrac{\sqrt{7}}{2}$

$\cot\theta = \dfrac{x}{y} = \dfrac{2}{\sqrt{3}} = \dfrac{2\sqrt{3}}{3}$

9. opposite side $= \sqrt{6^2 - 3^2}$

$= \sqrt{36 - 9}$

$= \sqrt{27} = 3\sqrt{3}$

$\sin\theta = \dfrac{\text{opp}}{\text{hyp}} = \dfrac{3\sqrt{3}}{6} = \dfrac{\sqrt{3}}{2}$

$\cos\theta = \dfrac{\text{adj}}{\text{hyp}} = \dfrac{3}{6} = \dfrac{1}{2}$

$\tan\theta = \dfrac{\text{opp}}{\text{adj}} = \dfrac{3\sqrt{3}}{3} = \sqrt{3}$

$\csc\theta = \dfrac{\text{hyp}}{\text{opp}} = \dfrac{6}{3\sqrt{3}} = \dfrac{2}{\sqrt{3}} = \dfrac{2\sqrt{3}}{3}$

$\sec\theta = \dfrac{\text{hyp}}{\text{adj}} = \dfrac{6}{3} = 2$

$\cot\theta = \dfrac{\text{adj}}{\text{opp}} = \dfrac{3}{3\sqrt{3}} = \dfrac{1}{\sqrt{3}} = \dfrac{\sqrt{3}}{3}$

11. hypotenuse $= \sqrt{5^2 + 6^2}$

$= \sqrt{25 + 36}$

$= \sqrt{61}$

$\sin\theta = \dfrac{\text{opp}}{\text{hyp}} = \dfrac{6}{\sqrt{61}} = \dfrac{6\sqrt{61}}{61}$

$\cos\theta = \dfrac{\text{adj}}{\text{hyp}} = \dfrac{5}{\sqrt{61}} = \dfrac{5\sqrt{61}}{61}$

$\tan\theta = \dfrac{\text{opp}}{\text{adj}} = \dfrac{6}{5}$

$\csc\theta = \dfrac{\text{hyp}}{\text{opp}} = \dfrac{\sqrt{61}}{6}$

$\sec\theta = \dfrac{\text{hyp}}{\text{adj}} = \dfrac{\sqrt{61}}{5}$

$\cot\theta = \dfrac{\text{adj}}{\text{opp}} = \dfrac{5}{6}$

For exercises 13 to 15, since $\sin\theta = \dfrac{y}{r} = \dfrac{3}{5}$, $y = 3$, $r = 5$, and $x = \sqrt{5^2 - 3^2} = 4$.

13. $\tan\theta = \dfrac{y}{x} = \dfrac{3}{4}$

15. $\cos\theta = \dfrac{x}{r} = \dfrac{4}{5}$

For exercises 16 to 18, since $\tan\theta = \dfrac{y}{x} = \dfrac{4}{3}$, $y = 4$, $x = 3$, and $r = \sqrt{3^2 + 4^2} = 5$.

17. $\cot\theta = \dfrac{x}{y} = \dfrac{3}{4}$

For exercises 19 to 21, since $\sec\beta = \dfrac{r}{x} = \dfrac{13}{12}$, $r = 13$, $x = 12$, and $y = \sqrt{13^2 - 12^2} = \sqrt{25} = 5$.

19. $\cos\beta = \dfrac{x}{r} = \dfrac{12}{13}$

21. $\csc\beta = \dfrac{r}{y} = \dfrac{13}{5}$

For exercises 22 to 24, since $\cos\theta \dfrac{x}{r} = \dfrac{2}{3}$, $x = 2$, $r = 3$, and $y = \sqrt{3^2 - 2^2} = \sqrt{9 - 4} = \sqrt{5}$.

23. $\sec\theta = \dfrac{r}{x} = \dfrac{3}{2}$

25. $\sin 45° + \cos 45° = \dfrac{\sqrt{2}}{2} + \dfrac{\sqrt{2}}{2} = \sqrt{2}$

27. $\sin 30° \cos 60° - \tan 45° = \dfrac{1}{2} \cdot \dfrac{1}{2} - 1$

$= \dfrac{1}{4} - 1$

$= -\dfrac{3}{4}$

29. $\sin 30° \cos 60° + \tan 45° = \dfrac{1}{2} \cdot \dfrac{1}{2} + 1 = \dfrac{1}{4} + 1 = \dfrac{5}{4}$

31. $\sin \dfrac{\pi}{3} + \cos \dfrac{\pi}{6} = \dfrac{\sqrt{3}}{2} + \dfrac{\sqrt{3}}{2} = 2 \cdot \dfrac{\sqrt{3}}{2} = \sqrt{3}$

33. $\sin \dfrac{\pi}{4} + \tan \dfrac{\pi}{6} = \dfrac{\sqrt{2}}{2} + \dfrac{\sqrt{3}}{3} = \dfrac{3\sqrt{2} + 2\sqrt{3}}{6}$

35. $\sec \dfrac{\pi}{3} \cos \dfrac{\pi}{3} - \tan \dfrac{\pi}{6} = 2 \cdot \dfrac{1}{2} - \dfrac{\sqrt{3}}{3} = 1 - \dfrac{\sqrt{3}}{3} = \dfrac{3 - \sqrt{3}}{3}$

37. $2 \csc \dfrac{\pi}{4} - \sec \dfrac{\pi}{3} \cos \dfrac{\pi}{6} = 2 \cdot \sqrt{2} - 2 \cdot \dfrac{\sqrt{3}}{2} = 2\sqrt{2} - \sqrt{3}$

39. $\tan 32° \approx 0.6249$ **41.** $\cos 63°20' \approx 0.4488$

43. $\cos 34.7° \approx 0.8221$ **45.** $\sec 5.9° \approx 1.0053$

47. $\tan \dfrac{\pi}{7} \approx 0.4816$

49. $\csc 1.2 \approx 1.0729$

51.

$\sin 52° = \dfrac{h}{12}$

$h = 12 \sin 52°$

$h \approx 9.5 \text{ ft}$

53.

$\tan 19° = \dfrac{32.0}{x}$

$x = \dfrac{32.0}{\tan 19°} = 92.9 \text{ in.}$

55.

$\cos 38° = \dfrac{d + 0.33}{6}$

$d = 6 \cos 38° + 0.33$

$d \approx 5.1 \text{ ft}$

57.

$d = \dfrac{240 \text{ mi}}{\text{hr}} \left(\dfrac{1 \text{ hr}}{60 \text{ min}} \right) 4 \text{ min}$

$d = 16 \text{ mi}$

$\sin 6° = \dfrac{h}{d}$

$h = 16 \sin 6°$

$h \approx 1.7 \text{ mi}$

59.

$$\tan 27.8° = \frac{x}{131}$$
$$x = 131\tan 27.8° = 69.1 \text{ ft}$$
$$\text{height} = 69.1 + 5.5 = 74.6 \text{ ft}$$

61.

$$\sin 0.056° = \frac{670,900}{d}$$
$$d = \frac{670,900}{\sin 0.056°} \text{ km}$$
$$\approx 686,000,000 \text{ km}$$

63.
$$A = \frac{1}{2}bh$$
$$= \frac{1}{2}(2a\sin\theta)(a\cos\theta)$$
$$= a^2\sin\theta\cos\theta$$

65.

$$\tan 36.4 = \frac{d}{350+x} \qquad \tan 51.9° = \frac{d}{x}$$
$$x = \frac{d}{\tan 51.9°}$$
$$\tan 36.4° = \frac{d}{350 + \dfrac{d}{\tan 51.9°}}$$
$$d = \frac{350\tan 36.4°}{1 - \dfrac{\tan 36.4°}{\tan 51.9°}}$$
$$d \approx 612 \text{ ft}$$

67.

$$\tan 42° = \frac{h}{x} \qquad\qquad \tan 37.8° = \frac{h}{100+x}$$
$$x = \frac{h}{\tan 42°} \qquad\qquad \tan 37.8° = \frac{h}{100 + \dfrac{h}{\tan 42°}}$$
$$h = \frac{100\tan 37.8°}{1 - \dfrac{\tan 37.8°}{\tan 42.0°}}$$
$$h \approx 5.60 \times 10^2 \text{ ft}$$
$$h \approx 560 \text{ ft}$$

69. a.

$$\tan 53.6° = \frac{h}{412}$$
$$h = 412 \tan 53.6°$$
$$h \approx 559 \text{ feet}$$

b. $(AC)^2 = 412^2 + 559^2$
$$AC = \sqrt{412^2 + 559^2}$$
$$AC = \sqrt{482,225}$$
$$AC \approx 694.4 \text{ feet}$$

$$\tan 15.5° = \frac{x}{694.4}$$
$$x = 694.4 \tan 15.5°$$
$$x \approx 193 \text{ feet}$$

••

71.

Consider the right triangle formed by A, B and the midpoint of AC.

$$r = \sqrt{6^2 - 3^2}$$
$$r = \sqrt{27}$$
$$r = 3\sqrt{3}$$
$$r \approx 5.2 \text{ m}$$

73.

if $\theta = 45°$ $d_1 = d_2$
$$\sin \theta = \frac{3}{d_1}$$
$$d_1 = \frac{3}{\sin 45°}$$
$$d = 2d_1 = \frac{6}{\sin 45°}$$
$$d \approx 8.5 \text{ ft}$$

•••••••••••••••••••••••••••••••••••

PS1. $-\dfrac{4}{3}$

PS2. $\dfrac{5}{2\sqrt{5}} \cdot \dfrac{\sqrt{5}}{\sqrt{5}} = \dfrac{5\sqrt{5}}{10} = \dfrac{\sqrt{5}}{2}$

PS3. $|120 - 180| = |-60| = 60$

PS4. $2\pi - \dfrac{9\pi}{5} = \dfrac{10\pi}{5} - \dfrac{9\pi}{5} = \dfrac{\pi}{5}$

PS5. $\dfrac{3}{2}\pi - \dfrac{1}{2}\pi = \dfrac{2}{2}\pi = \pi$

PS6. $\sqrt{(-3)^2 + (-5)^2} = \sqrt{9 + 25} = \sqrt{34}$

Section 2.3

1. $x = 2$, $y = 3$, $r = \sqrt{2^2 + 3^2} = \sqrt{13}$

$\sin\theta = \dfrac{y}{r} = \dfrac{3}{\sqrt{13}} = \dfrac{3\sqrt{13}}{13}$ $\csc\theta = \dfrac{\sqrt{13}}{3}$

$\cos\theta = \dfrac{x}{r} = \dfrac{2}{\sqrt{13}} = \dfrac{2\sqrt{13}}{13}$ $\sec\theta = \dfrac{\sqrt{13}}{2}$

$\tan\theta = \dfrac{y}{x} = \dfrac{3}{2}$ $\cot\theta = \dfrac{2}{3}$

3. $x = -2$, $y = 3$, $r = \sqrt{(-2)^2 + (3)^2} = \sqrt{13}$

$\sin\theta = \dfrac{y}{r} = \dfrac{3}{\sqrt{13}} = \dfrac{3\sqrt{13}}{13}$ $\csc\theta = \dfrac{\sqrt{13}}{3}$

$\cos\theta = \dfrac{x}{r} = \dfrac{-2}{\sqrt{13}} = -\dfrac{2\sqrt{13}}{13}$ $\sec\theta = -\dfrac{\sqrt{13}}{2}$

$\tan\theta = \dfrac{y}{x} = \dfrac{3}{-2} = -\dfrac{3}{2}$ $\cot\theta = -\dfrac{2}{3}$

5. $x = -8$, $y = -5$, $c = \sqrt{(-8)^2 + (-5)^2} = \sqrt{89}$

$\sin\theta = \dfrac{y}{r} = \dfrac{-5}{\sqrt{89}} = -\dfrac{5\sqrt{89}}{89}$ $\csc\theta = -\dfrac{\sqrt{89}}{5}$

$\cos\theta = \dfrac{x}{r} = \dfrac{-8}{\sqrt{89}} = -\dfrac{8\sqrt{89}}{89}$ $\sec\theta = -\dfrac{\sqrt{89}}{8}$

$\tan\theta = \dfrac{y}{x} = \dfrac{-5}{-8} = \dfrac{5}{8}$ $\cot\theta = \dfrac{8}{5}$

7. $x = -5$, $y = 0$, $r = \sqrt{(-5)^2 + (0)^2} = 5$

$\sin\theta = \dfrac{y}{r} = \dfrac{0}{5} = 0$ $\csc\theta$ is undefined

$\cos\theta = \dfrac{x}{r} = \dfrac{-5}{5} = -1$ $\sec\theta = -1$

$\tan\theta = \dfrac{y}{x} = \dfrac{0}{-5} = 0$ $\cot\theta$ is undefined

9. $\sin 180° = 0$

11. $\tan 180° = 0$

13. $\csc 90° = 1$

15. $\cos\dfrac{\pi}{2} = 0$

17. $\tan\dfrac{\pi}{2} =$ undefined

19. $\sin\dfrac{\pi}{2} = 1$

21. $\sin\theta > 0$ in quadrants I and II.
$\cos\theta > 0$ in quadrants I and IV.
quadrant I

23. $\cos\theta > 0$ in quadrants I and IV.
$\tan\theta < 0$ in quadrants II and IV.
quadrant IV

25. $\sin\theta < 0$ in quadrants III and IV.
$\cos\theta < 0$ in quadrants II and III.
quadrant III

27. $\sin\theta = -\dfrac{1}{2} = \dfrac{y}{r}$, $y = -1$, $r = 2$, $x = \pm\sqrt{2^2 - (-1)^2} = \pm\sqrt{3}$, $x = -\sqrt{3}$ in quadrant III, $\tan\theta = \dfrac{y}{x} = \dfrac{-y}{-\sqrt{3}} = \dfrac{\sqrt{3}}{3}$

29. $\csc\theta = \sqrt{2} = \dfrac{r}{y}$, $r = \sqrt{2}$, $y = 1$, $x = \pm\sqrt{\left(\sqrt{2}\right)^2 - 1^2} = \pm 1$, $x = -1$ in quadrant II, $\cot\theta = \dfrac{-1}{1} = -1$

31. θ is in quadrant IV, $\sin\theta = -\dfrac{1}{2} = \dfrac{y}{r}$, $y = -1$, $r = 2$, $x = \sqrt{2^2 - 1^2} = \sqrt{3}$, $\tan\theta = \dfrac{-1}{\sqrt{3}} = -\dfrac{\sqrt{3}}{3}$

33. $\cos\theta = \dfrac{1}{2}$, θ is in quadrant I or IV.

$\tan\theta = \sqrt{3}$, θ is in quadrant I or III.

θ is in quadrant I, $x = 1$, $y = \sqrt{3}$, $r = 2$

$\csc\theta = \dfrac{r}{y} = \dfrac{2}{\sqrt{3}} = \dfrac{2\sqrt{3}}{3}$

35. $\cos\theta = -\dfrac{1}{2}$, θ is in quadrant II or III.

$\sin\theta = \dfrac{\sqrt{3}}{2}$, θ is in quadrant I or II.

θ is in quadrant II, $x = -1$, $y = \sqrt{3}$, $r = 2$

$\cot\theta = \dfrac{x}{y} = \dfrac{-1}{\sqrt{3}} = -\dfrac{\sqrt{3}}{3}$

37. $\theta = 160°$

Since $90° < \theta < 180°$,

$\theta + \theta' = 180°$

$\theta' = 20°$

39. $\theta = 351°$

Since $270° < \theta < 360°$,

$\theta = \theta' = 360°$

$\theta' = 9°$

41. $\theta = \dfrac{11\pi}{5}$

$\theta > 2\pi = \dfrac{10\pi}{5}$,

θ is coterminal with $\alpha = \dfrac{11\pi}{5} - \dfrac{10\pi}{5} = \dfrac{\pi}{5}$.

Since $0 < \alpha < \dfrac{\pi}{2}$,

$\alpha' = \alpha = \theta'$

$\theta' = \dfrac{\pi}{5}$

43. $\theta = \dfrac{8}{3}$

Since $\dfrac{\pi}{2} < \theta < \pi$,

$\theta + \theta' = \pi$

$\theta' = \pi - \dfrac{8}{3}$

62   Chapter 2: Trigonometric Functions

45. $\theta = 1406° = 326° + 3 \cdot 360°$

θ is coterminal with $\alpha = 326°$.

Since $270° < \alpha < 360°$,

$\alpha + \alpha' = 360°$

$\alpha' = 34°$

$\theta' = 34°$

47. $\theta = -475° = 245° - 2 \cdot 360°$

θ is coterminal with $\alpha = 245°$

Since $180° < \alpha < 270°$,

$\alpha' + 180° = \alpha$

$\alpha' = 245° - 180°$

$\alpha' = 65°$

$\theta' = 65°$

49. $\theta = 225°$ is in quadrant III.

$225° - 180° = 45°$ so $\theta' = 45°$.

Thus, $\sin 225° = -\sin 45° = -\dfrac{\sqrt{2}}{2}$.

51. $\theta = 405°$ is in quadrant I.

$405° - 360° = 45°$ so $\theta' = 45°$.

Thus, $\tan 405° = \tan 45° = 1$.

53. $\theta = \dfrac{4}{3}\pi$ is in quadrant III.

$\dfrac{4}{3}\pi - \pi = \dfrac{\pi}{3}$ so $\theta' = \dfrac{\pi}{3}$.

Thus, $\csc \dfrac{4\pi}{3} = \dfrac{1}{\sin 4\pi/3} = \dfrac{1}{-\sin \pi/3}$

$= \dfrac{1}{-\sqrt{3}/2} = -\dfrac{2}{\sqrt{3}} = -\dfrac{2\sqrt{3}}{3}$.

55. $\theta = \dfrac{17\pi}{4} = \dfrac{16\pi}{4} + \dfrac{\pi}{4}$ is coterminal

with $\dfrac{\pi}{4}$ in quadrant I and $\theta' = \dfrac{\pi}{4}$,

so $\cos \dfrac{17\pi}{4} = \cos \dfrac{\pi}{4} = \dfrac{\sqrt{2}}{2}$.

57. $\theta = 765° = 720° + 45°$ is coterminal

with $45°$ in quadrant I and $\theta' = 45°$,

so $\sec 765° = \sec 45° = \dfrac{1}{\cos 45°} = \dfrac{1}{\sqrt{2}/2} = \sqrt{2}.$

59. $\theta = 540° = 360° + 180°$ is coterminal

with $180°$, so $\cot 540° = \cot 180° = \dfrac{\cos 180°}{\sin 180°} = \dfrac{-1}{0}$,

which is undefined.

61. $\sin 127° \approx 0.798636$

63. $\cos(-116°) \approx -0.438371$

65. $\sec 578° \approx -1.26902$

67. $\sin\left(-\dfrac{\pi}{5}\right) \approx -0.587785$

69. $\csc \dfrac{9\pi}{5} \approx -1.70130$

71. $\sec(-4.45) \approx -3.85522$

73. $\sin 210° - \cos 330° \tan 330° = -\dfrac{1}{2} - \dfrac{\sqrt{3}}{2}\left(-\dfrac{\sqrt{3}}{3}\right) = -\dfrac{1}{2} + \dfrac{1}{2} = 0$

75. $\sin^2 30° + \cos^2 30° = \left(\dfrac{1}{2}\right)^2 + \left(\dfrac{\sqrt{3}}{2}\right)^2 = \dfrac{1}{4} + \dfrac{3}{4} = 1$

77. $\sin \dfrac{3\pi}{2} \tan \dfrac{\pi}{4} - \cos \dfrac{\pi}{3} = (-1)(1) - \dfrac{1}{2} = -1 - \dfrac{1}{2} = -\dfrac{3}{2}$

79. $\sin^2 \dfrac{5\pi}{4} + \cos^2 \dfrac{5\pi}{4} = \left(-\dfrac{\sqrt{2}}{2}\right)^2 + \left(-\dfrac{\sqrt{2}}{2}\right)^2 = \dfrac{1}{2} + \dfrac{1}{2} = 1$

81. $\sin\theta = \dfrac{1}{2}$, θ is in quadrant I or quadrant II

$\theta = 30°, 150°$

83. $\cos\theta = \dfrac{-\sqrt{3}}{2}$, θ is in quadrant II or quadrant III

$\theta = 150°, 120°$

85. $\csc\theta = -\sqrt{2}$
θ is in quadrant III or IV

$\theta = 225°, 315°$

87. $\tan\theta = -1$
θ is in quadrant II or IV

$\theta = \dfrac{3\pi}{4}, \dfrac{7\pi}{4}$

89. $\tan\theta = \dfrac{-\sqrt{3}}{3}$

θ is in quadrant II or IV

$\theta = \dfrac{5\pi}{6}, \dfrac{11\pi}{6}$

91. $\sin\theta = \dfrac{\sqrt{3}}{2}$

θ is in quadrant I or II

$\theta = \dfrac{\pi}{3}, \dfrac{2\pi}{3}$

93. $1 + \tan^2\theta = \sec^2\theta$

$1 + \dfrac{y^2}{x^2} = \dfrac{x^2 + y^2}{x^2}$

$= \dfrac{r^2}{x^2}$

$= \sec^2\theta$

95.

$\cos(90° - \theta) = \sin\theta$

$\dfrac{y}{r} = \dfrac{y}{r}$

PS1. $x^2 + y^2 = 1$
$(0)^2 + (1)^2 = 1$
Yes

PS2. $x^2 + y^2 = 1$

$\left(\dfrac{1}{2}\right)^2 + \left(\dfrac{\sqrt{3}}{2}\right)^2 = 1$

$\dfrac{1}{4} + \dfrac{3}{4} = 1$

Yes

PS3. $x^2 + y^2 = 1$

$\left(\dfrac{\sqrt{2}}{2}\right)^2 + \left(\dfrac{\sqrt{3}}{2}\right)^2 = 1$

$\dfrac{2}{4} + \dfrac{3}{4} \neq 1$

No

PS4. $C = 2\pi r$
$= 2\pi(1)$
$= 2\pi$

PS5. even

PS6. neither

Section 2.4

1. $t = \dfrac{\pi}{6}$

$y = \sin t$ $x = \cos t$

$= \sin\dfrac{\pi}{6}$ $= \cos\dfrac{\pi}{6}$

$= \dfrac{1}{2}$ $= \dfrac{\sqrt{3}}{2}$

The point on the unit circle corresponding to $t = \dfrac{\pi}{6}$ is

$\left(\dfrac{\sqrt{3}}{2}, \dfrac{1}{2}\right)$.

3. $t = \dfrac{7\pi}{6}$

$y = \sin t$ $x = \cos t$

$= \sin\dfrac{7\pi}{6}$ $= \cos\dfrac{7\pi}{6}$

$= -\dfrac{1}{2}$ $= -\dfrac{\sqrt{3}}{2}$

The point on the unit circle corresponding to $t = \dfrac{7\pi}{6}$ is

$\left(-\dfrac{\sqrt{3}}{2}, -\dfrac{1}{2}\right)$.

5. $t = \dfrac{5\pi}{3}$

$y = \sin t$ $\qquad\qquad\qquad$ $x = \cos t$

$\quad = \sin\dfrac{5\pi}{3}$ $\qquad\qquad$ $= \cos\dfrac{5\pi}{3}$

$\quad = -\dfrac{\sqrt{3}}{2}$ $\qquad\qquad\;$ $= \dfrac{1}{2}$

The point on the unit circle corresponding to $t = \dfrac{5\pi}{3}$ is

$\left(\dfrac{1}{2}, -\dfrac{\sqrt{3}}{2}\right)$.

7. $t = \dfrac{11\pi}{6}$

$y = \sin t$ $\qquad\qquad\qquad$ $x = \cos t$

$\quad = \sin\dfrac{11\pi}{6}$ $\qquad\qquad$ $= \cos\dfrac{11\pi}{6}$

$\quad = -\dfrac{1}{2}$ $\qquad\qquad\;$ $= \dfrac{\sqrt{3}}{2}$

The point on the unit circle corresponding to $t = \dfrac{11\pi}{6}$ is

$\left(\dfrac{\sqrt{3}}{2}, -\dfrac{1}{2}\right)$.

9. $t = \pi$

$y = \sin t$ $\qquad\qquad$ $x = \cos t$

$\quad = \sin\pi$ $\qquad\qquad$ $= \cos\pi$

$\quad = 0$ $\qquad\qquad\;\;$ $= -1$

The point on the unit circle corresponding to $t = \pi$ is $(-1, 0)$.

11. $t = -\dfrac{2\pi}{3}$

$y = \sin t$ $\qquad\qquad$ $x = \cos t$

$\quad = \sin\left(-\dfrac{2\pi}{3}\right)$ $\quad$ $= \cos\left(-\dfrac{2\pi}{3}\right)$

$\quad = -\sin\dfrac{2\pi}{3}$ $\qquad$ $= \cos\dfrac{2\pi}{3}$

$\quad = -\dfrac{\sqrt{3}}{2}$ $\qquad\;\;$ $= -\dfrac{1}{2}$

The point on the unit circle corresponding to $t = -\dfrac{2\pi}{3}$ is

$\left(-\dfrac{1}{2}, -\dfrac{\sqrt{3}}{2}\right)$.

13. $\tan\dfrac{11\pi}{6} = -\tan\dfrac{\pi}{6} = -\dfrac{\sqrt{3}}{3}$

15. $\cos\left(-\dfrac{2\pi}{3}\right) = -\cos\dfrac{\pi}{3} = -\dfrac{1}{2}$

17. $\csc\left(-\dfrac{\pi}{3}\right) = -\csc\dfrac{\pi}{3} = -\dfrac{2\sqrt{3}}{3}$

19. $\sin\dfrac{3\pi}{2} = -\sin\dfrac{\pi}{2} = -1$

21. $\sec\left(-\dfrac{7\pi}{6}\right) = -\sec\dfrac{\pi}{6} = -\dfrac{2\sqrt{3}}{3}$

23. $\sin 1.22 \approx 0.9391$

25. $\csc(-1.05) \approx -1.1528$

27. $\tan\dfrac{11\pi}{12} \approx -0.2679$

29. $\cos\left(-\dfrac{\pi}{5}\right) \approx 0.8090$

31. $\sec 1.55 \approx 48.0889$

33. **a.** $\sin 2 \approx 0.9$

$\quad$ **b.** $\cos 2 \approx -0.4$

35. **a.** $\sin 5.4 \approx -0.8$

$\quad$ **b.** $\cos 5.4 \approx 0.6$

37. $\sin t = 0.4$ when $t = 0.4$ or 2.7

39. $\sin t = -0.3$ when $t = 3.4$ or 6.0

41. $f(-x) = -4\sin(-x)$

$\qquad\quad = 4\sin x$

$\qquad\quad = -f(x)$

The function defined by
$f(x) = -4\sin x$ is an odd function.

43. $G(-x) = \sin(-x) + \cos(-x)$

$\qquad\quad\; = -\sin x + \cos x$

The function defined by
$G(x) = \sin x + \cos x$ is neither
an even nor an odd function.

45. $S(-x) = \dfrac{\sin(-x)}{-x}$

$\qquad\quad = -\dfrac{\sin x}{-x} = \dfrac{\sin x}{x}$

$\qquad\quad = S(x)$

The function defined by

$S(x) = \dfrac{\sin(x)}{x}$ is an even function.

47. $v(-x) = 2\sin(-x)\cos(-x)$

$\qquad\quad = -2\sin x \cos x$

$\qquad\quad = -v(x)$

The function defined by
$v(x) = 2\sin x \cos x$ is an odd function.

49. 2π | **51.** π | **53.** 2π

55.

$$\cos t = x$$
$$\cos(-t) = x$$
$$\cos(-t) = \cos t$$

57.

$$\cos t = x$$
$$\cos(\pi + t) = -x$$
$$\cos t = -\cos(\pi + t)$$

59.

$$\sin(t - \pi) = -y$$
$$\sin t = y$$
$$\sin(t - \pi) = -\sin t$$

61. $\tan t \cos t = \dfrac{\sin t}{\cos t} \cdot \cos t$

$$= \sin t$$

63. $\dfrac{\csc t}{\cot t} = \dfrac{\dfrac{1}{\sin t}}{\dfrac{\cos t}{\sin t}}$

$$= \frac{1}{\sin t} \cdot \frac{\sin t}{\cos t}$$
$$= \frac{1}{\cos t} = \sec t$$

65. $1 - \sec^2 t = 1 - \dfrac{1}{\cos^2 t}$

$$= \frac{\cos^2 t - 1}{\cos^2 t}$$
$$= \frac{-\sin^2 t}{\cos^2 t} = -\tan^2 t$$

67. $\tan t - \dfrac{\sec^2 t}{\tan t} = \tan t - \dfrac{1 + \tan^2 t}{\tan t}$

$$= \tan t - \frac{1}{\tan t} - \frac{\tan^2 t}{\tan t}$$
$$= \tan t - \cot t - \tan t$$
$$= -\cot t$$

69. $\dfrac{1 - \cos^2 t}{\tan^2 t} = \dfrac{\sin^2 t}{\dfrac{\sin^2 t}{\cos^2 t}}$

$$= \sin^2 t \cdot \frac{\cos^2 t}{\sin^2 t}$$
$$= \cos^2 t$$

71. $\dfrac{1}{1 - \cos t} + \dfrac{1}{1 + \cos t} = \dfrac{1 + \cos t + 1 - \cos t}{(1 - \cos t)(1 + \cos t)}$

$$= \frac{2}{1 - \cos^2 t}$$
$$= \frac{2}{\sin^2 t} = 2 \csc^2 t$$

73. $\dfrac{\tan t + \cot t}{\tan t} = \dfrac{\dfrac{\sin t}{\cos t} + \dfrac{\cos t}{\sin t}}{\dfrac{\sin t}{\cos t}}$

$$= \left(\frac{\sin^2 t + \cos^2 t}{\sin t \cdot \cos t} \right) \frac{\cos t}{\sin t}$$
$$= \frac{\sin^2 t + \cos^2 t}{\sin^2 t}$$
$$= \frac{1}{\sin^2 t}$$
$$= \csc^2 t$$

75. $\sin^2 t (1 + \cot^2 t) = \sin^2 t (\csc^2 t)$

$$= \sin^2 t \cdot \frac{1}{\sin^2 t}$$
$$= 1$$

77. $\sin^2 t + \cos^2 t = 1$

$$\sin^2 t = 1 - \cos^2 t$$
$$\sin t = \pm \sqrt{1 - \cos^2 t}$$

Because $0 < t < \dfrac{\pi}{2}$, $\sin t$ is positive.

Thus, $\sin t = \sqrt{1 - \cos^2 t}$.

79. $\csc^2 t = 1 + \cot^2 t$

$\csc t = \pm\sqrt{1 + \cot^2 t}$

Because $\dfrac{\pi}{2} < t < \pi$, $\csc t$ is positive.

Thus, $\csc t = \sqrt{1 + \cot^2 t}$.

81. $d(t) = 1970\cos\left(\dfrac{\pi}{64}t\right)$

$d(24) = 1970\cos\left(\dfrac{\pi}{64} \cdot 24\right)$

$= 1970\cos\left(\dfrac{3\pi}{8}\right)$

≈ 750 miles

83. $\cos t - \dfrac{1}{\cos t} = \dfrac{\cos^2 t - 1}{\cos t} = -\dfrac{\sin^2 t}{\cos t}$

85. $\cot t + \dfrac{1}{\cot t} = \dfrac{\cot^2 t + 1}{\cot t} = \dfrac{\csc^2 t}{\cot t} = \dfrac{\dfrac{1}{\sin^2 t}}{\dfrac{\cos t}{\sin t}} = \dfrac{1}{\sin^2 t} \cdot \dfrac{\sin t}{\cos t} = \dfrac{1}{\sin t \cos t} = \dfrac{1}{\sin t} \cdot \dfrac{1}{\cos t} = \csc t \sec t$

87. $(1 - \sin t)^2 = 1 - 2\sin t + \sin^2 t$

89. $(\sin t - \cos t)^2 = \sin^2 t - 2\sin t \cos t + \cos^2 t$

$= 1 - 2\sin t \cos t$

91. $(1 - \sin t)(1 + \sin t) = 1 - \sin^2 t$

$= \cos^2 t$

93. $\dfrac{\sin t}{1 + \cos t} + \dfrac{1 + \cos t}{\sin t} = \dfrac{(\sin t)(\sin t) + (1 + \cos t)(1 + \cos t)}{\sin t(1 + \cos t)}$

$= \dfrac{\sin^2 t + 1 + 2\cos t + \cos^2 t}{\sin t(1 + \cos t)}$

$= \dfrac{2 + 2\cos t}{\sin t(1 + \cos t)} = \dfrac{2(1 + \cos t)}{\sin t(1 + \cos t)}$

$= \dfrac{2}{\sin t} = 2\csc t$

95. $\cos^2 t - \sin^2 t = (\cos t - \sin t)(\cos t + \sin t)$

97. $\tan^2 t - \tan t - 6 = (\tan t + 2)(\tan t - 3)$

99. $2\sin^2 t - \sin t - 1 = (2\sin t + 1)(\sin t - 1)$

• •

Connecting Concepts

101. $\csc t = \sqrt{2}, \quad 0 < t < \dfrac{\pi}{2}$

$\sin t = \dfrac{1}{\csc t} = \dfrac{\sqrt{2}}{2}$

$\cos^2 t + \sin^2 t = 1$

$\cos t = \pm\sqrt{1 - \sin^2 t}$

$\cos t$ is positive in quadrant I.

$\cos t = \sqrt{1 - \left(\dfrac{\sqrt{2}}{2}\right)^2} = \sqrt{\dfrac{1}{2}}$

$\cos t = \dfrac{\sqrt{2}}{2}$

103. $\sin t = \dfrac{1}{2}, \dfrac{\pi}{2} < t < \pi$

$\tan t = \dfrac{\sin t}{\cos t}$

$\tan t = \dfrac{\sin t}{\pm\sqrt{1 - \sin^2 t}}$

Because $\dfrac{\pi}{2} < t < \pi$, $\tan t$ is negative.

$\tan t = -\dfrac{\dfrac{1}{2}}{\sqrt{1 - \left(\dfrac{1}{2}\right)^2}}$

$\tan t = -\dfrac{\dfrac{1}{2}}{\dfrac{\sqrt{3}}{2}}$

$\tan t = -\dfrac{\sqrt{3}}{3}$

105. $\dfrac{\sin^2 t + \cos^2 t}{\sin^2 t} = \dfrac{1}{\sin^2 t}$

$\phantom{\dfrac{\sin^2 t + \cos^2 t}{\sin^2 t}} = \csc^2 t$

107. $(\cos t - 1)(\cos t + 1) = \cos^2 t - 1$

$ = -(1 - \cos^2 t)$

$ = -\sin^2 t$

●●

PS1. $\sin\dfrac{3\pi}{4} \approx 0.7$

PS2. $\cos\dfrac{5\pi}{4} \approx -0.7$

PS3. Reflect the graph of $y = f(x)$ across the x-axis to produce $y = -f(x)$.

PS4. Compress each point on the graph of $y = f(x)$ toward the y-axis by a factor of $\dfrac{1}{2}$.

PS5. $\dfrac{2\pi}{\frac{1}{3}} = \dfrac{2\pi}{1} \cdot \dfrac{3}{1} = 6\pi$

PS6. $\dfrac{2\pi}{\frac{2}{5}} = \dfrac{2\pi}{1} \cdot \dfrac{5}{2} = 5\pi$

Section 2.5

1. $y = 2\sin x$

$a = 2,\ p = 2\pi$

3. $y = \sin 2x$

$a = 1,\ p = \dfrac{2\pi}{2} = \pi$

5. $y = \dfrac{1}{2}\sin 2\pi x$

$a = \dfrac{1}{2},\ p = \dfrac{2\pi}{2\pi} = 1$

7. $y = -2\sin\dfrac{x}{2}$

$a = |-2| = 2,\ p = \dfrac{2\pi}{1/2} = 4\pi$

9. $y = \dfrac{1}{2}\cos x$

$a = \dfrac{1}{2},\ p = 2\pi$

11. $y = \cos\dfrac{x}{4}$

$a = 1,\ p = \dfrac{2\pi}{1/4} = 8\pi$

13. $y = 2\cos\dfrac{\pi x}{3}$

$a = 2,\ p = \dfrac{2\pi}{\pi/3} = 6$

15. $y = -3\cos\dfrac{2x}{3}$

$a = |-3| = 3,\ p = \dfrac{2\pi}{2/3} = 3\pi$

17. $y = 4.7\sin 0.8\pi t$

$a = 4.7,\ p = \dfrac{2\pi}{0.8\pi} = 2.5$

19. $y = \dfrac{1}{2}\sin x,\ a = \dfrac{1}{2},\ p = 2\pi$

21. $y = 3\cos x,\ a = 3,\ p = 2\pi$

23. $y = -\dfrac{7}{2}\cos x,\ a = \left|-\dfrac{7}{2}\right| = \dfrac{7}{2},\ p = 2\pi$

25. $y = -4\sin x,\ a = |-4| = 4,\ p = 2\pi$

27. $y = \cos 3x,\ a = 1,\ p = \dfrac{2\pi}{3}$

29. $y = \sin\dfrac{3x}{2},\ a = 1,\ p = \dfrac{4\pi}{3}$

31. $y = \cos\dfrac{\pi}{2}x,\ a = 1,\ p = 4$

33. $y = \sin 2\pi x,\ a = 1,\ p = 1$

35. $y = 4\cos\dfrac{x}{2},\ a = 4,\ p = \dfrac{2\pi}{1/2} = 4\pi$

37. $y = -2\cos\dfrac{x}{3}$, $a = |-2| = 2$, $p = \dfrac{2\pi}{1/3} = 6\pi$

39. $y = 2\sin\pi x$, $a = 2$, $p = \dfrac{2\pi}{\pi} = 2$

41. $y = \dfrac{3}{2}\cos\dfrac{\pi x}{2}$, $a = \dfrac{3}{2}$, $p = \dfrac{2\pi}{\pi/2} = 4$

43. $y = 4\sin\dfrac{2\pi}{3}x$, $a = |4| = 4$, $p = \dfrac{2\pi}{2\pi/3} = 3$

45. $y = 2\cos 2x$, $a = 2$, $p = \dfrac{2\pi}{2} = \pi$

47. $y = -2\sin 1.5x$, $a = |-2| = 2$, $p = \dfrac{2\pi}{1.5} = \dfrac{4\pi}{3}$

49. $y = \left|2\sin\dfrac{x}{2}\right|$

51. $y = |-2\cos 3x|$

53. $y = -\left|2\sin\dfrac{x}{3}\right|$

55. $y = -|3\cos\pi x|$

57. $\dfrac{2\pi}{b} = \pi$, $b = 2$, $a = 1$

$y = \cos 2x$

59. $\dfrac{2x}{b} = 3\pi$, $b = \dfrac{2}{3}$, $a = 2$

$y = 2\sin\dfrac{2}{3}x$

61. $\dfrac{2\pi}{b} = 2$, $b = \pi$, $a = 2$

$y = -2\cos\pi x$

63. a. Amplitude, $a = 4$, period, $b = \dfrac{2\pi}{2} = \pi$

$V = 4\sin\pi t$, $0 \le t \le 8$ ms

b. Frequency $= \dfrac{4\text{ cycles}}{8\text{ ms}} = \dfrac{1}{2}$ cycle/ms

65. $f(x) = 2\sin\dfrac{2x}{3}$, $a = 2$, $p = \dfrac{2\pi}{2/3} = 3\pi$

67. $y_1 = 2\cos\dfrac{x}{2}$, $a = 2$, $p = \dfrac{2\pi}{1/2} = 4\pi$

$y_2 = 2\cos x$, $a = 2$, $p = 2\pi$

69. $y = \frac{1}{2}x\sin x$

71. $y = -x\cos x$

73. $y = e^{\sin x}$

The maximum value of $e^{\sin x}$ is e.

The minimum value of $e^{\sin x}$ is $\frac{1}{e} \approx 0.3679$.

The function defined by $y = e^{\sin x}$ is periodic with a period of 2π.

• •

75. $a = 2$

$p = \dfrac{2\pi}{b} = 3\pi$

$b = \dfrac{2}{3}$

$y = 2\sin\dfrac{2}{3}x$

77. $a = 2.5$

$p = \dfrac{2\pi}{b} = 3.2$

$b = \dfrac{5\pi}{8}$

$y = 2.5\sin\dfrac{5\pi}{8}x$

79. $a = 3$

$p = \dfrac{2\pi}{b} = 2.5$

$b = \dfrac{4\pi}{5}$

$y = 3\cos\dfrac{4\pi}{5}x$

• •

PS1. $\tan\dfrac{\pi}{3} \approx 1.7$

PS2. $\cot\dfrac{\pi}{3} \approx 0.6$

PS3. Stretch each point on the graph of $y = f(x)$ away from the x-axis by a factor of 2 to produce $y = 2f(x)$.

PS4. Shift the graph of $y = f(x)$ 2 units to the right and up 3 units.

PS5. $\dfrac{\pi}{\frac{1}{2}} = \dfrac{\pi}{1} \cdot \dfrac{2}{1} = 2\pi$

PS6. $\dfrac{\pi}{|-3/4|} = \dfrac{\pi}{1} \cdot \left|-\dfrac{4}{3}\right| = \dfrac{\pi}{1} \cdot \dfrac{4}{3} = \dfrac{4}{3}\pi$

═══════════ Section 2.6 ═══════════

1. $y = \tan x$ is undefined for $\dfrac{\pi}{2} + k\pi$, k an integer.

3. $y = \sec x$ is undefined for $\dfrac{\pi}{2} + k\pi$, k an integer.

5. $p = 2\pi$

7. $p = \pi$

9. $p = \dfrac{\pi}{1/2} = 2\pi$

11. $p = \dfrac{2\pi}{3}$

13. $p = \dfrac{\pi}{3}$

15. $p = \dfrac{2\pi}{1/4} = 8\pi$

17. $p = \dfrac{\pi}{\pi} = 1$

19. $p = \dfrac{\pi}{\pi/5} = 5$

21. $p = \dfrac{2\pi}{\pi/4.25} = 8.5$

23. $y = 3\tan x,\ p = \pi$

25. $y = \dfrac{3}{2}\cot x,\ p = \pi$

27. $y(x) = 2\sec x,\ p = 2\pi$

29. $y = \dfrac{1}{2}\csc x,\ p = 2\pi$

31. $y = 2\tan\dfrac{x}{2},\ p = \dfrac{\pi}{1/2} = 2\pi$

33. $y = -3\cot\dfrac{x}{2},\ p = \dfrac{\pi}{1/2} = 2\pi$

35. $y = -2\csc\dfrac{x}{3},\ p = \dfrac{2\pi}{1/3} = 6\pi$

37. $y = \dfrac{1}{2}\sec 2x,\ p = \dfrac{2\pi}{2} = \pi$

39. $y = -2\sec\pi x,\ p = \dfrac{2\pi}{\pi} = 2$

41. $y = 3\tan 2\pi x,\ p = \dfrac{\pi}{2\pi} = \dfrac{1}{2}$

43. $y = 2\csc 3x,\ p = \dfrac{2\pi}{3}$

45. $y = 3\sec\pi x,\ p = \dfrac{2\pi}{\pi} = 2$

47. $y = 2\cot 2x,\ p = \dfrac{\pi}{2}$

49. $y = 3\tan\pi x,\ p = \dfrac{\pi}{\pi} = 1$

51. $\dfrac{\pi}{b} = \dfrac{2\pi}{3},\ b = \dfrac{3}{2}$

$y = \cot\dfrac{3}{2}x$

53. $\dfrac{2\pi}{b} = 3\pi,\ b = \dfrac{2}{3}$

$y = \csc\dfrac{2}{3}x$

55. $\dfrac{2\pi}{b}=\dfrac{8\pi}{3}, b=\dfrac{3}{4}$

$y=\sec\dfrac{3}{4}x$

57. $y=\tan|x|$

59. $y=|\csc x|$

61. **a.** $\tan x=\dfrac{h}{1.4}$

$h=1.4\tan x$

b. $\cos x=\dfrac{1.4}{d}$

$d=\dfrac{1.4}{\cos x}=1.4\sec x$

d. The graph of d is above the graph of h, however, the distance between the graphs approaches

0 as x approaches $\dfrac{\pi}{2}$.

c.

·· **Connecting Concepts**

63. $\dfrac{\pi}{b}=\dfrac{\pi}{3}, b=3$

$y=\tan 3x$

65. $\dfrac{2\pi}{b}=\dfrac{3\pi}{4}, b=\dfrac{8}{3}$

$y=\sec\dfrac{8}{3}x$

67. $\dfrac{\pi}{b}=2, b=\dfrac{\pi}{2}$

$y=\cot\dfrac{\pi}{2}x$

69. $\dfrac{2\pi}{b}=1.5, b=\dfrac{4}{3}\pi$

$y=\csc\dfrac{4\pi}{3}x$

·· **Prepare for Section 2.7**

PS1. $y=2\sin 2x$

amplitude = 2, period = π

PS2. $y=\dfrac{2}{3}\cos\dfrac{x}{3}$

amplitude = $\dfrac{2}{3}$, period = 6π

PS3. $y=-4\sin 2\pi x$

amplitude = 4, period = 1

PS4. maximum at 2

PS5. minimum at -3

PS6. $f(x)=\cos x$ is symmetric to y–axis.

Section 2.7

1. $a=2, p=2\pi$, phase shift $=\dfrac{\pi}{2}$

3. $a=1, p=\dfrac{2\pi}{2}=\pi$, phase shift $=\dfrac{\pi/4}{2}=\dfrac{\pi}{8}$

5. $a=|-4|=4, p=\dfrac{2\pi}{2/3}=3\pi$, phase shift $=\dfrac{-\pi/6}{2/3}=-\dfrac{\pi}{4}$

7. $a=\dfrac{5}{4}, p=\dfrac{2\pi}{3}$, phase shift $=\dfrac{2\pi}{3}$

9. $p=\dfrac{\pi}{2}$, phase shift $=\dfrac{\pi/4}{2}=\dfrac{\pi}{8}$

11. $p=\dfrac{2\pi}{1/3}=6\pi$, phase shift $=\dfrac{-\pi}{1/3}=-3\pi$

13. $p = \dfrac{2\pi}{2} = \pi$, phase shift $= \dfrac{\pi/8}{2} = \dfrac{\pi}{16}$

15. $p = \dfrac{\pi}{1/4} = 4\pi$, phase shift $= \dfrac{-3\pi}{1/4} = -12\pi$

17. $y = \sin\left(x - \dfrac{\pi}{2}\right)$

$0 \le x - \dfrac{\pi}{2} \le 2\pi$

$\dfrac{\pi}{2} \le x \le \dfrac{5\pi}{2}$

period $= 2\pi$, phase shift $\dfrac{\pi}{2}$

19. $y = \cos\left(\dfrac{x}{2} + \dfrac{\pi}{3}\right)$

$0 \le \dfrac{x}{2} + \dfrac{\pi}{3} \le 2\pi$

$-\dfrac{\pi}{3} \le \dfrac{x}{2} \le \dfrac{5\pi}{3}$

$-\dfrac{2\pi}{3} \le x \le \dfrac{10\pi}{3}$

period $= 4\pi$, phase shift $= -\dfrac{2\pi}{3}$

21. $y = \tan\left(x + \dfrac{\pi}{4}\right)$

$-\dfrac{\pi}{2} < x + \dfrac{\pi}{4} < \dfrac{\pi}{2}$

$-\dfrac{3\pi}{4} < x < \dfrac{\pi}{4}$

period $= \pi$, phase shift $= -\dfrac{\pi}{4}$

23. $y = 2\cot\left(\dfrac{x}{2} - \dfrac{\pi}{8}\right)$

$0 < \dfrac{x}{2} - \dfrac{\pi}{8} < \pi$

$\dfrac{\pi}{8} < \dfrac{x}{2} < \dfrac{9\pi}{8}$

$\dfrac{\pi}{4} < x < \dfrac{9\pi}{4}$

period $= 2\pi$, phase shift $= \dfrac{\pi}{4}$

25. $y = \sec\left(x + \dfrac{\pi}{4}\right)$

$0 \le x + \dfrac{\pi}{4} \le 2\pi$

$-\dfrac{\pi}{4} \le x \le \dfrac{7\pi}{4}$

period $= 2\pi$, phase shift $= -\dfrac{\pi}{4}$

27. $y = \csc\left(\dfrac{x}{3} - \dfrac{\pi}{2}\right)$

$0 \le \dfrac{x}{3} - \dfrac{\pi}{2} \le 2\pi$

$\dfrac{\pi}{2} \le \dfrac{x}{3} \le \dfrac{5\pi}{2}$

$\dfrac{3\pi}{2} \le x \le \dfrac{15\pi}{2}$

period $= 6\pi$, phase shift $= \dfrac{3\pi}{2}$

29. $y = -2\sin\left(\dfrac{x}{3} - \dfrac{2\pi}{3}\right)$

$0 \le \dfrac{x}{3} - \dfrac{2\pi}{3} \le 2\pi$

$\dfrac{2\pi}{3} \le \dfrac{x}{3} \le \dfrac{8\pi}{3}$

$2\pi \le x \le 8\pi$

period $= 6\pi$, phase shift $= 2\pi$

31. $y = -3\cos\left(3x + \dfrac{\pi}{4}\right)$

$0 \le 3x + \dfrac{\pi}{4} \le 2\pi$

$-\dfrac{\pi}{4} \le 3x \le \dfrac{7\pi}{4}$

$-\dfrac{\pi}{12} \le x \le \dfrac{7\pi}{12}$

period $= \dfrac{2\pi}{3}$, phase shift $= -\dfrac{\pi}{12}$

33. $y = \sin x + 1$, $p = 2\pi$

35. $y = -\cos x - 2$, $p = 2\pi$

37. $y = \sin 2x - 2$, $p = \pi$

39. $y = 4\cos(\pi x - 2) + 1,\ p = 2$

phase shift $= \dfrac{2}{\pi}$

41. $y = -\sin(\pi x + 1) - 2,\ p = 2$

phase shift $= -\dfrac{1}{\pi}$

43. $y = \sin\left(x - \dfrac{\pi}{2}\right) - \dfrac{1}{2},\ p = 2\pi$

phase shift $= \dfrac{\pi}{2}$

45. $y = \tan\dfrac{x}{2} - 4,\ p = 2\pi$

47. $y = \sec 2x - 2,\ p = \pi$

49. $y = \csc\dfrac{x}{2} - 1,\ p = 4\pi$

51. $y = x - \sin x$

53. $y = x + \sin 2x$

55. $y = \sin x + \cos x$

57. sine curve, $a = 1,\ \dfrac{2\pi}{b} = \pi,\ b = 2$

phase shift $= -\dfrac{c}{b} = \dfrac{\pi}{6},\ c = -\dfrac{\pi}{3}$

$y = \sin\left(2x - \dfrac{\pi}{3}\right)$

59. cosecant curve, $\dfrac{2\pi}{b} = 4\pi,\ b = \dfrac{1}{2}$

phase shift $= -\dfrac{c}{b} = 2\pi,\ c = -\pi$

$y = \csc\left(\dfrac{x}{2} - \pi\right)$

61. secant curve, $\dfrac{2\pi}{b} = 2\pi,\ b = 1,$ phase shift $= -\dfrac{c}{b} = \dfrac{\pi}{2},\ c = -\dfrac{\pi}{2},\ y = \sec\left(x - \dfrac{\pi}{2}\right)$

63. a. phase shift: $-\dfrac{c}{b} = -\dfrac{-1.25\pi}{\pi/6} = 1.25(6) = 7.5$ months

period: $\dfrac{2\pi}{b} = \dfrac{2\pi}{\pi/6} = 2(6) = 12$ months

b. First graph $y_1 = 4.1\cos\left(\dfrac{\pi}{6}t\right)$. Because the phase shift is 7.5 months, shift the graph of y_1 7.5 units to the right to produce the graph of y_2. Now shift the graph of y_2 upward 7 units to product the graph of S.

c. 7.5 months after January 1 is the middle of August.

65. $y = 2.3\sin 2\pi t + 1.25t + 315$

$t = 16$ between 1990 and 2006

$y = 2.3\sin 2\pi(16) + 1.25(16) + 315$

$y \approx 335$ ppm

difference $\approx 335 - 315$

≈ 20 ppm

67. 5 rpm $a = 7$

$p = 0.2$ min $p = \dfrac{2\pi}{b}$

$0.2 = \dfrac{2\pi}{b}$

$b = 10\pi$

$s = 7\cos 10\pi t + 5$

69. Change 6 rpm to radians/second.

$6 \text{ rpm} = \dfrac{6 \cdot 2\pi \text{ radians}}{1 \text{ minute}} \cdot \dfrac{1 \text{ minute}}{60 \text{ sec}}$

$= \dfrac{\pi}{5} \text{ radians/sec}$

in t seconds, θ increased by $\dfrac{\pi}{5}t$ radians.

$\tan \dfrac{\pi}{5}t = \dfrac{s}{400}$

$400\tan \dfrac{\pi}{5}t = s$ for $0 \le t < 2.5$ or $7.5 < t \le 10$

71. amplitude $A = 3$

$k = 9$

$24 = 2$ cycles

$12 = 1$ cycle

$\dfrac{2\pi}{B} = 12 \Rightarrow B = \dfrac{\pi}{6}$

$y = 3\cos \dfrac{\pi}{6}t + 9$

At 6:00 P.M., $t = 12$.

$y = 3\cos \dfrac{\pi}{6} \cdot 12 + 9$

$y = 3\cos 2\pi + 9$

$y = 3 + 9$

$y = 12$ ft

73. $y = \sin x - \cos \dfrac{x}{2}$

75. $y = 2\cos x + \sin \dfrac{x}{2}$

77. $y = \dfrac{x}{2}\sin x$

79. $y = x\sin \dfrac{x}{2}$

81.

83. sine function, $a = 2$, $p = \pi$

$\dfrac{2\pi}{b} = \pi$, $b = 2$

phase shift $= -\dfrac{c}{b} = \dfrac{\pi}{3}$, $c = -\dfrac{2\pi}{3}$

$y = 2\sin\left(2x - \dfrac{2\pi}{3}\right)$

85. tangent function, $p = 2\pi$

$\dfrac{\pi}{b} = 2\pi \Rightarrow b = \dfrac{1}{2}$

phase shift $= -\dfrac{c}{b} = \dfrac{\pi}{2} \Rightarrow c = -\dfrac{\pi}{4}$

$y = \tan\left(\dfrac{x}{2} - \dfrac{\pi}{4}\right)$

87. $g[h(x)] = (\cos x)^2 + 2$

$\qquad = \cos^2 x + 2$

89. $\dfrac{\sin x}{x} \to 1$ as $x \to 0$

91. $y = |x|\sin x$

PS1. $\dfrac{3}{2\pi}$

PS2. $\dfrac{5}{2}$

PS3. $4\cos 2\pi(0) = 4\cos 0 = 4$

PS4. $\sqrt{\dfrac{18}{2}} = \sqrt{9} = 3$

PS5. $4\cos\left(\sqrt{\dfrac{16}{4}} \cdot 2\pi\right) = 4\cos(4 \cdot 2\pi)$

$\qquad = 4\cos 8\pi = 4$

PS6. $y = 4\cos \pi x$

Section 2.8

1. $y = 2\sin 2t$

amplitude $= 2$

$p = \dfrac{2\pi}{2} = \pi$

frequency $= \dfrac{1}{p} = \dfrac{1}{\pi}$

3. $y = 3\cos\dfrac{2t}{3}$

amplitude $= 3$

$p = \dfrac{2\pi}{2/3} = 3\pi$

frequency $= \dfrac{1}{p} = \dfrac{1}{3\pi}$

5. $y = 4\cos \pi t$

amplitude $= 4$

$p = \dfrac{2\pi}{\pi} = 2$

frequency $= \dfrac{1}{p} = \dfrac{1}{2}$

7. $y = \dfrac{3}{4}\sin\dfrac{\pi t}{2}$

amplitude $= \dfrac{3}{4}$

$p = \dfrac{2\pi}{\pi/2} = 4$

frequency $= \dfrac{1}{p} = \dfrac{1}{4}$

9. $a = 4$

$p = \dfrac{1}{\text{frequency}} = \dfrac{1}{1.5} = \dfrac{2}{3}$

$\dfrac{2\pi}{b} = \dfrac{2}{3} \Rightarrow b = 3\pi$

$y = 4\cos 3\pi t$

11. $p = 1.5$

amplitude $= \dfrac{3}{2}$

$\dfrac{2\pi}{b} = 1.5 \Rightarrow b = \dfrac{4\pi}{3}$

$y = \dfrac{3}{2}\cos\dfrac{4\pi}{3}t$

13. amplitude $= 2$

$p = \pi$

$\dfrac{2\pi}{b} = \pi \Rightarrow b = 2$

$y = 2\sin 2t$

15. amplitude $= 1$

$p = 2$

$\dfrac{2\pi}{b} = 2 \Rightarrow b = \pi$

$y = \sin \pi t$

17. amplitude $= 2$

frequency $= 1$

$p = \dfrac{1}{\text{frequency}} = \dfrac{1}{1} = 1$

$\dfrac{2\pi}{b} = 1 \Rightarrow b = 2\pi$

$y = 2\sin 2\pi t$

19. amplitude $= \dfrac{1}{2}$

frequency $= \dfrac{2}{\pi} \Rightarrow p = \dfrac{\pi}{2}$

$\dfrac{2\pi}{b} = \dfrac{\pi}{2} \Rightarrow b = 4$

$y = \dfrac{1}{2}\cos 4t$

21. amplitude $= 2.5$

frequency $= 0.5 \Rightarrow p = 2$

$\dfrac{2\pi}{b} = 2 \Rightarrow b = \pi$

$y = 2.5\cos \pi t$

23. amplitude $= \dfrac{1}{2}$

$p = 3$

$\dfrac{2\pi}{b} = 3 \Rightarrow b = \dfrac{2}{3}\pi$

$y = \dfrac{1}{2}\cos\dfrac{2\pi}{3}t$

25. amplitude $= 4$

$p = \dfrac{\pi}{2}$

$\dfrac{2\pi}{b} = \dfrac{\pi}{2} \Rightarrow b = 4$

$y = 4\cos 4t$

27. amplitude $= 2$ feet, $a = -2$

frequency $= f = \dfrac{1}{2\pi}\sqrt{\dfrac{k}{m}} = \dfrac{1}{2\pi}\sqrt{\dfrac{8}{32}} = \dfrac{1}{2\pi}\cdot\dfrac{1}{2} = \dfrac{1}{4\pi}$

period $= p = \dfrac{1}{f} = 4\pi$

$y = a\cos 2\pi f t = -2\cos 2\pi\left(\dfrac{1}{4\pi}\right)t$

$= -2\cos\dfrac{1}{2}t$

29. **a.** frequency $= f = \dfrac{392\pi}{2\pi} = 196$ cycles/s

period $= p = \dfrac{1}{196}$ s

b. The amplitude needs to increase.

31. amplitude $= a = \dfrac{78-4}{2} = 37$

period $= p = \dfrac{2\pi}{45} = \dfrac{\pi}{22.5}$

$h = -37\cos\left(\dfrac{\pi}{22.5}t\right) + 41$

33. **a.** The pseudoperiod is $\dfrac{2\pi}{2} = \pi$. There are $\dfrac{10}{\pi} \approx 3$ complete oscillations of length π in $0 \le t \le 10$.

b. $\left|f(t)\right| < 0.01$ for all $t > 59.8$ (nearest tenth).

$\text{Xmin} = 56, \text{Xmax} = 65, \text{Xscl} = 1,$
$\text{Ymin} = -0.01, \text{Ymax} = 0.01, \text{Yscl} = 0.005$

35. **a.** The pseudoperiod is $\dfrac{2\pi}{2\pi} = 1$. There are 10 complete oscillations of length 1 in $0 \le t \le 10$.

b. $\left|f(t)\right| < 0.01$ for all $t > 71.0$ (nearest tenth).

$\text{Xmin} = 70, \text{Xmax} = 73, \text{Xscl} = 1$
$\text{Ymin} = -0.01, \text{Ymax} = 0.01, \text{Yscl} = 0.005$

37. **a.** The pseudoperiod is $\dfrac{2\pi}{2\pi} = 1$. There are 10 complete oscillations of length 1 in $0 \le t \le 10$.

b. $\left|f(t)\right| < 0.01$ for all $t > 9.1$ (nearest tenth).

$\text{Xmin} = 8, \text{Xmax} = 10, \text{Xscl} = 1,$
$\text{Ymin} = -0.01, \text{Ymax} = 0.01, \text{Yscl} = 0.005$

39. **a.** The pseudoperiod is $\dfrac{2\pi}{2\pi} = 1$. There are 10 complete oscillations of length 1 in $0 \le t \le 10$.

b. $\left|f(t)\right| < 0.01$ for all $t > 6.1$ (nearest tenth).

$\text{Xmin} = 5, \text{Xmax} = 7, \text{Xscl} = 1,$
$\text{Ymin} = -0.01, \text{Ymax} = .01, \text{Yscl} = 0.005$

•••

41. $p_1 = 2\pi\sqrt{\dfrac{m}{k}}, \ p_2 = 2\pi\sqrt{\dfrac{9m}{k}} = 3p_1$

Increasing the main mass to $9m$ will triple the period.

Connecting Concepts

43. yes

$\text{Xmin} = 0, \text{Xmax} = 15, \text{Xscl} = 1,$
$\text{Ymin} = -1, \text{Ymax} = 1.5, \text{Yscl} = 0.25$

45.   yes

Xmin = 0, Xmax = 10, Xscl = 1,
Ymin = −3, Ymax = 9, Yscl = 1

• •

Sinusoidal Families

1.

All three sine graphs have, a period of 2π, x-intercepts at $n\pi$, and no phase shift, but their amplitudes are 2, 4, and 6 respectively.

3.

All three sine graphs have a period of 2π and an amplitude of 1, but their phase shifts are $-\pi/4$, $-\pi/6$, and $-\pi/12$, respectively.

• •

1.   True

2.   False; $\sec^2\theta - \tan^2\theta = 1$ is an identity.

3.   False; 1 rad $\approx 57.3°$.

4.   True

5. $\dfrac{\pi}{4}$

6. $(0, 1)$

7.   The period is $\dfrac{2\pi}{3\pi/4} = \dfrac{8}{3}$.

8.   Shift the graph of y_1 to the left $\dfrac{\pi}{2}$ units.

9.   All real numbers except multiples of π.

10.   The vertical asymptotes are $x = \dfrac{\pi}{2}$ and $x = \dfrac{3\pi}{2}$.

1. complement: $90° - 65° = 25°$ [2.1]
    supplement: $180° - 65° = 115°$

2. $\theta = 980° = 260° + 2 \cdot 360°$ [2.3]
    θ is coterminal with $\alpha = 260°$ and
    $\theta' = \alpha'$.
    Since $180° < \alpha < 270°$,
    $180° + \alpha' = \alpha$
    $180° + a' = 260°$
    $\alpha' = 80°$
    $\theta = 80°$

3. $2 = 2\left(\dfrac{180°}{\pi}\right)$ [2.1]
    $= 114.59°$

4. $315° = 315°\left(\dfrac{\pi}{180°}\right)$ [2.1]
    $= \dfrac{7\pi}{4}$

5. $s = r\theta = 3(75°)\left(\dfrac{\pi}{180°}\right)$ [2.1]
    $= 3.93$ m

6. $\theta = \dfrac{s}{r} = \dfrac{12}{40}$ [2.1]
    $= 0.3$

7. $w = \dfrac{V}{r} = \dfrac{50}{16} \cdot \dfrac{63360}{3600}$ [2.1]
    ≈ 55 rad/sec

For exercises 8 to 11, $\csc\theta = \dfrac{3}{2} = \dfrac{r}{y}$, $r = 3$, $y = 2$, and $x = \sqrt{3^2 - 2^2} = \sqrt{5}$.

8. $\cos\theta = \dfrac{x}{r} = \dfrac{\sqrt{5}}{3}$ [2.2]

9. $\cot\theta = \dfrac{x}{y} = \dfrac{\sqrt{5}}{2}$ [2.2]

10. $\sin\theta = \dfrac{y}{r} = \dfrac{2}{3}$ [2.2]

11. $\sec\theta = \dfrac{r}{x} = \dfrac{3}{\sqrt{5}} = \dfrac{3\sqrt{5}}{5}$ [2.2]

12.

$x = 1$, $y = -3$, $r = \sqrt{1^2 + (-3)^2} = \sqrt{10}$

$\sin\theta = -\dfrac{3}{\sqrt{10}} = -\dfrac{3\sqrt{10}}{10}$ $\cos\theta = \dfrac{1}{\sqrt{10}} = \dfrac{\sqrt{10}}{10}$ $\tan\theta = \dfrac{-3}{1} = -3$

$\csc\theta = -\dfrac{\sqrt{10}}{3}$ $\sec\theta = \sqrt{10}$ $\cot\theta = -\dfrac{1}{3}$

13. a. $\sec 150° = \dfrac{2}{-\sqrt{3}} = -\dfrac{2\sqrt{3}}{3}$

     b. $\tan\left(-\dfrac{3\pi}{4}\right) = 1$

     c. $\cot(-225°) = -1$

     d. $\cos\dfrac{2\pi}{3} = -\dfrac{1}{2}$ [2.3]

14. a. $\cos 123° \approx -0.5446$

     b. $\cot 4.22 \approx 0.5365$

     c. $\sec 612° \approx -3.2361$

     d. $\tan\dfrac{2\pi}{5} \approx 3.0777$ [2.3]

15. $\cos\phi = -\dfrac{\sqrt{3}}{2} = \dfrac{x}{r},\ x = -\sqrt{3},\ r = 2,\ y = -\sqrt{2^2 - \left(-\sqrt{3}\right)^2} = -1$ [2.3]

     a. $\sin\phi = \dfrac{y}{r} = -\dfrac{1}{2}$

     b. $\tan\phi = \dfrac{y}{x} = \dfrac{-1}{-\sqrt{3}} = \dfrac{\sqrt{3}}{3}$

16. $\tan\phi = -\dfrac{\sqrt{3}}{3} = \dfrac{y}{x},\ y = \sqrt{3},\ x = -3,\ r = \sqrt{(-3)^2 + \left(\sqrt{3}\right)^2} = 2\sqrt{3}$ [2.3]

     a. $\sec\phi = \dfrac{r}{x} = \dfrac{2\sqrt{3}}{-3} = -\dfrac{2\sqrt{3}}{3}$

     b. $\csc\phi = \dfrac{r}{y} = \dfrac{2\sqrt{3}}{\sqrt{3}} = 2$

17. $\sin\phi = -\dfrac{\sqrt{2}}{2},\ y = -\sqrt{2},\ r = 2,\ x = -\sqrt{2^2 - \left(-\sqrt{2}\right)^2} = \sqrt{2}$ [2.3]

     a. $\cos\phi = \dfrac{x}{r} = \dfrac{\sqrt{2}}{2}$

     b. $\cot\phi = \dfrac{x}{y} = \dfrac{\sqrt{2}}{-\sqrt{2}} = -1$

18. a. $W(\pi) = (-1, 0)$ [2.4]

     b. $W\left(-\dfrac{\pi}{3}\right) = \left(\dfrac{1}{2}, -\dfrac{\sqrt{3}}{2}\right)$

     c. $W\left(\dfrac{5\pi}{4}\right) = \left(-\dfrac{\sqrt{2}}{2}, -\dfrac{\sqrt{2}}{2}\right)$

19. $f(x) = \sin(x)\tan(x)$ [2.4]

     $f(-x) = \sin(-x)\tan(-x) = (-\sin x)(-\tan x)$
     $\qquad = \sin x \tan x$
     $\qquad = f(x)$

The function defined by $f(x) = \sin(x)\tan(x)$ is an even function.

20.

$$\cos(\pi + t) = -x$$
$$\cos t = x$$
$$\cos(\pi + t) = -\cos t$$

21.

$$\tan(-t) = -\frac{y}{x}$$
$$\tan t = \frac{y}{x}$$
$$\tan(-t) = -\tan t$$

22.
$$1 + \frac{\sin^2 \phi}{\cos^2 \phi} = 1 + \tan^2 \phi \quad [2.4]$$
$$= \sec^2 \phi$$

23.
$$\frac{\tan \phi + 1}{\cot \phi + 1} = \frac{\dfrac{\sin \phi}{\cos \phi} + 1}{\dfrac{\cos \phi}{\sin \phi} + 1} \qquad [2.4]$$
$$= \frac{\dfrac{\sin \phi + \cos \phi}{\cos \phi}}{\dfrac{\cos \phi + \sin \phi}{\sin \phi}}$$
$$= \frac{\sin \phi (\sin \phi + \cos \phi)}{\cos \phi (\cos \phi + \sin \phi)}$$
$$= \tan \phi$$

24.
$$\frac{\cos^2 \phi + \sin^2 \phi}{\csc \phi} = \frac{1}{\csc \phi} \quad [2.4]$$
$$= \sin \phi$$

25.
$$\sin^2 \phi (\tan^2 \phi + 1) = \sin^2 \phi \sec^2 \phi \quad [2.4]$$
$$= \frac{\sin^2 \phi}{\cos^2 \phi}$$
$$= \tan^2 \phi$$

26.
$$1 + \frac{1}{\tan^2 \phi} = \frac{\tan^2 \phi + 1}{\tan^2 \phi} \quad [2.4]$$
$$= \frac{\sec^2 \phi}{\tan^2 \phi}$$
$$= \frac{\dfrac{1}{\cos^2 \phi}}{\dfrac{\sin^2 \phi}{\cos^2 \phi}}$$
$$= \frac{1}{\sin^2 \phi}$$
$$= \csc^2 \phi$$

27.
$$\frac{\cos^2 \phi}{1 - \sin^2 \phi} - 1 = \frac{1 - \sin^2 \phi}{1 - \sin^2 \phi} - 1 \quad [2.4]$$
$$= 1 - 1$$
$$= 0$$

28. $y = 3\cos(2x - \pi) \quad [2.5]$

$$a = |3| = 3; \text{ period} = \frac{2\pi}{b} = \frac{2\pi}{2} = \pi$$

$$\text{phase shift} = -\frac{c}{b} = -\frac{-\pi}{2} = \frac{\pi}{2}$$

29. $y = 2\tan 3x \quad [2.6]$

$$\text{no amplitude; period} = \frac{\pi}{b} = \frac{\pi}{3}$$

$$\text{phase shift} = 0$$

30. $y = -2\sin\left(3x + \dfrac{\pi}{3}\right)$ [2.5]

$a = |-2| = 2$; period $= \dfrac{2\pi}{b} = \dfrac{2\pi}{3}$

phase shift $= -\dfrac{c}{b} = -\dfrac{\pi/3}{3} = -\dfrac{\pi}{9}$

31. $y = \cos\left(2x - \dfrac{2\pi}{3}\right) + 2$ [2.5]

$a = |1| = 1$; period $= \dfrac{2\pi}{b} = \dfrac{2\pi}{2} = \pi$

phase shift $= -\dfrac{c}{b} = -\dfrac{-2\pi/3}{2} = \dfrac{\pi}{3}$

32. $y = -4\sec\left(4x - \dfrac{3\pi}{2}\right)$ [2.6]

no amplitude; period $= \dfrac{2\pi}{b} = \dfrac{2\pi}{4} = \dfrac{\pi}{2}$

phase shift $= -\dfrac{c}{b} = -\dfrac{-3\pi/2}{4} = \dfrac{3\pi}{8}$

33. $y = 2\csc\left(x - \dfrac{\pi}{4}\right) - 3$ [2.6]

no amplitude; period $= \dfrac{2\pi}{b} = \dfrac{2\pi}{1} = 2\pi$

phase shift $= -\dfrac{c}{b} = -\dfrac{-\pi/4}{1} = \dfrac{\pi}{4}$

34. $y = 2\cos\pi x,\ p = \dfrac{2\pi}{\pi} = 2$

35. $y = -\sin\dfrac{2x}{3},\ p = \dfrac{2\pi}{2/3} = 3\pi$

36. $y = 2\sin\dfrac{3x}{2},\ p = \dfrac{2\pi}{3/2} = \dfrac{4\pi}{3}$

37. $y = \cos\left(x - \dfrac{\pi}{2}\right),\ p = 2\pi$

phase shift $= \dfrac{\pi}{2}$

38. $y = \dfrac{1}{2}\sin\left(2x + \dfrac{\pi}{4}\right),\ p = \dfrac{2\pi}{2} = \pi$

phase shift $-\dfrac{\pi}{8}$

39. $y = 3\cos 3(x - \pi),\ p = \dfrac{2\pi}{3}$

phase shift $= \pi$

40. $y = -\tan\dfrac{x}{2},\ p = \dfrac{\pi}{1/2} = 2\pi$

41. $y = 2\cot 2x,\ p = \dfrac{\pi}{2}$

42. $y = \tan\left(x - \dfrac{\pi}{2}\right),\ p = \pi$

phase shift $= \dfrac{\pi}{2}$

43. $y = -\cot\left(2x + \dfrac{\pi}{4}\right),\ p = \dfrac{\pi}{2}$

phase shift $= -\dfrac{\pi}{8}$

44. $y = -2\csc\left(2x - \dfrac{\pi}{3}\right),\ p = \dfrac{2\pi}{2} = \pi$

phase shift $= \dfrac{\pi}{6}$

45. $y = 3\sec\left(x + \dfrac{\pi}{4}\right),\ p = 2\pi$

phase shift $= -\dfrac{\pi}{4}$

46. $y = 3\sin 2x - 3$

47. $y = 2\cos 3x + 3$

48. $y = -\cos\left(3x + \dfrac{\pi}{2}\right) + 2$

49. $y = 3\sin\left(4x - \dfrac{2\pi}{3}\right) - 3$

50. $y = 2 - \sin 2x$

51. $y = \sin x - \sqrt{3}\cos x$

52.

$\sin 4.5° = \dfrac{h}{1.14}$

$h = 1.14\sin 4.5° \approx 0.089$ mi

[2.2]

53.

$\tan 55.3° = \dfrac{h}{8.55}$

$h = 8.55\tan 55.3 \approx 12.3$ feet

[2.2]

54.

Speed for inner ring:

$v = \dfrac{s}{t}$

$= \dfrac{2\pi(14.5)}{24}$

≈ 3.79609 ft/s

Speed for outer ring:

$v = \dfrac{s}{t}$

$= \dfrac{2\pi(21)}{24}$

≈ 5.497787 ft/s

The outer swing has a greater speed of
$5.497787 - 3.79609 \approx 1.7$ ft/s. [2.1]

55.

(1) $\cot 18° = \dfrac{80 + x}{h} = \dfrac{80}{h} + \dfrac{x}{h}$

(2) $\cot 37° = \dfrac{x}{h}$

Substitute for $\dfrac{x}{h}$ in equation (1).

$\cot 18° = \dfrac{80}{h} + \cot 37°$

Solve for h. $\dfrac{80}{h} = \cot 18° - \cot 37°$

$\dfrac{h}{80} = \dfrac{1}{\cot 18° - \cot 37°}$

$h = \dfrac{80}{\cot 18° - \cot 37°} \approx 46$ ft

[2.2]

56. $y = 2.5 \sin 50t$ [2.8]

amplitude $= 2.5$

$p = \dfrac{2\pi}{b} = \dfrac{2\pi}{50} = \dfrac{\pi}{25}$

frequency $= \dfrac{1}{p} = \dfrac{25}{\pi}$

57. amplitude $= 0.5$ [2.8]

$f = \dfrac{1}{2\pi}\sqrt{\dfrac{k}{m}} = \dfrac{1}{2\pi}\sqrt{\dfrac{20}{5}} = \dfrac{1}{\pi}$

$p = \pi$

$y = -0.5 \cos 2\pi f t = -0.5 \cos 2\pi \left(\dfrac{1}{\pi}\right) t$

$y = -0.5 \cos 2t$

58. $\left| f(t) \right| < 0.01$ for all $t > 7.2$ [2.8]

$\text{Xmin} = 5, \text{Xmax} = 10, \text{Xcsl} = 1$

$\text{Ymin} = -.01, \text{Ymax} = .01, \text{Yscl} = .005$

· ·

Quantitative Reasoning

QR1. a. $\dfrac{2\pi}{3\pi} = \dfrac{m}{n} \Rightarrow 2\pi n = 3\pi m \Rightarrow$ period $= 6\pi$

$2\pi(3) = 3\pi(2)$

b. $\dfrac{2/3}{4} = \dfrac{m}{n} \Rightarrow \dfrac{2}{3} n = 4m \Rightarrow$ period $= 4$

$\dfrac{2}{3}(6) = 4(1)$

c. $\dfrac{\pi/2}{2\pi/3} = \dfrac{m}{n} \Rightarrow \dfrac{\pi}{2} n = \dfrac{2\pi}{3} m \Rightarrow$ period $= 2\pi$

$\dfrac{\pi}{2}(4) = \dfrac{2\pi}{3}(3)$

d. $\dfrac{3\pi/2}{8\pi/3} = \dfrac{m}{n} \Rightarrow \dfrac{3\pi}{2} n = \dfrac{8\pi}{3} m \Rightarrow$ period $= 24\pi$

$\dfrac{3\pi}{2}(16) = \dfrac{8\pi}{3}(9)$

e. $\dfrac{5/2}{3/2} = \dfrac{m}{n} \Rightarrow \dfrac{5}{2} n = \dfrac{3}{2} m \Rightarrow$ period $= 7.5$

$\dfrac{5}{2}(3) = \dfrac{3}{2}(5)$

f. $\dfrac{4\pi/5}{4\pi} = \dfrac{m}{n} \Rightarrow \dfrac{4\pi}{5} n = 4\pi m \Rightarrow$ period $= 4\pi$

$\dfrac{4\pi}{5}(5) = 4\pi(1)$

QR3. $\dfrac{1.25}{2.25} = \dfrac{m}{n} \Rightarrow 1.25n = 2.25m \Rightarrow$ period $= 11.25$ s

$1.25(9) = 2.25(5)$

· ·

Chapter Test

1. $150° = 150°\left(\dfrac{\pi}{180°}\right)$ [2.1]

$= \dfrac{5\pi}{6}$

2. $\pi - \dfrac{11}{12}\pi = \dfrac{\pi}{12}$ [2.1]

3. $s = r\theta$ [2.1]

$s = 10(75°)\left(\dfrac{\pi}{180°}\right)$

$s \approx 13.1$ cm

4. $w = 6\dfrac{rev}{sec}$ [2.1]

$w = 6\dfrac{rev}{sec}\left(\dfrac{2\pi \ rad}{rev}\right)$

$w = 12\pi$ rad/sec

5. $v = rw$ [2.1]

$= 8 \cdot 10$

$= 80$ cm/sec

6.

$r = \sqrt{7^2 + 3^2}$

$r = \sqrt{58}$

$\sec\theta = \dfrac{\sqrt{58}}{7}$ [2.2]

7. $\csc 67° \approx 1.0864$ [2.2]

8. $\tan\dfrac{\pi}{6}\cos\dfrac{\pi}{3} - \sin\dfrac{\pi}{2} = \dfrac{1}{\sqrt{3}}\cdot\dfrac{1}{2} - 1$ [2.3]

$$= \dfrac{1}{2\sqrt{3}} - 1$$

$$= \dfrac{\sqrt{3}}{6} - 1$$

$$= \dfrac{\sqrt{3}-6}{6}$$

9. $t = \dfrac{11\pi}{6}$ [2.4]

$x = \cos t \qquad\quad y = \sin t$

$\quad = \dfrac{\sqrt{3}}{2} \qquad\quad = -\dfrac{1}{2}$

$W(x,y) = W\left(\dfrac{\sqrt{3}}{2}, -\dfrac{1}{2}\right)$

10. $\dfrac{\sec^2 t - 1}{\sec^2 t} = \dfrac{\dfrac{1}{\cos^2 t} - 1}{\dfrac{1}{\cos^2 t}}$ [2.4]

$$= \dfrac{\dfrac{1-\cos^2 t}{\cos^2 t}}{\dfrac{1}{\cos^2 t}}$$

$$= 1 - \cos^2 t$$

$$= \sin^2 t$$

11. $\text{period} = \dfrac{\pi}{b} = \dfrac{\pi}{3}$ [2.6]

12. $a = |-3| = 3; \ \text{period} = \dfrac{2\pi}{b} = \dfrac{2\pi}{2} = \pi$ [2.7]

$\text{phase shift} = -\dfrac{\pi}{4}$

13. $\text{period} = \dfrac{\pi}{\pi/3} = 3$ [2.7]

$\text{phase shift} = -\dfrac{c}{b} = -\dfrac{\pi/6}{\pi/3} = -\dfrac{1}{2}$

14. $y = 3\cos\dfrac{1}{2}x, \ p = 4\pi$

15. $y = -2\sec\dfrac{1}{2}x, \qquad p = 4\pi$

16. Shift the graph [of $y = 2\sin(2x)$]   [2.7]

$\dfrac{\pi}{4}$ units to the right and down 1 unit.

17. $y = 2 - \sin\dfrac{x}{2}$

18. $y = \sin x - \cos 2x$

19.

$$\tan 42.2° = \frac{h}{x}$$

$$x = \frac{h}{\tan 42.2°}$$

$$= h \cot 422.2°$$

$$\tan 37.4° = \frac{h}{5.24x}$$

$$= \frac{h}{5.24 + h \cot 42.2°}$$

Solve for h.

$$\tan 37.4° = \frac{h}{5.24 + h \cot 42.2°}$$

$$\tan 37.4°(5.24 + h \cot 42.2°) = h$$

$$5.24 \tan 37.4° + h \tan 37.4° \cot 42.2° = h$$

$$h - h \tan 37.4° \cot 42.2° = 5.24 \tan 37.4°$$

$$h(1 - \tan 37.4° \cot 42.2°) = 5.24 \tan 37.4°$$

$$h = \frac{5.24 \tan 37.4°}{1 - \tan 37.4° \cot 42.2°}$$

$$h \approx 25.5 \text{ meters}$$

The height of the tree is approximately 25.5 meters. [2.2]

20. $p = 5$ $\quad 5 = \frac{2\pi}{b}, b = \frac{2\pi}{5}$

$a = 13$

$$y = 13 \cos \frac{2\pi}{5} t \text{ or } y = 13 \sin \frac{2\pi}{5} t \quad [2.8]$$

Cumulative Review

1. $d = \sqrt{(4-(-3))^2 + (1-2)^2} = \sqrt{49+1} = \sqrt{50}$ [1.1]

2. $c^2 = a^2 + b^2$ [1.1]

$$1^2 = \left(\frac{1}{2}\right)^2 + b^2$$

$$\frac{3}{4} = b^2$$

$$\frac{\sqrt{3}}{2} = b$$

3. Intercepts: (–3, 0), (3, 0), (–9, 0) [1.2]

4. $f(-x) = \frac{-x}{(-x)^2+1} = \frac{-x}{x^2+1} = -f(x)$ [1.4]

Odd function

5.
$$f(x) = \frac{x}{2x-3} \qquad [1.5]$$

$$x = \frac{y}{2y-3}$$

$$x(2y-3) = 2xy - 3x = y$$

$$2xy - y = y(2x-1) = 3x$$

$$y = \frac{3x}{2x-1}$$

$$f^{-1}(x) = \frac{3x}{2x-1}$$

6. Domain: $(-\infty, 4) \cup (4, \infty)$ [1.3]

7.
$$x^2 + x - 6 = 0$$

$$(x+3)(x-2) = 0$$

$$x+3 = 0 \qquad x - 2 = 0$$

$$x = -3 \qquad x = 2$$

The solutions are −3 and 2.   [1.1]

8. Shift the graph of $y = f(x)$ horizontally 3 units to the right. [1.4]

9. Reflect the graph of $y = f(x)$ across the y-axis. [1.4[

10.
$$300° = 300° \left(\frac{\pi}{180°} \right) = \frac{5\pi}{3} \quad [2.1]$$

11.
$$\frac{5\pi}{4} = \frac{5\pi}{4} \left(\frac{180°}{\pi} \right) = 225° \quad [2.1]$$

12.
$$f\left(\frac{\pi}{3} \right) = \sin\left(\frac{\pi}{3} + \frac{\pi}{6} \right) = \sin\left(\frac{\pi}{2} \right) = 1 \quad [2.3]$$

13.
$$f\left(\frac{\pi}{3} \right) = \sin\left(\frac{\pi}{3} \right) + \sin\left(\frac{\pi}{6} \right) = \frac{\sqrt{3}}{2} + \frac{1}{2} = \frac{\sqrt{3}+1}{2} \quad [2.2]$$

14.
$$\cos^2 45° + \sin^2 60° + = \left(\frac{\sqrt{2}}{2} \right)^2 + \left(\frac{\sqrt{3}}{2} \right)^2 = \frac{2}{4} + \frac{3}{4} = \frac{5}{4} \quad [2.2]$$

15. negative [2.3]

16.
$$\theta = 210° \quad [2.3]$$
Since $180° < \theta < 270°$,
$$\theta' + 180° = \theta$$
$$\theta' = 30°$$

17.
$$\theta = \frac{2\pi}{3} \quad [2.3]$$
Since $\frac{\pi}{2} < \theta < \pi$,
$$\theta + \theta' = \pi$$
$$\theta' = \frac{\pi}{3}$$

18. Domain: $(-\infty, \infty)$ [2.4]

19. Range: $[-1, 1]$ [2.4]

20.
$$\tan \theta = \frac{\text{opp}}{\text{adj}} = \frac{3}{4}$$

$$\text{hypotenuse} = \sqrt{3^2 + 4^2}$$
$$= \sqrt{9 + 16}$$
$$= \sqrt{25}$$
$$= 5$$

$$\sin \theta = \frac{\text{opp}}{\text{hyp}} = \frac{3}{5} \quad [2.2]$$

Chapter 3
Trigonometric Identities and Equations

Section 3.1

1. $\tan x \csc x \cos x = \dfrac{\sin x}{\cos x} \cdot \dfrac{1}{\sin x} \cdot \cos x = 1$

3. $\dfrac{4\sin^2 x - 1}{2\sin x + 1} = \dfrac{(2\sin x - 1)(2\sin x + 1)}{2\sin x + 1} = 2\sin x - 1$

5. $(\sin x - \cos x)(\sin x + \cos x) = \sin^2 x - \cos^2 x$
$$= 1 - \cos^2 x - \cos^2 x$$
$$= 1 - 2\cos^2 x$$

7. $\dfrac{1}{\sin x} - \dfrac{1}{\cos x} = \dfrac{\cos x}{\sin x \cos x} - \dfrac{\sin x}{\sin x \cos x}$
$$= \dfrac{\cos x - \sin x}{\sin x \cos x}$$

9. $\dfrac{\cos x}{1 - \sin x} = \dfrac{\cos x(1 + \sin x)}{(1 - \sin x)(1 + \sin x)}$
$$= \dfrac{\cos x(1 + \sin x)}{1 - \sin^2 x}$$
$$= \dfrac{\cos x(1 + \sin x)}{\cos^2 x}$$
$$= \dfrac{(1 + \sin x)}{\cos x} = \dfrac{1}{\cos x} + \dfrac{\sin x}{\cos x}$$
$$= \sec x + \tan x$$

11. $\dfrac{1 - \tan^4 x}{\sec^2 x} = \dfrac{(1 + \tan^2 x)(1 - \tan^2 x)}{\sec^2 x}$
$$= \dfrac{\sec^2 x(1 - \tan^2 x)}{\sec^2 x}$$
$$= 1 - \tan^2 x$$

13. $\dfrac{1 + \tan^3 x}{1 + \tan x} = \dfrac{(1 + \tan x)(1 - \tan x + \tan^2 x)}{1 + \tan x}$
$$= 1 - \tan x + \tan^2 x$$

15. $\dfrac{\sin x - 2 + \dfrac{1}{\sin x}}{\sin x - \dfrac{1}{\sin x}} = \dfrac{\sin x - 2 + \dfrac{1}{\sin x}}{\sin x - \dfrac{1}{\sin x}} \cdot \dfrac{\sin x}{\sin x}$
$$= \dfrac{\sin^2 x - 2\sin x + 1}{\sin^2 x - 1}$$
$$= \dfrac{(\sin x - 1)(\sin x - 1)}{(\sin x - 1)(\sin x + 1)}$$
$$= \dfrac{\sin x - 1}{\sin x + 1}$$

17. $(\sin x + \cos x)^2 = \sin^2 x + 2\sin x \cos x + \cos^2 x$
$$= \sin^2 x + \cos^2 x + 2\sin x \cos x$$
$$= 1 + 2\sin x \cos x$$

19. $\dfrac{\cos x}{1 + \sin x} = \dfrac{\cos x(1 - \sin x)}{(1 + \sin x)(1 - \sin x)}$
$$= \dfrac{\cos x(1 - \sin x)}{1 - \sin^2 x}$$
$$= \dfrac{\cos x(1 - \sin x)}{\cos^2 x}$$
$$= \dfrac{1 - \sin x}{\cos x}$$
$$= \dfrac{1}{\cos x} - \dfrac{\sin x}{\cos x}$$
$$= \sec x - \tan x$$

21. $\dfrac{\cot x + \tan x}{\sec x} = \dfrac{\dfrac{\cos x}{\sin x} + \dfrac{\sin x}{\cos x}}{\dfrac{1}{\cos x}}$

$\qquad = \dfrac{\dfrac{\cos x}{\sin x} + \dfrac{\sin x}{\cos x}}{\dfrac{1}{\cos x}} \cdot \dfrac{\sin x \cos x}{\sin x \cos x}$

$\qquad = \dfrac{\cos^2 x + \sin^2 x}{\sin x}$

$\qquad = \dfrac{1}{\sin x}$

$\qquad = \csc x$

23. $\dfrac{\cos x \tan x + 2\cos x - \tan x - 2}{\tan x + 2} = \dfrac{\cos x(\tan x + 2) - (\tan x + 2)}{\tan x + 2}$

$\qquad = \dfrac{(\tan x + 2)(\cos x - 1)}{\tan x + 2}$

$\qquad = \cos x - 1$

25. $\dfrac{1 - \sin x}{\cos x} = \dfrac{1}{\cos x} - \dfrac{\sin x}{\cos x} = \sec x - \tan x$

27. $\sin^2 x - \cos^2 x = \sin^2 x - (1 - \sin^2 x)$

$\qquad = \sin^2 x - 1 + \sin^2 x$

$\qquad = 2\sin^2 x - 1$

29. $\dfrac{1}{\sin^2 x} + \dfrac{1}{\cos^2 x} = \dfrac{\cos^2 x + \sin^2 x}{\sin^2 x \cos^2 x}$

$\qquad = \dfrac{1}{\sin^2 x \cos^2 x}$

$\qquad = \csc^2 x \sec^2 x$

31. $\sec x - \cos x = \dfrac{1}{\cos x} - \cos x$

$\qquad = \dfrac{1 - \cos^2 x}{\cos x}$

$\qquad = \dfrac{\sin^2 x}{\cos x}$

$\qquad = \sin x \tan x$

33. $\dfrac{\dfrac{1}{\sin x} + 1}{\dfrac{1}{\sin x} - 1} = \dfrac{\dfrac{1}{\sin x} + 1}{\dfrac{1}{\sin x} - 1} \cdot \dfrac{\sin x}{\sin x}$

$\qquad = \dfrac{1 + \sin x}{1 - \sin x}$

$\qquad = \dfrac{(1 + \sin x)}{1 - \sin x} \cdot \dfrac{(1 + \sin x)}{1 + \sin x}$

$\qquad = \dfrac{1 + 2\sin x + \sin^2 x}{1 - \sin^2 x}$

$\qquad = \dfrac{1 + 2\sin x + \sin^2 x}{\cos^2 x}$

$\qquad = \dfrac{1}{\cos^2 x} + \dfrac{2\sin x}{\cos^2 x} + \dfrac{\sin^2 x}{\cos^2 x}$

$\qquad = \sec^2 x + 2\tan x \sec x + \tan^2 x$

35. $\sin^4 x - \cos^4 x = (\sin^2 x + \cos^2 x)(\sin^2 x - \cos^2 x)$

$\qquad = 1(\sin^2 x - \cos^2 x)$

$\qquad = \sin^2 x - (1 - \sin^2 x)$

$\qquad = \sin^2 x - 1 + \sin^2 x$

$\qquad = 2\sin^2 x - 1$

37. $\dfrac{1}{1 - \cos x} = \dfrac{1}{1 - \cos x} \cdot \dfrac{1 + \cos x}{1 + \cos x}$

$\qquad = \dfrac{1 + \cos x}{1 - \cos^2 x}$

$\qquad = \dfrac{1 + \cos x}{\sin^2 x}$

39. $\dfrac{\sin x}{1 - \sin x} - \dfrac{\cos x}{1 - \sin x} = \dfrac{\sin x - \cos x}{1 - \sin x}$

$\qquad = \dfrac{\dfrac{\sin x}{\sin x} - \dfrac{\cos x}{\sin x}}{\dfrac{1}{\sin x} - \dfrac{\sin x}{\sin x}}$

$\qquad = \dfrac{1 - \cot x}{\csc x - 1}$

41. $\dfrac{1}{1+\cos x} - \dfrac{1}{1-\cos x} = \dfrac{(1-\cos x)-(1+\cos x)}{(1+\cos x)(1-\cos x)}$

$$= \dfrac{1-\cos x - 1 - \cos x}{1-\cos^2 x}$$

$$= \dfrac{-2\cos x}{\sin^2 x}$$

$$= -2\cot x \csc x$$

43. $\dfrac{\dfrac{1}{\sin x}+\csc x}{\dfrac{1}{\sin x}-\sin x} = \dfrac{\dfrac{1}{\sin x}+\csc x}{\dfrac{1}{\sin x}-\sin x} \cdot \dfrac{\sin x}{\sin x}$

$$= \dfrac{1+1}{1-\sin^2 x}$$

$$= \dfrac{2}{\cos^2 x}$$

45. $\dfrac{\cot x}{1+\csc x} + \dfrac{1+\csc x}{\cot x} = \dfrac{\dfrac{\cos x}{\sin x}}{1+\dfrac{1}{\sin x}} + \dfrac{1+\dfrac{1}{\sin x}}{\dfrac{\cos x}{\sin x}}$

$$= \dfrac{\cos x}{\sin x + 1} + \dfrac{\sin x + 1}{\cos x}$$

$$= \dfrac{\cos^2 x + (\sin x + 1)^2}{\cos x (\sin x + 1)}$$

$$= \dfrac{\cos^2 x + \sin^2 x + 2\sin x + 1}{\cos x(\sin x + 1)}$$

$$= \dfrac{1 + 2\sin x + 1}{\cos x(\sin x + 1)}$$

$$= \dfrac{2(1+\sin x)}{\cos x(\sin x + 1)}$$

$$= 2\sec x$$

47. $\sqrt{\dfrac{1+\sin x}{1-\sin x}} = \sqrt{\dfrac{1+\sin x}{1-\sin x} \cdot \dfrac{1+\sin x}{1+\sin x}}$

$$= \sqrt{\dfrac{(1+\sin x)^2}{1-\sin^2 x}}$$

$$= \sqrt{\dfrac{(1+\sin x)^2}{\cos^2 x}}$$

$$= \dfrac{1+\sin x}{\cos x}, \quad \cos x > 0$$

49. $\dfrac{\sin^3 x + \cos^3 x}{\sin x + \cos x} = \dfrac{(\sin x + \cos x)(\sin^2 x - \sin x \cos x + \cos^2 x)}{\sin x + \cos x}$

$$= \sin^2 x - \sin x \cos x + \cos^2 x$$

$$= 1 - \sin x \cos x$$

51. $\dfrac{\sec x - 1}{\sec x + 1} - \dfrac{\sec x + 1}{\sec x - 1} = \dfrac{(\sec x - 1)^2 - (\sec x + 1)^2}{(\sec x - 1)(\sec x + 1)}$

$$= \dfrac{\sec^2 x - 2\sec x + 1 - \sec^2 x - 2\sec x - 1}{\sec^2 x - 1}$$

$$= \dfrac{-4\sec x}{\tan^2 x}$$

$$= \dfrac{-4}{\cos x} \cdot \dfrac{\cos^2 x}{\sin^2 x}$$

$$= -4\csc x \cot x$$

53. $\dfrac{1+\sin x}{\cos x} - \dfrac{\cos x}{1-\sin x} = \dfrac{(1+\sin x)(1-\sin x) - \cos x(\cos x)}{\cos x(1-\sin x)} = \dfrac{1 - \sin^2 x - \cos^2 x}{\cos x(1-\sin x)} = \dfrac{\cos^2 x - \cos^2 x}{\cos x(1-\sin x)} = 0$

55. $\dfrac{\sec x + \tan x}{\sec x - \tan x} = \dfrac{\dfrac{1}{\cos x}+\dfrac{\sin x}{\cos x}}{\dfrac{1}{\cos x}-\dfrac{\sin x}{\cos x}} \cdot \dfrac{\cos x}{\cos x}$

$$= \dfrac{1+\sin x}{1-\sin x}$$

$$= \dfrac{1+\sin x}{1-\sin x} \cdot \dfrac{1+\sin x}{1+\sin x}$$

$$= \dfrac{(1+\sin x)^2}{1-\sin^2 x}$$

$$= \dfrac{(1+\sin x)^2}{\cos^2 x}$$

57. Identity

59. Identity

61. Identity

63. Not an identity

65. Not an identity. If $x = \pi/4$, the left side is 2 and the right side is 1.

67. Not an identity. If $x = 0°$, the left side is $\sqrt{3}/2$ and the right side is $(2+\sqrt{3})/2$.

69. Not an identity. If $x = 0$, the left side is -1 and the right side is 1.

••

Connecting Concepts

71.
$$\frac{1-\sin x+\cos x}{1+\sin x+\cos x} = \frac{1-\sin x+\cos x}{1+\sin x+\cos x} \cdot \frac{1+\sin x+\cos x}{1+\sin x-\cos x}$$
$$= \frac{1-\sin^2 x+2\sin x\cos x-\cos^2 x}{1+2\sin x+\sin^2 x-\cos^2 x}$$
$$= \frac{1-(\sin^2 x+\cos^2 x)+2\sin x\cos x}{1+2\sin x+\sin^2 x-(1-\sin^2 x)}$$
$$= \frac{1-1+2\sin x\cos x}{1+2\sin x+\sin^2 x-1+\sin^2 x}$$
$$= \frac{2\sin x\cos x}{2\sin x+2\sin^2 x} = \frac{2\sin x\cos x}{2\sin x(1+\sin x)}$$
$$= \frac{\cos x}{1+\sin x}$$

73.
$$\frac{2\sin^4 x+2\sin^2 x\cos^2 x-3\sin^2 x-3\cos^2 x}{2\sin^2 x} = \frac{2\sin^2 x(\sin^2 x+\cos^2 x)-3(\sin^2 x+\cos^2 x)}{2\sin^2 x}$$
$$= \frac{(2\sin^2 x-3)(\sin^2 x+\cos^2 x)}{2\sin^2 x}$$
$$= \frac{2\sin^2 x-3}{2\sin^2 x}$$
$$= \frac{2\sin^2 x}{2\sin^2 x}-\frac{3}{2\sin^2 x}$$
$$= 1-\frac{3}{2}\csc^2 x$$

75.
$$\frac{\sin x(\tan x+1)-2\tan x\cos x}{\sin x-\cos x}=\frac{\sin x\tan x+\sin x-2\dfrac{\sin x}{\cos x}\cos x}{\sin x-\cos x}$$

$$=\frac{\sin x\tan x+\sin x-2\sin x}{\sin x-\cos x}$$

$$=\frac{\sin x(\tan x-1)}{\sin x-\cos x}$$

$$=\frac{\dfrac{\sin x(\tan x-1)}{\cos x}}{\dfrac{\sin x}{\cos x}-\dfrac{\cos x}{\cos x}}$$

$$=\frac{\tan x(\tan x-1)}{\tan x-1}$$

$$=\tan x$$

77.
$$\sin^4 x+\cos^4 x=\sin^4 x+2\sin^2 x\cos^2 x+\cos^4 x-2\sin^2 x\cos^2 x$$

$$=(\sin^2 x+\cos^2 x)^2-2\sin^2 x\cos^2 x$$

$$=1-2\sin^2 x\cos^2 x$$

•••   **Prepare for Section 3.2**

PS1. $\cos\left(\dfrac{\pi}{2}-\dfrac{\pi}{6}\right)=\cos\left(\dfrac{\pi}{3}\right)=\dfrac{1}{2}$

$\cos\left(\dfrac{\pi}{2}\right)\cos\left(\dfrac{\pi}{6}\right)+\sin\left(\dfrac{\pi}{2}\right)\sin\left(\dfrac{\pi}{6}\right)=0\cdot\dfrac{\sqrt{3}}{2}+1\cdot\dfrac{1}{2}=\dfrac{1}{2}$

Both functional values equal $\dfrac{1}{2}$.

PS2. $\sin\left(\dfrac{\pi}{2}+\dfrac{\pi}{3}\right)=\sin\left(\dfrac{5\pi}{6}\right)=\dfrac{1}{2}$

$\sin\left(\dfrac{\pi}{2}\right)\cos\left(\dfrac{\pi}{3}\right)+\cos\left(\dfrac{\pi}{2}\right)\sin\left(\dfrac{\pi}{3}\right)=1\cdot\dfrac{1}{2}+0\cdot\dfrac{\sqrt{3}}{2}=\dfrac{1}{2}$

Both functional values equal to $\dfrac{1}{2}$.

PS3. $\sin(90°-30°)=\sin(60°)=\dfrac{\sqrt{3}}{2}=\cos(30°)$

$\sin(90°-45°)=\sin(45°)=\dfrac{\sqrt{2}}{2}=\cos(45°)$

$\sin(90°-120°)=\sin(-30°)=-\dfrac{1}{2}=\cos(120°)$

For each of the given values of θ, the functional values are equal.

PS4. $\tan\left(\dfrac{\pi}{2}-\dfrac{\pi}{6}\right)=\tan\left(\dfrac{\pi}{3}\right)=\cot\left(\dfrac{\pi}{6}\right)$

$\tan\left(\dfrac{\pi}{2}-\dfrac{\pi}{4}\right)=\tan\left(\dfrac{\pi}{4}\right)=\cot\left(\dfrac{\pi}{4}\right)$

$\tan\left(\dfrac{\pi}{2}-\dfrac{4\pi}{3}\right)=\tan\left(-\dfrac{5\pi}{6}\right)=\cot\left(\dfrac{4\pi}{3}\right)$

For each of the given values of θ, the functional values are equal.

PS5. $\tan\left(\dfrac{\pi}{3}-\dfrac{\pi}{6}\right)=\tan\left(\dfrac{\pi}{6}\right)=\dfrac{\sqrt{3}}{3}$

$\dfrac{\tan\left(\dfrac{\pi}{3}\right)-\tan\left(\dfrac{\pi}{6}\right)}{1+\tan\left(\dfrac{\pi}{3}\right)\tan\left(\dfrac{\pi}{6}\right)}=\dfrac{\sqrt{3}-\dfrac{\sqrt{3}}{3}}{1+\sqrt{3}\cdot\dfrac{\sqrt{3}}{3}}=\dfrac{\dfrac{3\sqrt{3}}{3}-\dfrac{\sqrt{3}}{3}}{1+1}=\dfrac{\dfrac{2\sqrt{3}}{3}}{2}=\dfrac{\sqrt{3}}{3}$

Both functional values equal $\dfrac{\sqrt{3}}{3}$.

PS6. For k is any integer, the value of $(2k+1)$ will result in odd integers.

Thus $\sin[(2k+1)\pi]$ will be 0.

Section 3.2

1. $\sin(45° + 30°) = \sin 45° \cos 30° + \cos 45° \sin 30°$

$$= \frac{\sqrt{2}}{2} \cdot \frac{\sqrt{3}}{2} + \frac{\sqrt{2}}{2} \cdot \frac{1}{2}$$

$$= \frac{\sqrt{6}}{4} + \frac{\sqrt{2}}{4} = \frac{\sqrt{6} + \sqrt{2}}{4}$$

3. $\cos(45° - 30°) = \cos 45° \cos 30° + \sin 45° \sin 30°$

$$= \frac{\sqrt{2}}{2} \cdot \frac{\sqrt{3}}{2} + \frac{\sqrt{2}}{2} \cdot \frac{1}{2}$$

$$= \frac{\sqrt{6}}{4} + \frac{\sqrt{2}}{4} = \frac{\sqrt{6} + \sqrt{2}}{4}$$

5. $\tan(45° - 30°) = \dfrac{\tan 45° - \tan 30°}{1 + \tan 45° \tan 30°}$

$$= \frac{1 - \dfrac{\sqrt{3}}{3}}{1 + 1\left(\dfrac{\sqrt{3}}{3}\right)} = \frac{\dfrac{3 - \sqrt{3}}{3}}{\dfrac{3 + \sqrt{3}}{3}} = \frac{3 - \sqrt{3}}{3 + \sqrt{3}} = 2 - \sqrt{3}$$

7. $\sin\left(\dfrac{5\pi}{4} - \dfrac{\pi}{6}\right) = \sin\dfrac{5\pi}{4}\cos\dfrac{\pi}{6} - \cos\dfrac{5\pi}{4}\sin\dfrac{\pi}{6}$

$$= -\frac{\sqrt{2}}{2} \cdot \frac{\sqrt{3}}{2} - \left(-\frac{\sqrt{2}}{2}\right) \cdot \frac{1}{2}$$

$$= -\frac{\sqrt{6}}{4} + \frac{\sqrt{2}}{4} = \frac{-\sqrt{6} + \sqrt{2}}{4}$$

9. $\cos\left(\dfrac{3\pi}{4} + \dfrac{\pi}{6}\right) = \cos\dfrac{3\pi}{4}\cos\dfrac{\pi}{6} - \sin\dfrac{3\pi}{4}\sin\dfrac{\pi}{6}$

$$= -\frac{\sqrt{2}}{2} \cdot \frac{\sqrt{3}}{2} - \frac{\sqrt{2}}{2} \cdot \frac{1}{2}$$

$$= -\frac{\sqrt{6}}{4} - \frac{\sqrt{2}}{4} = -\frac{\sqrt{6} + \sqrt{2}}{4}$$

11. $\tan\left(\dfrac{\pi}{6} + \dfrac{\pi}{4}\right) = \dfrac{\tan\dfrac{\pi}{6} + \tan\dfrac{\pi}{4}}{1 - \tan\dfrac{\pi}{6}\tan\dfrac{\pi}{4}}$

$$= \frac{\dfrac{\sqrt{3}}{3} + 1}{1 - \dfrac{\sqrt{3}}{3} \cdot 1} = \frac{\dfrac{\sqrt{3} + 3}{3}}{\dfrac{3 - \sqrt{3}}{3}}$$

$$= \frac{\sqrt{3} + 3}{3 - \sqrt{3}} \cdot \frac{3 + \sqrt{3}}{3 + \sqrt{3}} = \frac{9 + 6\sqrt{3} + 3}{9 - 3}$$

$$= \frac{12 + 6\sqrt{3}}{6} = 2 + \sqrt{3}$$

13. $\cos 212° \cos 122° + \sin 212° \sin 122° = \cos(212° - 122°) = \cos 90° = 0$

15. $\sin\dfrac{5\pi}{12}\cos\dfrac{\pi}{4} - \cos\dfrac{5\pi}{12}\sin\dfrac{\pi}{4} = \sin\left(\dfrac{5\pi}{12} - \dfrac{\pi}{4}\right) = \sin\dfrac{\pi}{6} = \dfrac{1}{2}$

17. $\dfrac{\tan\dfrac{7\pi}{12} - \tan\dfrac{\pi}{4}}{1 + \tan\dfrac{7\pi}{12}\tan\dfrac{\pi}{4}} = \tan\left(\dfrac{7\pi}{12} - \dfrac{\pi}{4}\right) = \tan\dfrac{\pi}{3} = \sqrt{3}$

19. $\sin 42° = \cos(90° - 42°)$
$ = \cos 48°$

21. $\tan 15° = \cot(90° - 15°)$
$ = \cot 75°$

23. $\sec 25° = \csc(90° - 25°)$
$ = \csc 65°$

25. $\sin 7x \cos 2x - \cos 7x \sin 2x = \sin(7x - 2x) = \sin 5x$

27. $\cos x \cos 2x + \sin x \sin 2x = \cos(x - 2x) = \cos(-x) = \cos x$

29. $\sin 7x \cos 3x - \cos 7x \sin 3x = \sin(7x - 3x) = \sin 4x$

31. $\cos 4x \cos(-2x) - \sin 4x \sin(-2x) = \cos 4x \cos 2x + \sin 4x \sin 2x$

$$= \cos(4x - 2x)$$

$$= \cos 2x$$

33. $\sin\dfrac{x}{3}\cos\dfrac{2x}{3} + \cos\dfrac{x}{3}\sin\dfrac{2x}{3} = \sin\left(\dfrac{x}{3} + \dfrac{2x}{3}\right) = \sin x$

35. $\dfrac{\tan 3x + \tan 4x}{1 - \tan 3x \tan 4x} = \tan(3x + 4x) = \tan 7x$

37. $\tan\alpha = -\dfrac{4}{3}$, $\sin\alpha = \dfrac{4}{5}$, $\cos\alpha = -\dfrac{3}{5}$,

$\tan\beta = \dfrac{15}{8}$, $\sin\beta = -\dfrac{15}{17}$, $\cos\beta = -\dfrac{8}{17}$

a. $\sin(\alpha - \beta) = \sin\alpha\cos\beta - \cos\alpha\sin\beta$

$$= \left(\dfrac{4}{5}\right)\left(-\dfrac{8}{17}\right) - \left(-\dfrac{3}{5}\right)\left(-\dfrac{15}{17}\right)$$

$$= -\dfrac{32}{85} - \dfrac{45}{85} = -\dfrac{77}{85}$$

b. $\cos(\alpha + \beta) = \cos\alpha\cos\beta - \sin\alpha\sin\beta$

$$= \left(-\dfrac{3}{5}\right)\left(-\dfrac{8}{17}\right) - \left(\dfrac{4}{5}\right)\left(-\dfrac{15}{17}\right)$$

$$= \dfrac{24}{85} + \dfrac{60}{85} = \dfrac{84}{85}$$

c. $\tan(\alpha - \beta) = \dfrac{\tan\alpha - \tan\beta}{1 + \tan\alpha\tan\beta}$

$$= \dfrac{-\dfrac{4}{3} - \dfrac{15}{8}}{1 + \left(-\dfrac{4}{3}\right)\left(\dfrac{15}{8}\right)} = \dfrac{-\dfrac{4}{3} - \dfrac{15}{8}}{1 - \dfrac{60}{24}} \cdot \dfrac{24}{24}$$

$$= \dfrac{-32 - 45}{24 - 60} = \dfrac{77}{36}$$

39. $\sin\alpha = \dfrac{3}{5}$, $\cos\alpha = \dfrac{4}{5}$, $\tan\alpha = \dfrac{3}{4}$,

$\cos\beta = -\dfrac{5}{13}$, $\sin\beta = \dfrac{12}{13}$, $\tan\beta = -\dfrac{12}{5}$

a. $\sin(\alpha - \beta) = \sin\alpha\cos\beta - \cos\alpha\sin\beta$

$$= \dfrac{3}{5}\left(-\dfrac{5}{13}\right) - \dfrac{4}{5}\left(\dfrac{12}{13}\right)$$

$$= -\dfrac{15}{65} - \dfrac{48}{65}$$

$$= -\dfrac{63}{65}$$

b. $\cos(\alpha + \beta) = \cos\alpha\cos\beta - \sin\alpha\sin\beta$

$$= \dfrac{4}{5}\left(-\dfrac{5}{13}\right) - \dfrac{3}{5}\left(\dfrac{12}{13}\right)$$

$$= -\dfrac{20}{65} - \dfrac{36}{65}$$

$$= -\dfrac{56}{65}$$

c. $\tan(\alpha - \beta) = \dfrac{\tan\alpha - \tan\beta}{1 + \tan\alpha\tan\beta}$

$$= \dfrac{\dfrac{3}{4} - \left(-\dfrac{12}{5}\right)}{1 + \left(\dfrac{3}{4}\right)\left(-\dfrac{12}{5}\right)}$$

$$= \dfrac{\dfrac{3}{4} + \dfrac{12}{5}}{1 - \dfrac{36}{20}} \cdot \dfrac{20}{20}$$

$$= \dfrac{15 + 48}{20 - 36} = -\dfrac{63}{16}$$

41. $\sin\alpha = -\dfrac{4}{5}$, $\cos\alpha = -\dfrac{3}{5}$, $\tan\alpha = \dfrac{4}{3}$,

$\cos\beta = -\dfrac{12}{13}$, $\sin\beta = \dfrac{5}{13}$, $\tan\beta = -\dfrac{5}{12}$

a. $\sin(\alpha - \beta) = \sin\alpha\cos\beta - \cos\alpha\sin\beta$

$$= \left(-\dfrac{4}{5}\right)\left(-\dfrac{12}{13}\right) - \left(-\dfrac{3}{5}\right)\dfrac{5}{13}$$

$$= \dfrac{48}{65} + \dfrac{15}{65} = \dfrac{63}{65}$$

b. $\cos(\alpha + \beta) = \cos\alpha\cos\beta - \sin\alpha\sin\beta$

$$= \left(-\dfrac{3}{5}\right)\left(-\dfrac{12}{13}\right) - \left(-\dfrac{4}{5}\right)\dfrac{5}{13}$$

$$= \dfrac{36}{65} + \dfrac{20}{65} = \dfrac{56}{65}$$

c. $\tan(\alpha + \beta) = \dfrac{\tan\alpha + \tan\beta}{1 - \tan\alpha\tan\beta}$

$$= \dfrac{\dfrac{4}{3} + \left(-\dfrac{5}{12}\right)}{1 - \left(\dfrac{4}{3}\right)\left(-\dfrac{5}{12}\right)}$$

$$= \dfrac{\dfrac{4}{3} - \dfrac{5}{12}}{1 + \dfrac{29}{36}} = \dfrac{\dfrac{4}{3} - \dfrac{5}{12}}{1 + \dfrac{20}{36}} \cdot \dfrac{36}{36}$$

$$= \dfrac{48 - 15}{36 + 20} = \dfrac{33}{56}$$

43. $\cos\alpha = \dfrac{15}{17}$, $\sin\alpha = \dfrac{8}{17}$, $\tan\alpha = \dfrac{8}{15}$,

$\sin\beta = -\dfrac{3}{5}$, $\cos\beta = -\dfrac{4}{5}$, $\tan\beta = \dfrac{3}{4}$

a. $\sin(\alpha + \beta) = \sin\alpha\cos\beta + \cos\alpha\sin\beta$

$$= \dfrac{8}{17}\left(-\dfrac{4}{5}\right) + \dfrac{15}{17}\left(-\dfrac{3}{5}\right)$$

$$= -\dfrac{32}{85} - \dfrac{45}{85} = -\dfrac{77}{85}$$

b. $\cos(\alpha - \beta) = \cos\alpha\cos\beta + \sin\alpha\sin\beta$

$$= \dfrac{15}{17}\left(-\dfrac{4}{5}\right) + \dfrac{8}{17}\left(-\dfrac{3}{5}\right)$$

$$= -\dfrac{60}{85} - \dfrac{24}{85} = -\dfrac{84}{85}$$

c. $\tan(\alpha - \beta) = \dfrac{\tan\alpha - \tan\beta}{1 + \tan\alpha\tan\beta}$

$$= \dfrac{\dfrac{8}{15} - \dfrac{3}{4}}{1 + \dfrac{8}{15}\left(\dfrac{3}{4}\right)} = \dfrac{\dfrac{8}{15} - \dfrac{3}{4}}{1 + \dfrac{24}{60}} = \dfrac{\dfrac{8}{15} - \dfrac{3}{4}}{1 + \dfrac{24}{60}} \cdot \dfrac{60}{60}$$

$$= \dfrac{32 - 45}{60 + 24} = -\dfrac{13}{84}$$

45. $\cos\alpha = -\dfrac{3}{5}$, $\sin\alpha = -\dfrac{4}{5}$, $\tan\alpha = \dfrac{4}{3}$,

$\sin\beta = \dfrac{5}{13}$, $\cos\beta = \dfrac{12}{13}$, $\tan\beta = \dfrac{5}{12}$

a. $\sin(\alpha - \beta) = \sin\alpha\cos\beta - \cos\alpha\sin\beta$

$$= \left(-\dfrac{4}{5}\right)\dfrac{12}{13} - \left(-\dfrac{3}{5}\right)\left(\dfrac{5}{13}\right)$$

$$= -\dfrac{48}{65} + \dfrac{15}{65} = -\dfrac{33}{65}$$

b. $\cos(\alpha + \beta) = \cos\alpha\cos\beta - \sin\alpha\sin\beta$

$$= \left(-\dfrac{3}{5}\right)\dfrac{12}{13} - \left(-\dfrac{4}{5}\right)\dfrac{5}{13}$$

$$= -\dfrac{36}{65} + \dfrac{20}{65} = -\dfrac{16}{65}$$

c. $\tan(\alpha + \beta) = \dfrac{\tan\alpha + \tan\beta}{1 - \tan\alpha\tan\beta}$

$$= \dfrac{\dfrac{4}{3} + \dfrac{5}{12}}{1 - \dfrac{4}{3}\left(\dfrac{5}{12}\right)} = \dfrac{\dfrac{4}{3} + \dfrac{5}{12}}{1 - \dfrac{20}{36}}$$

$$= \dfrac{\dfrac{4}{3} + \dfrac{5}{12}}{1 - \dfrac{20}{36}} \cdot \dfrac{36}{36} = \dfrac{48 + 15}{36 - 20} = \dfrac{63}{16}$$

47. $\sin\alpha = \dfrac{3}{5}$, $\cos\alpha = \dfrac{4}{5}$, $\tan\alpha = \dfrac{3}{4}$,

$\tan\beta = \dfrac{5}{12}$, $\sin\beta = -\dfrac{5}{13}$, $\cos\beta = -\dfrac{12}{13}$

a. $\sin(\alpha + \beta) = \sin\alpha\cos\beta + \cos\alpha\sin\beta$

$$= \left(\dfrac{3}{5}\right)\left(-\dfrac{12}{13}\right) + \left(\dfrac{4}{5}\right)\left(-\dfrac{5}{13}\right)$$

$$= -\dfrac{36}{65} - \dfrac{20}{65} = -\dfrac{56}{65}$$

b. $\cos(\alpha - \beta) = \cos\alpha\cos\beta + \sin\alpha\sin\beta$

$$= \left(\dfrac{4}{5}\right)\left(-\dfrac{12}{13}\right) + \left(\dfrac{3}{5}\right)\left(-\dfrac{5}{13}\right)$$

$$= -\dfrac{48}{65} - \dfrac{15}{65} = -\dfrac{63}{65}$$

c. $\tan(\alpha - \beta) = \dfrac{\tan\alpha - \tan\beta}{1 + \tan\alpha\tan\beta}$

$$= \dfrac{\dfrac{3}{4} - \dfrac{5}{12}}{1 + \left(\dfrac{3}{4}\right)\left(\dfrac{5}{12}\right)} = \dfrac{\dfrac{3}{4} - \dfrac{5}{12}}{1 + \dfrac{15}{48}} \cdot \dfrac{48}{48}$$

$$= \dfrac{36 - 20}{48 + 15} = \dfrac{16}{63}$$

49. $\cos\left(\dfrac{\pi}{2} - \theta\right) = \cos\dfrac{\pi}{2}\cos\theta + \sin\dfrac{\pi}{2}\sin\theta$

$$= 0 \cdot \cos\theta + 1 \cdot \sin\theta$$

$$= \sin\theta$$

51. $\sin\left(\theta + \dfrac{\pi}{2}\right) = \sin\theta\cos\dfrac{\pi}{2} + \cos\theta\sin\dfrac{\pi}{2}$

$$= \sin\theta(0) + \cos\theta(1)$$

$$= \cos\theta$$

53. $\tan\left(\theta + \dfrac{\pi}{4}\right) = \dfrac{\tan\theta + \tan\dfrac{\pi}{4}}{1 - \tan\theta\tan\dfrac{\pi}{4}}$

$$= \dfrac{\tan\theta + 1}{1 - \tan\theta}$$

55. $\cos\left(\dfrac{3\pi}{2} - \theta\right) = \cos\dfrac{3\pi}{2}\cos\theta + \sin\dfrac{3\pi}{2}\sin\theta$

$$= 0(\cos\theta) + (-1)\sin\theta$$

$$= -\sin\theta$$

57. $\cot\left(\dfrac{\pi}{2} - \theta\right) = \dfrac{\cos(\pi/2 - \theta)}{\sin(\pi/2 - \theta)}$

$$= \dfrac{\left(\cos\dfrac{\pi}{2}\right)\cos\theta + \left(\sin\dfrac{\pi}{2}\right)\sin\theta}{\left(\sin\dfrac{\pi}{2}\right)\cos\theta - \left(\cos\dfrac{\pi}{2}\right)\sin\theta}$$

$$= \dfrac{(0)\cos\theta + (1)\sin\theta}{(1)\cos\theta - (0)\sin\theta}$$

$$= \dfrac{\sin\theta}{\cos\theta}$$

$$= \tan\theta$$

59. $\csc(\pi - \theta) = \dfrac{1}{\sin(\pi - \theta)}$

$$= \dfrac{1}{\sin\pi\cos\theta - \cos\pi\sin\theta}$$

$$= \dfrac{1}{(0)\cos\theta - (-1)\sin\theta}$$

$$= \dfrac{1}{\sin\theta}$$

$$= \csc\theta$$

61. $\sin 6x\cos 2x - \cos 6x\sin 2x = \sin(6x - 2x)$

$$= \sin 4x$$

$$= \sin(2x + 2x)$$

$$= \sin 2x\cos 2x + \cos 2x\sin 2x$$

$$= 2\sin 2x\cos 2x$$

63. $\cos(\alpha + \beta) + \cos(\alpha - \beta) = \cos\alpha\cos\beta - \sin\alpha\sin\beta + \cos\alpha\cos\beta + \sin\alpha\sin\beta$

$$= 2\cos\alpha\cos\beta$$

65. $\sin(\alpha+\beta)+\sin(\alpha-\beta)=\sin\alpha\cos\beta+\cos\alpha\sin\beta+\sin\alpha\cos\beta-\cos\alpha\sin\beta$
$$=2\sin\alpha\cos\beta$$

67. $\dfrac{\cos(\alpha-\beta)}{\sin(\alpha+\beta)}=\dfrac{\cos\alpha\cos\beta+\sin\alpha\sin\beta}{\sin\alpha\cos\beta+\cos\alpha\sin\beta}$

$$=\dfrac{\dfrac{\cos\alpha\cos\beta}{\sin\alpha\cos\beta}+\dfrac{\sin\alpha\sin\beta}{\sin\alpha\cos\beta}}{\dfrac{\sin\alpha\cos\beta}{\sin\alpha\cos\beta}+\dfrac{\cos\alpha\sin\beta}{\sin\alpha\cos\beta}}$$

$$=\dfrac{\cot\alpha+\tan\beta}{1+\cot\alpha\tan\beta}$$

69. $\dfrac{\sin(x+h)-\sin x}{h}=\dfrac{\sin x\cos h+\cos x\sin h-\sin x}{h}$

$$=\dfrac{\sin x(\cos h-1)}{h}+\dfrac{\cos x\sin h}{h}$$

$$=\sin x\dfrac{\cos h-1}{h}+\cos x\dfrac{\sin h}{h}$$

71. $\sin\left(\dfrac{\pi}{2}+\alpha-\beta\right)=\sin\left[\dfrac{\pi}{2}+(\alpha-\beta)\right]$

$$=\sin\dfrac{\pi}{2}\cos(\alpha-\beta)+\cos\dfrac{\pi}{2}\sin(\alpha-\beta)$$

$$=(1)\cos(\alpha-\beta)+(0)\sin(\alpha-\beta)$$

$$=\cos(\alpha-\beta)$$

$$=\cos\alpha\cos\beta+\sin\alpha\sin\beta$$

73. $\sin 3x=\sin(2x+x)$

$$=\sin 2x\cos x+\cos 2x\sin x$$

$$=\sin(x+x)\cos x+\cos(x+x)\sin x$$

$$=(\sin x\cos x+\cos x\sin x)\cos x+(\cos x\cos x-\sin x\sin x)\sin x$$

$$=2\sin x\cos^2 x+\sin x\cos^2 x-\sin^3 x$$

$$=3\sin x\cos^2 x-\sin^3 x$$

$$=3\sin x(1-\sin^2 x)-\sin^3 x$$

$$=3\sin x-3\sin^3 x-\sin^3 x$$

$$=3\sin x-4\sin^3 x$$

75. $\cos(\theta+3\pi)=\cos\theta\cos 3\pi-\sin\theta\sin 3\pi$
$$=(\cos\theta)(-1)-(\sin\theta)(0)$$
$$=-\cos\theta$$

77. $\tan(\theta+\pi)=\dfrac{\tan\theta+\tan\pi}{1-\tan\theta\tan\pi}$

$$=\dfrac{\tan\theta+0}{1-(\tan\theta)(0)}$$

$$=\tan\theta$$

79. $\sin(\theta+2k\pi)=\sin\theta\cos(2k\pi)+\cos\theta\sin 2k\pi$
$$=(\sin\theta)(1)+(\cos\theta)(0)$$
$$=\sin\theta$$

81. $y=\sin\left(\dfrac{\pi}{2}-x\right)$ and $y=\cos x$ both have the following graph.

83. $y=\sin 7x\,\cos 2x-\cos 7x\,\sin 2x$ and $y=\sin 5x$ both have the following graph.

Connecting Concepts

85. $\sin(x-y) \cdot \sin(x+y) = (\sin x \cos y - \cos x \sin y)(\sin x \cos y + \cos x \sin y)$

$$= \sin^2 x \cos^2 y - \cos^2 x \sin^2 y$$

87. $\cos(x+y+z) = \cos[x+(y+z)]$

$$= \cos x \cos(y+z) - \sin x \sin(y+z)$$

$$= \cos x[\cos y \cos z - \sin y \sin z] - \sin x[\sin y \cos z + \cos y \sin z]$$

$$= \cos x \cos y \cos z - \cos x \sin y \sin z - \sin x \sin y \cos z - \sin x \cos y \sin z$$

89. $\dfrac{\cos(x-y)}{\cos x \sin y} = \dfrac{\cos x \cos y + \sin x \sin y}{\cos x \sin y}$

$$= \dfrac{\cos x \cos y}{\cos x \sin y} + \dfrac{\sin x \sin y}{\cos x \sin y}$$

$$= \cot y + \tan x$$

Prepare for Section 3.3

PS1. $\sin 2\alpha = \sin(\alpha + \alpha)$

$$= \sin\alpha\cos\alpha + \cos\alpha\sin\alpha$$

$$= 2\sin\alpha\cos\alpha$$

PS2. $\cos 2\alpha = \cos(\alpha + \alpha)$

$$= \cos\alpha\cos\alpha - \sin\alpha\sin\alpha$$

$$= \cos^2\alpha - \sin^2\alpha$$

PS3. $\tan 2\alpha = \tan(\alpha + \alpha)$

$$= \dfrac{\tan\alpha + \tan\alpha}{1 - \tan\alpha\tan\alpha}$$

$$= \dfrac{2\tan\alpha}{1 - \tan^2\alpha}$$

PS4. $\tan\left(\dfrac{60°}{2}\right) = \tan(30°) = \dfrac{\sqrt{3}}{3}$ $\dfrac{\sin(60°)}{1 + \cos(60°)} = \dfrac{\frac{\sqrt{3}}{2}}{1 + \frac{1}{2}} = \dfrac{\frac{\sqrt{3}}{2}}{\frac{3}{2}} = \dfrac{\sqrt{3}}{3}$

$\tan\left(\dfrac{90°}{2}\right) = \tan(45°) = 1$ $\dfrac{\sin(90°)}{1 + \cos(90°)} = \dfrac{1}{1 + 0} = 1$

$\tan\left(\dfrac{120°}{2}\right) = \tan(60°) = \sqrt{3}$ $\dfrac{\sin(120°)}{1 + \cos(120°)} = \dfrac{\frac{\sqrt{3}}{2}}{1 - \frac{1}{2}} = \dfrac{\frac{\sqrt{3}}{2}}{\frac{1}{2}} = \sqrt{3}$

For each of the given values of α, the functional values are equal.

PS5. Let $\alpha = 45°$; then the left side of the equation is 1, and the right side of the equation is $\sqrt{2}$.

PS6. Let $\alpha = 60°$; then the left side of the equation is $\dfrac{\sqrt{3}}{2}$, and the right side of the equation is $\dfrac{1}{4}$.

Section 3.3

1. $2\sin 2\alpha\cos 2\alpha = \sin 2(2\alpha)$

$$= \sin 4\alpha$$

3. $1 - 2\sin^2 5\beta = \cos 2(5\beta)$

$$= \cos 10\beta$$

5. $\cos^2 3\alpha - \sin^2 3\alpha = \cos 2(3\alpha)$

$$= \cos 6\alpha$$

7. $\dfrac{2\tan 3\alpha}{1 - \tan^2 3\alpha} = \tan 2(3\alpha)$

$$= \tan 6\alpha$$

9. $\cos\alpha = -\dfrac{4}{5}$, $\sin\alpha = \sqrt{1 - \left(\dfrac{4}{5}\right)^2} = \dfrac{3}{5}$, $\tan\alpha = \dfrac{\sin\alpha}{\cos\alpha} = \dfrac{3/5}{-4/5} = -\dfrac{3}{4}$

$$\sin 2\alpha = 2\sin\alpha\cos\alpha$$
$$= 2\left(\dfrac{3}{5}\right)\left(-\dfrac{4}{5}\right)$$
$$= -\dfrac{24}{25}$$

$$\cos 2\alpha = \cos^2\alpha - \sin^2\alpha$$
$$= \left(-\dfrac{4}{5}\right)^2 - \left(\dfrac{3}{5}\right)^2$$
$$= \dfrac{16}{25} - \dfrac{9}{25}$$
$$= \dfrac{7}{25}$$

$$\tan 2\alpha = \dfrac{2\tan\alpha}{1 - \tan^2\alpha}$$
$$= \dfrac{2\left(-\dfrac{3}{4}\right)}{1 - \left(-\dfrac{3}{4}\right)^2} = \dfrac{-\dfrac{6}{4}}{1 - \dfrac{9}{16}}$$
$$= \dfrac{-\dfrac{6}{4}}{1 - \dfrac{9}{16}} \cdot \dfrac{16}{16} = \dfrac{-24}{16 - 9}$$
$$= -\dfrac{24}{7}$$

11. $\sin\alpha = \dfrac{8}{17}$, $\cos\alpha = -\sqrt{1 - \left(\dfrac{8}{17}\right)^2} = -\dfrac{15}{17}$, $\tan\alpha = \dfrac{\sin\alpha}{\cos\alpha} = \dfrac{8/17}{-15/17} = -\dfrac{8}{15}$

$$\sin 2\alpha = 2\sin\alpha\cos\alpha$$
$$= 2\left(\dfrac{8}{17}\right)\left(-\dfrac{15}{17}\right)$$
$$= -\dfrac{240}{289}$$

$$\cos 2\alpha = \cos^2\alpha - \sin^2\alpha$$
$$= \left(-\dfrac{15}{17}\right)^2 - \left(\dfrac{8}{17}\right)^2$$
$$= \dfrac{225}{289} - \dfrac{64}{289}$$
$$= \dfrac{161}{289}$$

$$\tan 2\alpha = \dfrac{2\tan\alpha}{1 - \tan^2\alpha}$$
$$= \dfrac{2\left(-\dfrac{8}{15}\right)}{1 - \left(-\dfrac{8}{15}\right)^2} = \dfrac{-\dfrac{16}{15}}{1 - \dfrac{64}{225}}$$
$$= \dfrac{-\dfrac{16}{15}}{1 - \dfrac{64}{225}} \cdot \dfrac{225}{225} = \dfrac{-240}{225 - 64}$$
$$= -\dfrac{240}{161}$$

13. $\tan\alpha = -\dfrac{24}{7}$, $r = \sqrt{24^2 + 7^2} = \sqrt{576 + 49} = \sqrt{625} = 25$, $\sin\alpha = -\dfrac{24}{25}$, $\cos\alpha = \dfrac{7}{25}$

$$\sin 2\alpha = 2\sin\alpha\cos\alpha$$
$$= 2\left(-\dfrac{24}{25}\right)\left(\dfrac{7}{25}\right)$$
$$= -\dfrac{336}{625}$$

$$\cos 2\alpha = \cos^2\alpha - \sin^2\alpha$$
$$= \left(\dfrac{7}{25}\right)^2 - \left(-\dfrac{24}{25}\right)^2$$
$$= \dfrac{49}{625} - \dfrac{576}{625}$$
$$= -\dfrac{527}{625}$$

$$\tan 2\alpha = \dfrac{2\tan\alpha}{1 - \tan^2\alpha}$$
$$= \dfrac{2\left(-\dfrac{24}{7}\right)}{1 - \left(-\dfrac{24}{7}\right)^2}$$
$$= \dfrac{-\dfrac{48}{7}}{1 - \dfrac{576}{49}} \cdot \dfrac{49}{49}$$
$$= \dfrac{-336}{49 - 576} = \dfrac{336}{527}$$

15. $\sin\alpha = \dfrac{15}{17}$, $\cos\alpha = \sqrt{1 - \left(\dfrac{15}{17}\right)^2} = \sqrt{1 - \dfrac{225}{289}} = \sqrt{\dfrac{289 - 225}{289}} = \sqrt{\dfrac{64}{289}} = \dfrac{8}{17}$, $\tan\alpha = \dfrac{15/17}{8/17} = \dfrac{15}{8}$

$$\sin 2\alpha = 2\sin\alpha\cos\alpha$$
$$= 2\left(\dfrac{15}{17}\right)\left(\dfrac{8}{17}\right)$$
$$= \dfrac{240}{289}$$

$$\cos 2\alpha = \cos^2\alpha - \sin^2\alpha$$
$$= \left(\dfrac{8}{17}\right)^2 - \left(\dfrac{15}{17}\right)^2$$
$$= \dfrac{64}{289} - \dfrac{225}{289}$$
$$= -\dfrac{161}{289}$$

$$\tan 2\alpha = \dfrac{2\tan\alpha}{1 - \tan^2\alpha}$$
$$= \dfrac{2\left(\dfrac{15}{8}\right)}{1 - \left(\dfrac{15}{8}\right)^2} = \dfrac{\dfrac{15}{4}}{1 - \dfrac{225}{64}}$$
$$= \dfrac{\dfrac{15}{4}}{1 - \dfrac{225}{64}} \cdot \dfrac{64}{64} = \dfrac{240}{64 - 225}$$
$$= -\dfrac{240}{161}$$

17. $\cos\alpha = \dfrac{40}{41}$, $\sin\alpha = -\sqrt{1-\left(\dfrac{40}{41}\right)^2} = -\sqrt{1-\dfrac{1600}{1681}} = -\dfrac{9}{41}$, $\tan\alpha = \dfrac{-9/41}{40/41} = -\dfrac{9}{40}$

$$\sin 2\alpha = 2\sin\alpha\cos\alpha$$
$$= 2\left(-\dfrac{9}{41}\right)\left(\dfrac{40}{41}\right)$$
$$= -\dfrac{720}{1681}$$

$$\cos 2\alpha = \cos^2\alpha - \sin^2\alpha$$
$$= \left(\dfrac{40}{41}\right)^2 - \left(-\dfrac{9}{41}\right)^2$$
$$= \dfrac{1600}{1681} - \dfrac{81}{1681}$$
$$= \dfrac{1519}{1681}$$

$$\tan 2\alpha = \dfrac{2\tan\alpha}{1-\tan^2\alpha}$$
$$= \dfrac{2\left(-\dfrac{9}{40}\right)}{1-\left(-\dfrac{9}{40}\right)^2} = \dfrac{-\dfrac{9}{20}}{1-\dfrac{81}{1600}}$$
$$= \dfrac{-\dfrac{9}{20}}{1-\dfrac{81}{1600}} \cdot \dfrac{1600}{1600}$$
$$= \dfrac{-720}{1600-81} = -\dfrac{720}{1519}$$

19. $6\cos^2 x = 6\left(\dfrac{1+\cos 2x}{2}\right) = 3(1+\cos 2x)$

21. $\cos^4 x = \left(\dfrac{1+\cos 2x}{2}\right)^2$
$$= \dfrac{1}{4}(1+2\cos 2x + \cos^2 2x)$$
$$= \dfrac{1}{4}\left(1+2\cos 2x + \dfrac{1+\cos 4x}{2}\right)$$
$$= \dfrac{1}{8}(3+4\cos 2x + \cos 4x)$$

23. $\sin^4 x\,\cos^2 x = \left(\dfrac{1-\cos 2x}{2}\right)^2\left(\dfrac{1+\cos 2x}{2}\right)$
$$= \dfrac{1}{8}(1-\cos^2 2x)(1-\cos 2x)$$
$$= \dfrac{1}{8}\left(1-\dfrac{1+\cos 4x}{2}\right)(1-\cos 2x)$$
$$= \dfrac{1}{8}\left(\dfrac{1}{2}-\dfrac{1}{2}\cos 4x\right)(1-\cos 2x)$$
$$= \dfrac{1}{16}(1-\cos 2x - \cos 4x + \cos 4x\cos 2x)$$

25. $\sin 75° = \sin\dfrac{150°}{2}$
$$= +\sqrt{\dfrac{1-\cos 150°}{2}}$$
$$= \sqrt{\dfrac{1-(-\sqrt{3}/2)}{2}}$$
$$= \sqrt{\dfrac{2+\sqrt{3}}{4}}$$
$$= \dfrac{\sqrt{2+\sqrt{3}}}{2}$$

27. $\tan 67.5° = \tan\dfrac{135°}{2}$
$$= \dfrac{1-\cos 135°}{\sin 135°}$$
$$= \dfrac{1-(-\sqrt{2}/2)}{\sqrt{2}/2}$$
$$= \dfrac{2+\sqrt{2}}{\sqrt{2}}\cdot\dfrac{\sqrt{2}}{\sqrt{2}}$$
$$= \dfrac{2\sqrt{2}+2}{2}$$
$$= \sqrt{2}+1$$

29. $\cos 157.5° = \cos\dfrac{315°}{2}$
$$= -\sqrt{\dfrac{1+\cos 315°}{2}}$$
$$= -\sqrt{\dfrac{1+(\sqrt{2}/2)}{2}}$$
$$= -\sqrt{\dfrac{2+\sqrt{2}}{4}}$$
$$= -\dfrac{\sqrt{2+\sqrt{2}}}{2}$$

31. $\sin 22.5° = \sin\dfrac{45°}{2}$
$$= +\sqrt{\dfrac{1-\cos 45°}{2}}$$
$$= \sqrt{\dfrac{1-\sqrt{2}/2}{2}}$$
$$= \sqrt{\dfrac{2-\sqrt{2}}{4}}$$
$$= \dfrac{\sqrt{2-\sqrt{2}}}{2}$$

33. $\sin\dfrac{7\pi}{8} = \sin\dfrac{7\pi/4}{2}$
$$= +\sqrt{\dfrac{1-\cos 7\pi/4}{2}}$$
$$= \sqrt{\dfrac{1-\sqrt{2}/2}{2}}$$
$$= \sqrt{\dfrac{2-\sqrt{2}}{4}}$$
$$= \dfrac{\sqrt{2-\sqrt{2}}}{2}$$

35. $\cos\dfrac{5\pi}{12} = \cos\dfrac{5\pi/6}{2}$
$$= +\sqrt{\dfrac{1+\cos 5\pi/6}{2}}$$
$$= \sqrt{\dfrac{1-\sqrt{3}/2}{2}}$$
$$= \sqrt{\dfrac{2-\sqrt{3}}{4}}$$
$$= \dfrac{\sqrt{2-\sqrt{3}}}{2}$$

37. $\sin\alpha = \dfrac{5}{13}$, $\cos\alpha = -\sqrt{1-\left(\dfrac{5}{13}\right)^2} = -\sqrt{1-\dfrac{25}{169}} = -\dfrac{12}{13}$.

$\sin\dfrac{\alpha}{2} = \sqrt{\dfrac{1-\cos\alpha}{2}}$

$\quad = \sqrt{\dfrac{1-(-12/13)}{2}}$

$\quad = \sqrt{\dfrac{13+12}{26}}$

$\quad = \sqrt{\dfrac{25}{26}} = \dfrac{5}{\sqrt{26}}$

$\quad = \dfrac{5\sqrt{26}}{26}$

$\cos\dfrac{\alpha}{2} = \sqrt{\dfrac{1+\cos\alpha}{2}}$

$\quad = \sqrt{\dfrac{1-12/13}{2}}$

$\quad = \sqrt{\dfrac{13-12}{26}}$

$\quad = \sqrt{\dfrac{1}{26}} = \dfrac{1}{\sqrt{26}}$

$\quad = \dfrac{\sqrt{26}}{26}$

$\tan\dfrac{\alpha}{2} = \dfrac{1-\cos\alpha}{\sin\alpha}$

$\quad = \dfrac{1+\dfrac{12}{13}}{\dfrac{5}{13}}$

$\quad = \dfrac{13+12}{5}$

$\quad = 5$

39. $\cos\alpha = -\dfrac{8}{17}$, $\sin\alpha = -\sqrt{1-\left(-\dfrac{8}{17}\right)^2} = -\sqrt{1-\dfrac{64}{289}} = -\dfrac{15}{17}$

$\sin\dfrac{\alpha}{2} = \sqrt{\dfrac{1-\cos\alpha}{2}}$

$\quad = \sqrt{\dfrac{1-(-8/17)}{2}}$

$\quad = \sqrt{\dfrac{17+8}{34}}$

$\quad = \sqrt{\dfrac{25}{34}}$

$\quad = \dfrac{5\sqrt{34}}{34}$

$\cos\dfrac{\alpha}{2} = -\sqrt{\dfrac{1+\cos\alpha}{2}}$

$\quad = -\sqrt{\dfrac{1-8/17}{2}}$

$\quad = -\sqrt{\dfrac{17-8}{34}}$

$\quad = -\sqrt{\dfrac{9}{34}}$

$\quad = -\dfrac{3\sqrt{34}}{34}$

$\tan\dfrac{\alpha}{2} = \dfrac{1-\cos\alpha}{\sin\alpha}$

$\quad = \dfrac{1+\dfrac{18}{17}}{-\dfrac{15}{17}}$

$\quad = \dfrac{17+8}{-15}$

$\quad = -\dfrac{5}{3}$

41. $\tan\alpha = \dfrac{4}{3}$, $r = \sqrt{3^2+4^2} = \sqrt{25} = 5$, $\sin\alpha = \dfrac{4}{5}$, $\cos = \dfrac{3}{5}$

$\sin\dfrac{\alpha}{2} = \sqrt{\dfrac{1-\cos\alpha}{2}}$

$\quad = \sqrt{\dfrac{1-3/5}{2}}$

$\quad = \sqrt{\dfrac{5-3}{10}}$

$\quad = \sqrt{\dfrac{1}{5}}$

$\quad = \dfrac{\sqrt{5}}{5}$

$\cos\dfrac{\alpha}{2} = \sqrt{\dfrac{1+\cos\alpha}{2}}$

$\quad = \sqrt{\dfrac{1+3/5}{2}}$

$\quad = \sqrt{\dfrac{5+3}{10}}$

$\quad = \sqrt{\dfrac{4}{5}}$

$\quad = \dfrac{2\sqrt{5}}{5}$

$\tan\dfrac{\alpha}{2} = \dfrac{1-\cos\alpha}{\sin\alpha}$

$\quad = \dfrac{1-\dfrac{3}{5}}{\dfrac{4}{5}}$

$\quad = \dfrac{5-3}{4}$

$\quad = \dfrac{1}{2}$

43. $\cos\alpha = \dfrac{24}{25}$, $\sin\alpha = -\sqrt{1-\left(\dfrac{24}{25}\right)^2} = -\sqrt{1-\dfrac{576}{625}} = -\dfrac{7}{25}$

$\sin\dfrac{\alpha}{2} = \sqrt{\dfrac{1-\cos\alpha}{2}}$

$\quad = \sqrt{\dfrac{1-24/25}{2}}$

$\quad = \sqrt{\dfrac{25-24}{50}} = \sqrt{\dfrac{1}{50}}$

$\quad = \dfrac{\sqrt{2}}{10}$

$\cos\dfrac{\alpha}{2} = -\sqrt{\dfrac{1+\cos\alpha}{2}}$

$\quad = -\sqrt{\dfrac{1+24/25}{2}}$

$\quad = -\sqrt{\dfrac{25+24}{50}} = -\sqrt{\dfrac{49}{50}}$

$\quad = -\dfrac{7\sqrt{2}}{10}$

$\tan\dfrac{\alpha}{2} = \dfrac{1-\cos\alpha}{\sin\alpha}$

$\quad = \dfrac{1-\dfrac{24}{25}}{-\dfrac{7}{25}} = \dfrac{25-24}{-7}$

$\quad = -\dfrac{1}{7}$

45.
$$\sin 3x \cos 3x = \frac{1}{2}(2\sin 3x \cos 3x)$$
$$= \frac{1}{2}\sin 2(3x)$$
$$= \frac{1}{2}\sin 6x$$

47.
$$\sin^2 x + \cos 2x = \sin^2 x + \cos^2 x - \sin^2 x$$
$$= \cos^2 x$$

49.
$$\frac{1+\cos 2x}{\sin 2x} = \frac{1+2\cos^2 x-1}{2\sin x\cos x}$$
$$= \frac{2\cos^2 x}{2\sin x\cos x}$$
$$= \cot x$$

51.
$$\frac{\sin 2x}{1-\sin^2 x} = \frac{2\sin x\cos x}{\cos^2 x}$$
$$= 2\tan x$$

53.
$$\frac{\cos 2x}{\cos^2 x} = \frac{\cos^2 x - \sin^2 x}{\cos^2 x}$$
$$= \frac{\cos^2 x}{\cos^2 x} - \frac{\sin^2 x}{\cos^2 x}$$
$$= 1 - \tan^2 x$$

55.
$$\sin 2x - \tan x = 2\sin x\cos x - \frac{\sin x}{\cos x}$$
$$= \frac{2\sin x\cos^2 x - \sin x}{\cos x}$$
$$= \frac{\sin x(2\cos^2 x - 1)}{\cos x}$$
$$= \tan x\cos 2x$$

57.
$$\cos^4 x - \sin^4 x = (\cos^2 x + \sin^2 x)(\cos^2 x - \sin^2 x)$$
$$= \cos^2 x - \sin^2 x$$
$$= \cos 2x$$

59.
$$\cos^2 x - 2\sin^2 x\cos^2 x - \sin^2 x + 2\sin^4 x = \cos^2 x(1-2\sin^2 x) - \sin^2 x(1-2\sin^2 x)$$
$$= (1-2\sin^2 x)(\cos^2 x - \sin^2 x)$$
$$= \cos 2x\cos 2x$$
$$= \cos^2 2x$$

61.
$$\cos 4x = \cos 2(2x)$$
$$= 2\cos^2 2x - 1$$
$$= 2(2\cos^2 x - 1)^2 - 1$$
$$= 2(4\cos^4 x - 4\cos^2 x + 1) - 1$$
$$= 8\cos^4 x - 8\cos^2 x + 1$$

63.
$$\cos 3x - \cos x = \cos(2x+x) - \cos x$$
$$= \cos 2x\cos x - \sin 2x\sin x - \cos x$$
$$= (2\cos^2 x - 1)\cos x - 2\sin x\cos x \cdot \sin x - \cos x$$
$$= 2\cos^3 x - \cos x - 2\sin^2 x\cos x - \cos x$$
$$= 2\cos^3 x - 2\cos x - 2\sin^2 x\cos x$$
$$= 2\cos^3 x - 2\cos x - 2(1-\cos^2 x)\cos x$$
$$= 2\cos^3 x - 2\cos x - 2\cos x + 2\cos^3 x$$
$$= 4\cos^3 x - 4\cos x$$

65. $\sin^3 x + \cos^3 x = (\sin x + \cos x)(\sin^2 x - \sin x \cos x + \cos^2 x)$

$$= (\sin x + \cos x)\left(\sin^2 x + \cos^2 x - \frac{2\sin x \cos x}{2}\right)$$

$$= (\sin x + \cos x)\left(1 - \frac{1}{2}\sin 2x\right)$$

67. $\sin^2 \dfrac{x}{2} = \left[\pm\sqrt{\dfrac{1-\cos x}{2}}\,\right]^2$

$$= \frac{1-\cos x}{2}$$

$$= \frac{1-\cos x}{2} \cdot \frac{\sec x}{\sec x}$$

$$= \frac{\sec x - 1}{2\sec x}$$

69. $\tan\dfrac{x}{2} = \dfrac{1-\cos x}{\sin x}$

$$= \frac{1}{\sin x} - \frac{\cos x}{\sin x}$$

$$= \csc x - \cot x$$

71. $2\sin\dfrac{x}{2}\cos\dfrac{x}{2} = \sin 2\left(\dfrac{x}{2}\right)$

$$= \sin x$$

73. $\left(\cos\dfrac{x}{2} + \sin\dfrac{x}{2}\right)^2 = \cos^2\dfrac{x}{2} + 2\sin\dfrac{x}{2}\cos\dfrac{x}{2} + \sin^2\dfrac{x}{2}$

$$= \cos^2\frac{x}{2} + \sin^2\frac{x}{2} + \sin 2\left(\frac{x}{2}\right)$$

$$= 1 + \sin x$$

75. $\sin^2\dfrac{x}{2}\sec x = \left(\pm\sqrt{\dfrac{1-\cos x}{2}}\right)^2 \sec x$

$$= \frac{1-\cos x}{2} \cdot \sec x$$

$$= \frac{1}{2}(\sec x - 1)$$

77. $\cos^2\dfrac{x}{2} - \cos x = \left(\pm\sqrt{\dfrac{1+\cos x}{2}}\right)^2 - \cos x$

$$= \frac{1+\cos x}{2} - \cos x$$

$$= \frac{1+\cos x - 2\cos x}{2}$$

$$= \frac{1-\cos x}{2}$$

$$= \sin^2\frac{x}{2}$$

79. $\sin^2\dfrac{x}{2} - \cos^2\dfrac{x}{2} = -\left(\cos^2\dfrac{x}{2} - \sin^2\dfrac{x}{2}\right)$

$$= -\cos 2\left(\frac{x}{2}\right)$$

$$= -\cos x$$

81. $\sin 2x - \cos x = 2\sin x \cos x - \cos x$

$$= (\cos x)(2\sin x - 1)$$

83. $\tan 2x = \dfrac{2\tan x}{1 - \tan^2 x}$

$$= \frac{\dfrac{2\tan x}{\tan x}}{\dfrac{1}{\tan x} - \dfrac{\tan^2 x}{\tan x}}$$

$$= \frac{2}{\cot x - \tan x}$$

85. $\dfrac{\sin^2 x + 1 - \cos^2 x}{\sin x(1+\cos x)} = \dfrac{1 - \cos^2 x + 1 - \cos^2 x}{\sin x(1+\cos x)}$

$$= \frac{2(1-\cos^2 x)}{\sin x(1+\cos x)}$$

$$= \frac{2(1-\cos x)(1+\cos x)}{\sin x(1+\cos x)}$$

$$= \frac{2(1-\cos x)}{\sin x}$$

$$= 2\tan\frac{x}{2}$$

87. $\csc 2x = \dfrac{1}{\sin 2x}$

$\qquad = \dfrac{1}{2\sin x \cos x}$

$\qquad = \dfrac{1}{2}\csc x \sec x$

89. $\cos \dfrac{x}{5} = \cos 2\left(\dfrac{x}{10}\right)$

$\qquad = 1 - 2\sin^2 \dfrac{x}{10}$

91. **a.** $M = \dfrac{1}{\sin \dfrac{(\pi/4)}{2}} = 1 + \sqrt{\dfrac{1 - \cos\dfrac{\pi}{4}}{2}}$

$\qquad = 1 + \sqrt{\dfrac{1 - \dfrac{\sqrt{2}}{2}}{2}}$

$\qquad = 1 + \sqrt{\dfrac{2 - \sqrt{2}}{4}}$

$\qquad = \dfrac{2}{\sqrt{2 - \sqrt{2}}}$

$\qquad \approx 2.61$

b. $M \sin \dfrac{\alpha}{2} = 1$

$\qquad \sin \dfrac{\alpha}{2} = \dfrac{1}{M}$

$\qquad \dfrac{\alpha}{2} = \sin^{-1}\left(\dfrac{1}{M}\right)$

$\qquad \alpha = 2\sin^{-1}\left(\dfrac{1}{M}\right)$

c. As M increases $\dfrac{1}{M}$ decreases. So if $\sin^{-1}\left(\dfrac{1}{M}\right)$ decreases, then α decreases.

Connecting Concepts

93. $y = \dfrac{\sin 2x}{1 - \sin^2 x}$ and $y = 2\tan x$ both have the following graph.

95. $y = \left(\cos \dfrac{x}{2} + \sin \dfrac{x}{2}\right)^2$ and $y = 1 + \sin x$ both have the following graph.

97. $\cos^4 x = \cos^2 x \cdot \cos^2 x$

$\qquad = \dfrac{\cos 2x + 1}{2} \cdot \dfrac{\cos 2x + 1}{2}$

$\qquad = \dfrac{1}{4}\left(\cos^2 2x + 2\cos 2x + 1\right)$

$\qquad = \dfrac{1}{4}\left(\dfrac{\cos 4x + 1}{2} + 2\cos 2x + 1\right)$

$\qquad = \dfrac{1}{8}\cos 4x + \dfrac{1}{8} + \dfrac{1}{2}\cos 2x + \dfrac{1}{4}$

$\qquad = \dfrac{1}{8}\cos 4x + \dfrac{1}{2}\cos 2x + \dfrac{3}{8}$

PS1. $\dfrac{1}{2}[\sin(\alpha+\beta)+\sin(\alpha-\beta)]$

$=\dfrac{1}{2}[\sin\alpha\cos\beta+\cos\alpha\sin\beta+\sin\alpha\cos\beta-\cos\alpha\sin\beta]$

$=\sin\alpha\cos\beta$

PS2. $\dfrac{1}{2}[\cos(\alpha+\beta)+\cos(\alpha-\beta)]$

$=\dfrac{1}{2}[\cos\alpha\cos\beta-\sin\alpha\sin\beta+\cos\alpha\cos\beta-\sin\alpha\sin\beta]$

$=\cos\alpha\cos\beta$

PS3. $\sin\pi-\sin\dfrac{\pi}{6}=0-\dfrac{1}{2}=-\dfrac{1}{2}$

$2\cos\left(\dfrac{\overset{1}{\cancel{\dfrac{\pi+\dfrac{\pi}{6}}{2}}}}{}\right)\sin\left(\dfrac{\pi+\dfrac{\pi}{6}}{2}\right)=2\cos\left(\dfrac{7\pi}{6}\right)\cos\left(\dfrac{5\pi}{6}\right)$

$2\left(-\sqrt{\dfrac{1+\cos\left(\dfrac{7\pi}{6}\right)}{2}}+\sqrt{\dfrac{1-\cos\left(\dfrac{5\pi}{6}\right)}{2}}\right)$

$=-2\sqrt{\dfrac{1-\dfrac{\sqrt{3}}{2}}{2}}\sqrt{\dfrac{1+\dfrac{\sqrt{3}}{2}}{2}}$

$=-\sqrt{1-\dfrac{3}{4}}=-\sqrt{\dfrac{1}{4}}$

$=-\dfrac{1}{2}$

Both functional values equal $-\dfrac{1}{2}$.

PS4. $\sqrt{2}\sin\left(x+\dfrac{\pi}{4}\right)=\sqrt{2}\left[\sin x\cos\left(\dfrac{\pi}{4}\right)+\cos x\sin\left(\dfrac{\pi}{4}\right)\right]$

$=\sqrt{2}\left[(\sin x)\left(\dfrac{\sqrt{2}}{2}\right)+(\cos x)\left(\dfrac{\sqrt{2}}{2}\right)\right]$

$=\sin x+\cos x$

PS5. Answers will vary.

PS6. $\sqrt{(-1)^2+\left(\sqrt{3}\right)^2}=\sqrt{1+3}=\sqrt{4}=2$

Section 3.4

1. $2\sin x\cos 2x=2\cdot\dfrac{1}{2}[\sin(x+2x)+\sin(x-2x)]$

$=\sin 3x+\sin(-x)$

$=\sin 3x-\sin x$

3. $\cos 6x\sin 2x=\dfrac{1}{2}[\sin(6x+2x)-\sin(6x-2x)]$

$=\dfrac{1}{2}[\sin 8x-\sin 4x]$

5. $2\sin 5x\cos 3x=\sin(5x+3x)+\sin(5x-3x)$

$=\sin 8x+\sin 2x$

7. $\sin x\cos 5x=\dfrac{1}{2}[\cos(x-5x)-\cos(x+5x)]$

$=\dfrac{1}{2}[\cos(-4x)-\cos 6x]$

$=\dfrac{1}{2}(\cos 4x-\cos 6x)$

9. $\cos 75°\cos 15°=\dfrac{1}{2}[\cos(75°+15°)+\cos(75°-15°)]$

$=\dfrac{1}{2}(\cos 90°+\cos 60°)$

$=\dfrac{1}{2}\left(0+\dfrac{1}{2}\right)$

$=\dfrac{1}{4}$

11.
$$\cos 157.5° \sin 22.5° = \frac{1}{2}\left[\sin(157.5° + 22.5°) - \sin(157.5° - 22.5°)\right]$$
$$= \frac{1}{2}(\sin 180° - \sin 135°)$$
$$= \frac{1}{2}\left(0 - \frac{\sqrt{2}}{2}\right)$$
$$= -\frac{\sqrt{2}}{4}$$

13.
$$\sin\frac{13\pi}{12}\cos\frac{\pi}{12} = \frac{1}{2}\left[\sin\left(\frac{13\pi}{12} + \frac{\pi}{12}\right) + \sin\left(\frac{13\pi}{12} - \frac{\pi}{12}\right)\right]$$
$$= \frac{1}{2}\left(\sin\frac{7\pi}{6} + \sin\pi\right)$$
$$= \frac{1}{2}\left(-\frac{1}{2} + 0\right)$$
$$= -\frac{1}{4}$$

15.
$$\sin\frac{\pi}{12}\cos\frac{7\pi}{12} = \frac{1}{2}\left[\sin\left(\frac{\pi}{12} + \frac{7\pi}{12}\right) + \sin\left(\frac{\pi}{12} - \frac{7\pi}{12}\right)\right]$$
$$= \frac{1}{2}\left[\sin\frac{2\pi}{3} + \sin\left(-\frac{\pi}{2}\right)\right]$$
$$= \frac{1}{2}\left(\sin\frac{2\pi}{3} - \sin\frac{\pi}{2}\right)$$
$$= \frac{1}{2}\left(\frac{\sqrt{3}}{2} - 1\right)$$
$$= \frac{\sqrt{3} - 2}{4}$$

17.
$$\sin 4\theta + \sin 2\theta = 2\sin\frac{4\theta + 2\theta}{2}\cos\frac{4\theta - 2\theta}{2}$$
$$= 2\sin 3\theta\cos\theta$$

19.
$$\cos 3\theta + \cos\theta = 2\cos\frac{3\theta + \theta}{2}\cos\frac{3\theta - \theta}{2}$$
$$= 2\cos 2\theta\cos\theta$$

21.
$$\cos 6\theta - \cos 2\theta = -2\sin\frac{6\theta + 2\theta}{2}\sin\frac{6\theta - 2\theta}{2}$$
$$= -2\sin 4\theta\sin 2\theta$$

23.
$$\cos\theta + \cos 7\theta = 2\cos\frac{\theta + 7\theta}{2}\cos\frac{\theta - 7\theta}{2}$$
$$= 2\cos 4\theta\cos(-3\theta)$$
$$= 2\cos 4\theta\cos 3\theta$$

25.
$$\sin 5\theta + \sin 9\theta = 2\sin\frac{5\theta + 9\theta}{2}\cos\frac{5\theta - 9\theta}{2}$$
$$= 2\sin 7\theta\cos(-2\theta)$$
$$= 2\sin 7\theta\cos 2\theta$$

27.
$$\cos 2\theta - \cos\theta = -2\sin\frac{2\theta + \theta}{2}\sin\frac{2\theta - \theta}{2}$$
$$= -2\sin\frac{3}{2}\theta\sin\frac{1}{2}\theta$$

29.
$$\cos\frac{\theta}{2} - \cos\theta = -2\sin\frac{\frac{\theta}{2} + \theta}{2}\sin\frac{\frac{\theta}{2} - \theta}{2}$$
$$= -2\sin\frac{3}{4}\theta\sin\left(-\frac{1}{4}\theta\right)$$
$$= 2\sin\frac{3}{4}\theta\sin\frac{1}{4}\theta$$

31.
$$\sin\frac{\theta}{2} - \sin\frac{\theta}{3} = 2\cos\frac{\frac{\theta}{2} + \frac{\theta}{3}}{2}\sin\frac{\frac{\theta}{2} - \frac{\theta}{3}}{2}$$
$$= 2\cos\frac{5}{12}\theta\sin\frac{1}{12}\theta$$

33.
$$\cos(\alpha + \beta) + \cos(\alpha - \beta) = \cos\alpha\cos\beta - \sin\alpha\sin\beta + \cos\alpha\cos\beta + \sin\alpha\sin\beta$$
$$= 2\cos\alpha\cos\beta$$

35.
$$2\cos 3x\sin x = 2\cdot\frac{1}{2}\left[\sin(3x + x) - \sin(3x - x)\right]$$
$$= \sin 4x - \sin 2x$$
$$= 2\sin 2x\cos 2x - \sin 2x$$
$$= \sin 2x(2\cos 2x - 1)$$
$$= 2\sin x\cos x\left[2(1 - 2\sin^2 x) - 1\right]$$
$$= 4\sin x\cos x - 8\sin^3 x\cos x - 2\sin x\cos x$$
$$= 2\sin x\cos x - 8\cos x\sin^3 x$$

37.
$$2\cos 5x\cos 7x = 2\cdot\frac{1}{2}\left[\cos(5x + 7x) + \cos(5x - 7x)\right]$$
$$= \cos 12x + \cos(-2x)$$
$$= \cos 12x + \cos 2x$$
$$= \cos^2 6x - \sin^2 6x + 2\cos^2 x - 1$$

39. $\quad \sin 3x - \sin x = 2\cos\dfrac{3x+x}{2}\sin\dfrac{3x-x}{2}$

$\qquad\qquad\qquad = 2\cos 2x\sin x$

$\qquad\qquad\qquad = 2(1 - 2\sin^2 x)\sin x$

$\qquad\qquad\qquad = 2\sin x - 4\sin^3 x$

41. $\quad \sin 2x + \sin 4x = 2\cos\dfrac{2x+4}{2}\cos\dfrac{2x-x4}{2}$

$\qquad\qquad\qquad = 2\cos 3x\cos(-x)$

$\qquad\qquad\qquad = 2\cos 3x\cos x$

$\qquad\qquad\qquad = 2\cos x\sin(2x+x)$

$\qquad\qquad\qquad = 2\cos x(\sin 2x\cos x + \cos 2x\sin x)$

$\qquad\qquad\qquad = 2\cos x[(2\sin x\cos x)\cos x + (2\cos^2 x - 1)\sin x]$

$\qquad\qquad\qquad = 2\cos x\sin x(4\cos^2 x - 1)$

43. $\quad \dfrac{\sin 3x - \sin x}{\cos 3x - \cos x} = \dfrac{2\cos\frac{3x+x}{2}\sin\frac{3x-x}{2}}{-2\sin\frac{3x+x}{2}\sin\frac{3x-x}{2}}$

$\qquad\qquad\qquad = -\dfrac{\cos 2x}{\sin 2x}$

$\qquad\qquad\qquad = -\cot 2x$

45. $\quad \dfrac{\sin 5x + \sin 3x}{4\sin x\cos^3 x - 4\sin^3 x\cos x} = \dfrac{2\sin\frac{5x+3x}{2}\cos\frac{5x-3x}{2}}{4\sin x\cos x(\cos^2 x - \sin^2 x)}$

$\qquad\qquad\qquad\qquad = \dfrac{\sin 4x\cos x}{2\sin x\cos x\cos 2x}$

$\qquad\qquad\qquad\qquad = \dfrac{2\sin 2x\cos 2x\cos x}{\sin 2x\cos 2x}$

$\qquad\qquad\qquad\qquad = 2\cos x$

47. $\quad \sin(x+y)\cos(x-y) = \dfrac{1}{2}\big[\sin(x+y+x-y) + \sin(x+y-x+y)\big]$

$\qquad\qquad\qquad\qquad = \dfrac{1}{2}\big[\sin 2x + \sin 2y\big]$

$\qquad\qquad\qquad\qquad = \dfrac{1}{2}\big[2\sin x\cos x + 2\sin y\cos y\big]$

$\qquad\qquad\qquad\qquad = \sin x\cos x + \sin y\cos y$

49. $\quad a = -1,\ b = -1,\ k = \sqrt{(-1)^2 + (-1)^2} = \sqrt{2},$

α is a third quadrant angle.

$\sin\beta = \left|\dfrac{-1}{\sqrt{2}}\right| = \dfrac{1}{\sqrt{2}}$

$\qquad \beta = 45°$

$\qquad \alpha = -180° + 45° = -135°$

$\qquad y = \sqrt{2}\sin(x - 135°)$

51. $\quad a = \dfrac{1}{2},\ b = -\dfrac{\sqrt{3}}{2},\ k = \sqrt{\left(\dfrac{1}{2}\right)^2 + \left(-\dfrac{\sqrt{3}}{2}\right)^2} = 1,$

α is a fourth quadrant angle.

$\sin\beta = \left|\dfrac{-\frac{\sqrt{3}}{2}}{1}\right| = \dfrac{\sqrt{3}}{2}$

$\qquad \beta = 60°$

$\qquad \alpha = -60°$

$\qquad y = \sin(x - 60°)$

53. $a = \frac{1}{2}$, $b = -\frac{1}{2}$, $k = \sqrt{\left(\frac{1}{2}\right)^2 + \left(\frac{1}{2}\right)^2} = \frac{\sqrt{2}}{2}$,

α is a fourth quadrant angle.

$\sin\beta = \left|\dfrac{-1/2}{\sqrt{2}/2}\right| = \dfrac{\sqrt{2}}{2}$

$\beta = 45°$

$\alpha = -45°$

$y = \dfrac{\sqrt{2}}{2}\sin(x - 45°)$

55. $a = -3$, $b = 3$, $k = \sqrt{(-3)^2 + 3^2} = 3\sqrt{2}$,

α is a second quadrant angle.

$\sin\beta = \left|\dfrac{3}{3\sqrt{2}}\right| = \dfrac{1}{\sqrt{2}} - \dfrac{\sqrt{2}}{2}$

$\beta = 45°$

$\alpha = 180° - 45° = 135°$

$y = 3\sqrt{2}\sin(x + 135°)$

57. $a = \pi$, $b = -\pi$, $k = \sqrt{\pi^2 + (-\pi)^2} = \pi\sqrt{2}$,

α is a fourth quadrant angle.

$\sin\beta = \left|\dfrac{-\pi}{\pi\sqrt{2}}\right| = \dfrac{1}{\sqrt{2}} = \dfrac{\sqrt{2}}{2}$

$\beta = 45°$

$\alpha = -45°$

$y = \pi\sqrt{2}\sin(x - 45°)$

59. $a = -1$, $b = 1$, $k = \sqrt{(-1)^2 + 1^2} = \sqrt{2}$,

α is a second quadrant angle.

$\sin\beta = \left|\dfrac{1}{\sqrt{2}}\right| = \dfrac{1}{\sqrt{2}} = \dfrac{\sqrt{2}}{2}$

$\beta = \dfrac{\pi}{4}$

$\alpha = \pi - \dfrac{\pi}{4} = \dfrac{3\pi}{4}$

$y = \sqrt{2}\sin\left(x + \dfrac{3\pi}{4}\right)$

61. $a = \dfrac{\sqrt{3}}{2}$, $b = \dfrac{1}{2}$, $k = \sqrt{\left(\dfrac{\sqrt{3}}{2}\right)^2 + \left(\dfrac{1}{2}\right)^2} = 1$,

α is a first quadrant angle.

$\sin\beta = \left|\dfrac{1/2}{1}\right| = \dfrac{1}{2}$

$\beta = \dfrac{\pi}{6}$

$\alpha = \dfrac{\pi}{6}$

$y = \sin\left(x + \dfrac{\pi}{6}\right)$

63. $a = -10$, $b = 10\sqrt{3}$, $k = \sqrt{(-10)^2 + (10\sqrt{3})^2} = 20$,

α is a second quadrant angle.

$\sin\beta = \dfrac{10\sqrt{3}}{20} = \dfrac{\sqrt{3}}{2}$

$\beta = \dfrac{\pi}{3}$

$\alpha = \pi - \dfrac{\pi}{3} = \dfrac{2\pi}{3}$

$y = 20\sin\left(x + \dfrac{2\pi}{3}\right)$

65. $a = -5$, $b = 5$, $k = \sqrt{(-5)^2 + 5^2} = 5\sqrt{2}$,

α is a second quadrant angle.

$\sin\beta = \left|\dfrac{5}{5\sqrt{2}}\right| = \dfrac{\sqrt{2}}{2}$

$\beta = \dfrac{\pi}{4}$

$\alpha = \pi - \dfrac{\pi}{4} = \dfrac{3\pi}{4}$

$y = 5\sqrt{2}\sin\left(x + \dfrac{3\pi}{4}\right)$

67. $y = -\sin x - \sqrt{3}\cos x$

$y = 2\sin\left(x - \dfrac{2\pi}{3}\right)$

69. $y = 2\sin x + 2\cos x$

$y = 2\sqrt{2}\sin\left(x + \dfrac{\pi}{4}\right)$

71. $y = -\sqrt{3}\sin x - \cos x$

$y = 2\sin\left(x - \dfrac{5\pi}{6}\right)$

73. $y = -5\sin x + 5\sqrt{3}\cos x$

$y = 10\sin\left(x + \dfrac{2\pi}{3}\right)$

75. $y = 6\sqrt{3}\sin x - 6\cos x$

$y = 12\sin\left(x - \dfrac{\pi}{6}\right)$

77. **a.** $p(t) = \sin(2\pi \cdot 1336t) + \sin(2\pi \cdot 770t)$

b. $p(t) = 2\sin\left(\dfrac{2\pi \cdot 1336t + 2\pi \cdot 770t}{2}\right)\sin\left(\dfrac{2\pi \cdot 1336t - 2\pi \cdot 770t}{2}\right)$

$= 2\sin(2106\pi t)\sin(556\pi t)$

c. $\dfrac{1336 + 770}{2} = \dfrac{2106}{2} = 1053$ cycles per second

79. Identity

81. Identity

83. Identity

• •

Connecting Concepts

85. Let $x = \alpha + \beta$ and $y = \alpha - \beta$.

$x + y = \alpha + \beta + \alpha - \beta$ and $x - y = \alpha + \beta - (\alpha - \beta)$

$x + y = 2\alpha \qquad\qquad x - y = 2\beta$

$\alpha = \dfrac{x + y}{2} \qquad\qquad 2\beta = \dfrac{x - y}{2}$

$\cos(\alpha - \beta) + \cos(\alpha + \beta) = 2\cos\alpha\cos\beta$

$\cos\left[\dfrac{x + y}{2} - \dfrac{x - y}{2}\right] + \cos\left[\dfrac{x + y}{2} + \dfrac{x - y}{2}\right] = 2\cos\dfrac{x + y}{2}\cos\dfrac{x - y}{2}$

$\cos y + \cos x = 2\cos\dfrac{x + y}{2}\cos\dfrac{x - y}{2}$

87. $x + y = 180°$

$y = 180° - x$

$\sin x + \sin y = \sin x + \sin(180° - x)$

$= \sin x + \sin 180°\cos x - \cos 180°\sin x$

$= \sin x + 0(\cos x) - (-1)\sin x$

$= 2\sin x$

89. $\sin 2x + \sin 4x + \sin 6x = 2\sin\dfrac{2x + 4x}{2}\cos\dfrac{2x - 4x}{2} + 2\sin 3x\cos 3x$

$= 2\sin 3x\cos x + 2\sin 3x\cos 3x$

$= 2\sin 3x(\cos x + \cos 3x)$

$= 2\sin 3x\left(2\cos\dfrac{x + 3x}{2}\cos\dfrac{x - 3x}{2}\right)$

$= 4\sin 3x\cos 2x\cos x$

91. $\dfrac{\cos 10x + \cos 8x}{\sin 10x - \sin 8x} = \dfrac{2\cos\frac{10x+8x}{2}\cos\frac{10x-8x}{2}}{2\cos\frac{10x+8x}{2}\sin\frac{10x-8x}{2}}$

$\phantom{\dfrac{\cos 10x + \cos 8x}{\sin 10x - \sin 8x}} = \dfrac{2\cos 9x\cos x}{2\cos 9x\sin x}$

$\phantom{\dfrac{\cos 10x + \cos 8x}{\sin 10x - \sin 8x}} = \cot x$

93. $\dfrac{\sin 2x + \sin 4x + \sin 6x}{\cos 2x + \cos 4x + \cos 6x} = \dfrac{\sin 2x + \sin 6x + \sin 4x}{\cos 2x + \cos 6x + \cos 4x}$

$ = \dfrac{2\sin\frac{2x+6x}{2}\cos\frac{2x-6x}{2} + \sin 4x}{2\cos\frac{2x+6x}{2}\cos\frac{2x-6x}{2} + \cos 4x}$

$ = \dfrac{2\sin 4x\cos 2x + \sin 4x}{2\cos 4x\cos 2x + \cos 4x}$

$ = \dfrac{\sin 4x(2\cos 2x + 1)}{\cos 4x(2\cos 2x + 1)}$

$ = \dfrac{\sin 4x}{\cos 4x}$

$ = \tan 4x$

95. $\cos^2 x - \sin^2 x = \cos x\cdot\cos x - \sin x\cdot\sin x$

$ = \frac{1}{2}[\cos(x+x)+\cos(x-x)] - \frac{1}{2}[\cos(x-x)-\cos(x+x)]$

$ = \frac{1}{2}\cos 2x + \frac{1}{2}\cos 0 - \frac{1}{2}\cos 0 + \frac{1}{2}\cos 2x$

$ = \cos 2x$

97. Let $k = \sqrt{a^2+b^2}$, $\tan\alpha = \dfrac{a}{b}$

$a\sin x + b\cos x = \dfrac{\sqrt{a^2+b^2}}{\sqrt{a^2+b^2}}(a\sin x + b\cos x)$

$ = \sqrt{a^2+b^2}\left(\dfrac{a}{\sqrt{a^2+b^2}}\sin x + \dfrac{b}{\sqrt{a^2+b^2}}\cos x\right)$

$ = k(\sin\alpha\sin x + \cos\alpha\cos x)$ because $\sin\alpha = \dfrac{a}{\sqrt{a^2+b^2}}$ and $\cos\alpha = \dfrac{b}{\sqrt{a^2+b^2}}$

$ = k(\cos x\cos\alpha + \sin x\sin\alpha) = k\cos(x-\alpha)$

Prepare for Section 3.5

PS1. A one-to-one function is a function for which each range value (y-value) is paired with one and only one domain value (x-value).

PS2. If every horizontal line intersects the graph of a function at most once, then the function is a one-to-one function.

PS3. $f[g(x)] = f\left[\frac{1}{2}x - 2\right]$

$ = 2\left(\frac{1}{2}x - 2\right) + 4$

$ = x - 4 + 4$

$ = x$

PS4. $f[f^{-1}(x)] = x$

PS5. The graph of f^{-1} is the reflection of the graph of f across the line given by $y = x$.

PS6. No, it does not pass the horizontal line test.

Section 3.5

1. $y = \sin^{-1} 1$

$\sin y = 1$ with $-\dfrac{\pi}{2} \le y \le \dfrac{\pi}{2}$

$y = \dfrac{\pi}{2}$

3. $y = \cos^{-1}\left(-\dfrac{\sqrt{3}}{2}\right)$

$\cos y = -\dfrac{\sqrt{3}}{2} \qquad 0 \le y \le \pi$

$y = \dfrac{5\pi}{6}$

5. $y = \tan^{-1}(1)$

$\tan y = -1 \qquad -\dfrac{\pi}{2} < y < \dfrac{\pi}{2}$

$y = -\dfrac{\pi}{4}$

7. $y = \cot^{-1}\dfrac{\sqrt{3}}{3}$

$\cot y = \dfrac{\sqrt{3}}{3} \qquad 0 < y < \pi$

$y = \dfrac{\pi}{3}$

9. $y = \sec^{-1} 2$

$\sec y = 2 \qquad 0 \le y \le \pi$

$y = \dfrac{\pi}{3}$

11. $y = \csc^{-1}\left(-\sqrt{2}\right)$

$\csc y = -\sqrt{2} \qquad -\dfrac{\pi}{2} \le y \le \dfrac{\pi}{2}$

$y = -\dfrac{\pi}{4}$

13. $y = \sin^{-1}\left(-\dfrac{\sqrt{3}}{2}\right)$

$\sin y = -\dfrac{\sqrt{3}}{2} \qquad -\dfrac{\pi}{2} \le y \le \dfrac{\pi}{2}$

$y = -\dfrac{\pi}{3}$

15. $y = \cos^{-1}\left(-\dfrac{1}{2}\right)$

$\cos y = -\dfrac{1}{2} \qquad 0 \le y \le \pi$

$y = \dfrac{2\pi}{3}$

17. $y = \tan^{-1}\dfrac{\sqrt{3}}{3}$

$\tan y = \dfrac{\sqrt{3}}{3} \qquad -\dfrac{\pi}{2} < y < \dfrac{\pi}{2}$

$y = \dfrac{\pi}{6}$

19. a. $\sin^{-1}(0.8422) \approx 1.0014$

b. $\tan^{-1}(0.2385) \approx 0.2341$

21. a. $\sec^{-1}(2.2500) = \cos^{-1}\left(\dfrac{1}{2.2500}\right) \approx 1.1102$

b. $\cot^{-1}(3.4545) = \tan^{-1}\left(\dfrac{1}{3.4545}\right) \approx 0.2818$

23. $\cos\theta = \dfrac{x}{7}$ or $\theta = \cos^{-1}\left(\dfrac{x}{7}\right)$

25. $y = \cos\left(\cos^{-1}\dfrac{1}{2}\right)$

$y = \cos\dfrac{\pi}{3}$

$y = \dfrac{1}{2}$

27. $y = \tan(\tan^{-1} 2)$

$y = 2$

29. $y = \sin\left(\tan^{-1}\dfrac{3}{4}\right)$

$y = \dfrac{3}{5}$

31. $y = \tan\left(\sin^{-1}\dfrac{\sqrt{2}}{2}\right)$

$y = 1$

33. $y = \cos(\sec^{-1} 2)$

$y = \dfrac{1}{2}$

35. $y = \sin^{-1}\left(\sin\dfrac{\pi}{6}\right)$

$= \sin^{-1}\dfrac{1}{2}$

$y = \dfrac{\pi}{6}$

37. $y = \cos^{-1}\left(\sin\dfrac{\pi}{4}\right)$

$= \cos^{-1}\dfrac{\sqrt{2}}{2}$

$y = \dfrac{\pi}{4}$

39. $y = \sin^{-1}\left(\tan\dfrac{\pi}{3}\right)$

$= \sin^{-1}\sqrt{3}$

y is not defined.

41. $y = \tan^{-1}\left(\sin\dfrac{\pi}{6}\right)$

$\quad = \tan^{-1}\dfrac{1}{2}$

$y \approx 0.4636$

43. $y = \sin^{-1}\left(\cos\left[-\dfrac{2\pi}{3}\right]\right)$

$\quad = \sin^{-1}\left(-\dfrac{1}{2}\right)$

$y = -\dfrac{\pi}{6}$

45. Let $\theta = \sin^{-1}\dfrac{1}{2}$ and find $y = \tan\theta$.

Then $\sin\theta = \dfrac{1}{2}$ and $-\dfrac{\pi}{2} \le \theta \le \dfrac{\pi}{2}$.

Thus $\tan\theta = \dfrac{1}{\sqrt{3}} = \dfrac{\sqrt{3}}{3}$.

$y = \dfrac{\sqrt{3}}{3}$

47. Let $\theta = \sin^{-1}\dfrac{1}{4}$ and find $y = \sec\theta$.

Then $\sin\theta = \dfrac{1}{4}$, and $-\dfrac{\pi}{2} \le \theta \le \dfrac{\pi}{2}$.

Thus $\sec\theta = \dfrac{4}{\sqrt{15}} = \dfrac{4\sqrt{15}}{15}$.

$y = \dfrac{4\sqrt{15}}{15}$

49. Let $\theta = \sin^{-1}\dfrac{7}{25}$ and find $y = \cos\theta$.

Then $\sin\theta = \dfrac{7}{25}$, and $-\dfrac{\pi}{2} \le \theta \le \dfrac{\pi}{2}$.

Thus $\cos\theta = \dfrac{24}{25}$.

$y = \dfrac{24}{25}$

51. Let $\alpha = \sin^{-1}\dfrac{\sqrt{2}}{2}$, $\alpha = \dfrac{\pi}{4}$, $\sin\alpha = \dfrac{\sqrt{2}}{2}$, $\cos\alpha = \dfrac{\sqrt{2}}{2}$.

$y = \cos\left(2\sin^{-1}\dfrac{\sqrt{2}}{2}\right)$

$\quad = \cos 2\alpha$

$\quad = \cos^2\alpha - \sin^2\alpha$

$\quad = \left(\dfrac{\sqrt{2}}{2}\right)^2 - \left(\dfrac{\sqrt{2}}{2}\right)^2$

$\quad = 0$

53. Let $\alpha = \sin^{-1}\dfrac{4}{5}$, $\sin\alpha = \dfrac{4}{5}$, $\cos\alpha = \sqrt{1 - \left(\dfrac{4}{5}\right)^2} = \dfrac{3}{5}$.

$y = \sin\left(2\sin^{-1}\dfrac{4}{5}\right)$

$\quad = \sin 2\alpha = 2\sin\alpha\cos\alpha$

$\quad = 2\left(\dfrac{4}{5}\right)\left(\dfrac{3}{5}\right) = \dfrac{24}{25}$

55. $y = \sin\left(\sin^{-1}\dfrac{2}{3} + \cos^{-1}\dfrac{1}{2}\right)$

Let $\alpha = \sin^{-1}\dfrac{2}{3}$, $\sin\alpha = \dfrac{2}{3}$, $\cos\alpha = \sqrt{1 - \left(\dfrac{2}{3}\right)^2} = \dfrac{\sqrt{5}}{3}$.

$\beta = \cos^{-1}\dfrac{1}{2}$, $\cos\beta = \dfrac{1}{2}$, $\sin\beta = \sqrt{1 - \left(\dfrac{1}{2}\right)^2} = \dfrac{\sqrt{3}}{2}$.

$y = \sin(\alpha + \beta)$

$\quad = \sin\alpha\cos\beta + \cos\alpha\sin\beta$

$\quad = \dfrac{2}{3}\left(\dfrac{1}{2}\right) + \dfrac{\sqrt{5}}{3}\left(\dfrac{\sqrt{3}}{2}\right)$

$\quad = \dfrac{1}{3} + \dfrac{\sqrt{15}}{6} = \dfrac{2 + \sqrt{15}}{6}$

57. $y = \tan\left(\cos^{-1}\frac{1}{2} - \sin^{-1}\frac{3}{4}\right)$

Let $\alpha = \cos^{-1}\frac{1}{2}$, $\cos\alpha = \frac{1}{2}$, $\sin\alpha = \sqrt{1-\left(\frac{1}{2}\right)^2} = \frac{\sqrt{3}}{2}$, $\tan\alpha = \frac{\frac{\sqrt{3}}{2}}{\frac{1}{2}} = \sqrt{3}$.

$\beta = \sin^{-1}\frac{3}{4}$, $\sin\beta = \frac{3}{4}$, $\cos\beta = \sqrt{1-\left(\frac{3}{4}\right)^2} = \frac{\sqrt{7}}{4}$, $\tan\beta = \frac{\frac{3}{4}}{\frac{\sqrt{7}}{4}} = \frac{3}{\sqrt{7}} = \frac{3\sqrt{7}}{7}$.

$y = \tan(\alpha - \beta)$

$= \frac{\tan\alpha - \tan\beta}{1 + \tan\alpha\tan\beta}$

$= \frac{\sqrt{3} - \frac{3\sqrt{7}}{7}}{1+\sqrt{3}\cdot\frac{3\sqrt{7}}{7}} = \frac{\sqrt{3}-\frac{3\sqrt{7}}{7}}{1+\sqrt{3}\cdot\frac{3\sqrt{7}}{7}}\cdot\frac{7}{7} = \frac{7\sqrt{3}-3\sqrt{7}}{7+3\sqrt{21}} = \frac{7\sqrt{3}-3\sqrt{7}}{7+3\sqrt{21}}\cdot\frac{7-3\sqrt{21}}{7-3\sqrt{21}} = \frac{112\sqrt{3}-84\sqrt{7}}{-140} = \frac{3\sqrt{7}-4\sqrt{3}}{5} = \frac{1}{5}\left(3\sqrt{7}-4\sqrt{3}\right)$

59. $\sin^{-1}x = \cos^{-1}\frac{5}{13}$

$\sin(\sin^{-1}x) = \sin\left(\cos^{-1}\frac{5}{13}\right)$

$x = \frac{12}{13}$

61. $\sin^{-1}(x-1) = \frac{\pi}{2}$

$(x-1) = \sin\frac{\pi}{2}$

$(x-1) = 1$

$x = 2$

63. $\tan^{-1}\left(x+\frac{\sqrt{2}}{2}\right) = \frac{\pi}{4}$

$\left(x+\frac{\sqrt{2}}{2}\right) = \tan\frac{\pi}{4}$

$x = 1 - \frac{\sqrt{2}}{2}$

$= \frac{2-\sqrt{2}}{2}$

65. $\sin^{-1}\frac{3}{5} + \cos^{-1}x = \frac{\pi}{4}$

$\cos^{-1}x = \frac{\pi}{4} - \sin^{-1}\frac{3}{5}$

$x = \cos\left(\frac{\pi}{4} - \sin^{-1}\frac{3}{5}\right)$

Let $\alpha = \sin^{-1}\frac{3}{5}$, $\sin\alpha = \frac{3}{5}$, $\cos\alpha = \frac{4}{5}$.

$x = \cos\left(\frac{\pi}{4} - \alpha\right)$

$x = \cos\frac{\pi}{4}\cos\alpha + \sin\frac{\pi}{4}\sin\alpha$

$x = \frac{\sqrt{2}}{2}\cdot\frac{4}{5} + \frac{\sqrt{2}}{2}\cdot\frac{3}{5}$

$= \frac{4\sqrt{2}}{10} + \frac{3\sqrt{2}}{10} = \frac{7\sqrt{2}}{10}$

67. $\sin^{-1}\frac{\sqrt{2}}{2} + \cos^{-1}x = \frac{2\pi}{3}$

$\cos^{-1}x = \frac{2\pi}{3} - \sin^{-1}\frac{\sqrt{2}}{2}$

$x = \cos\left(\frac{2\pi}{3} - \sin^{-1}\frac{\sqrt{2}}{2}\right)$

Let $\alpha = \sin^{-1}\frac{\sqrt{2}}{2}$, $\sin\alpha = \frac{\sqrt{2}}{2}$, $\cos\alpha = \frac{\sqrt{2}}{2}$.

$x = \cos\left(\frac{2\pi}{3} - \alpha\right)$

$x = \cos\frac{2\pi}{3}\cos\alpha + \sin\frac{2\pi}{3}\sin\alpha$

$x = -\frac{1}{2}\cdot\frac{\sqrt{2}}{2} + \frac{\sqrt{3}}{2}\cdot\frac{\sqrt{2}}{2}$

$= -\frac{\sqrt{2}}{4} + \frac{\sqrt{6}}{4} = \frac{\sqrt{6}-\sqrt{2}}{4} \approx 0.2588$

Note: Since $\sin\alpha = \frac{\sqrt{2}}{2}$ and $\cos\alpha = \frac{\sqrt{2}}{2}$, then $\alpha = \frac{\pi}{4}$.

Thus, $\cos\left(\frac{2\pi}{3} - \alpha\right) = \cos\left(\frac{2\pi}{3} - \frac{\pi}{4}\right) = \cos\left(\frac{5\pi}{12}\right) \approx 0.2588$.

69. $\tan(\cos^{-1}x) = \dfrac{\sqrt{1-x^2}}{x}$

71. Let $\alpha = \sin^{-1}x$, $\sin\alpha = x$, $\cos\alpha = \sqrt{1-x^2}$.

Let $\beta = \sin^{-1}(-x)$, $\sin\beta = -x$, $\cos\beta = \sqrt{1-x^2}$.

$$\sin^{-1}x + \sin^{-1}(-x) = \alpha + \beta$$
$$= \sin^{-1}[\sin(\alpha+\beta)]$$
$$= \sin^{-1}(\sin\alpha\cos\beta + \cos\alpha\sin\beta)$$
$$= \sin^{-1}\left[x\sqrt{1-x^2} + \sqrt{1-x^2}(-x)\right]$$
$$= \sin^{-1}0$$
$$= 0$$

73. Let $\alpha = \tan^{-1}x$, $\tan\alpha = x$, $\beta = \tan^{-1}\dfrac{1}{x}$, $\tan\beta = \dfrac{1}{x}$.

$$\tan^{-1}x + \tan^{-1}\dfrac{1}{x} = \alpha + \beta$$
$$= \tan^{-1}[\tan(\alpha+\beta)]$$
$$= \tan^{-1}\left[\dfrac{\tan\alpha + \tan\beta}{1 - \tan\alpha\tan\beta}\right]$$
$$= \tan^{-1}\left[\dfrac{x+\dfrac{1}{x}}{1-x\cdot\dfrac{1}{x}}\right]$$
$$= \tan^{-1}\dfrac{\dfrac{x^2+1}{x}}{1-1}, \text{ which is undefined}$$

Thus $x = \dfrac{\pi}{2}$

75. The graph of $y = \sin^{-1}(x) + 2$ (shown as a black graph) is the graph of $y = \sin^{-1}x$ (shown as a gray graph) moved two units up.

77. The graph of $y = \sin^{-1}(x+1) - 2$ (shown as a black graph) is the graph of $y = \sin^{-1}x$ (shown as a gray graph) moved one unit to the left and two units down.

79. The graph of $y = 2\cos^{-1}x$ (shown as a black graph) is the graph of $y = \cos^{-1}x$ (shown as a gray graph) stretched.

81. The graph of $y = \tan^{-1}(x+1) - 2$ (shown as a black graph) is the graph of $y = \tan^{-1}x$ (shown as a gray graph) moved one unit to the left and two units down.

83.   **a.** $s = 3960\theta$

$$\cos\theta = \frac{3960}{a+3960}$$

$$\theta = \cos^{-1}\left(\frac{3960}{a+3960}\right)$$

$$s = 3960\cos^{-1}\left(\frac{3960}{a+3960}\right)$$

b.

$$5500 = 3960\cos^{-1}\left(\frac{3960}{a+3960}\right)$$

$$\frac{5500}{3960} = \cos^{-1}\left(\frac{3960}{a+3960}\right)$$

$$\cos\left(\frac{5500}{3960}\right) = \frac{3960}{a+3960}$$

$$a+3960 = \frac{3960}{\cos\left(\frac{5500}{3960}\right)}$$

$$a = \frac{3960}{\cos\left(\frac{5500}{3960}\right)} - 3960$$

$$a \approx 17{,}930 \text{ mi}$$

85.

f and *g* have the same graph in Quadrant I

[Note: $f(x) = \cos^{-1} x$ is neither odd nor even.

$g(x) = \sin^{-1}\sqrt{1-x^2}$ is an even function.]

No, $f(x) \neq g(x)$ on the interval $[-1, 1]$.

87. $y = \csc^{-1} 2x$

$\csc y = 2x$

$-\dfrac{\pi}{2} \le y \le \dfrac{\pi}{2},\ y \neq 0$

$2x \le -1\ $ or $\ 2x \ge 1$

$x \le -\dfrac{1}{2}\ $ or $\ x \ge \dfrac{1}{2}$

89. $y = \sec^{-1}(x-1)$

$\sec y = x-1$

$0 < y < \pi$

$y \neq \dfrac{\pi}{2}$

$x-1 \le -1 \qquad x-1 \ge 1$

$\quad x \le 0 \qquad\quad x \ge 2$

91. $y = 2\tan^{-1} 2x$

$\dfrac{y}{2} = \tan^{-1} 2x$

$\tan y = 2x$

$-\dfrac{\pi}{2} < \dfrac{y}{2} < \dfrac{\pi}{2}$

$-\pi < y < \pi$

$-\infty < 2x < \infty$

93. Let $\alpha = \sin^{-1} x$
$\sin \alpha = x$

$\cos\left(\sin^{-1} x\right) = \cos \alpha$

$\quad = \dfrac{\sqrt{1-x^2}}{1}$

$\quad = \sqrt{1-x^2}$

95. Let $\alpha = \csc^{-1} x$
$\csc \alpha = x$

$\tan(\csc^{-1} x) = \tan \alpha$

$\quad = \dfrac{1}{\sqrt{x^2-1}}$

$\quad = \dfrac{\sqrt{x^2-1}}{x^2-1}$

97. $5x = \tan^{-1} 3y$

$\tan 5x = 3y$

$y = \dfrac{1}{3} \tan 5x$

99. $x - \dfrac{\pi}{3} = \cos^{-1}(y-3)$

$\cos\left(x - \dfrac{\pi}{3}\right) = y - 3$

$y = 3 + \cos\left(x - \dfrac{\pi}{3}\right)$

PS1. $x = \dfrac{5 \pm \sqrt{(-5)^2 - 4(3)(-4)}}{2(3)} = \dfrac{5 \pm \sqrt{73}}{6}$

PS2. $\sin^2 x + \cos^2 x = 1$
$\qquad \sin^2 x = 1 - \cos^2 x$

PS3. $\dfrac{\pi}{2} + 2(1)\pi = \dfrac{5}{2}\pi$

$\dfrac{\pi}{2} + 2(2)\pi = \dfrac{9}{2}\pi$

$\dfrac{\pi}{2} + 2(3)\pi = \dfrac{13}{2}\pi$

PS4. $x^2 - \dfrac{\sqrt{3}}{2}x + x - \dfrac{\sqrt{3}}{2} = x\left(x - \dfrac{\sqrt{3}}{2}\right) + 1\left(x - \dfrac{\sqrt{3}}{2}\right)$

$\qquad\qquad\qquad\quad = (x+1)\left(x - \dfrac{\sqrt{3}}{2}\right)$

PS5.

PS6. $2x^2 - 2x = 0$

$2x(x-1) = 0$

$x = 0 \quad x - 1 = 0$

$\qquad\qquad x = 1$

The solutions are 0, 1.

Section 3.6

1. $\sec x - \sqrt{2} = 0$

$\sec x = \sqrt{2}$

$x = \dfrac{\pi}{4}, \dfrac{7\pi}{4}$

3. $\tan x - \sqrt{3} = 0$

$\tan x = \sqrt{3}$

$x = \dfrac{\pi}{3}, \dfrac{4\pi}{3}$

5.

$2\sin x \cos x = \sqrt{2}\cos x$

$2\sin x \cos x - \sqrt{2}\cos x = 0$

$\cos x(2\sin x - \sqrt{2}) = 0$

$\cos x = 0 \qquad 2\sin x - \sqrt{2} = 0$

$$\sin x = \frac{\sqrt{2}}{2}$$

$x = \dfrac{\pi}{2}, \dfrac{3\pi}{2} \qquad\qquad x = \dfrac{\pi}{4}, \dfrac{3\pi}{4}$

The solutions are $\dfrac{\pi}{4}, \dfrac{\pi}{2}, \dfrac{3\pi}{4}, \dfrac{3\pi}{2}$.

7.

$\sin^2 x - 1 = 0$

$\sin^2 x = 1$

$\sin x = \pm\sqrt{1}$

$\sin x = \pm 1$

$x = \dfrac{\pi}{2}, \dfrac{3\pi}{2}$

9.

$4\sin x \cos x - 2\sqrt{3}\sin x - 2\sqrt{2}\cos x + \sqrt{6} = 0$

$2\sin x(2\cos x - \sqrt{3}) - \sqrt{2}(2\cos x - \sqrt{3}) = 0$

$(2\cos x - \sqrt{3})(2\sin x - \sqrt{2}) = 0$

$2\cos x - \sqrt{3} = 0 \qquad\qquad 2\sin x - \sqrt{2} = 0$

$$\cos x = \frac{\sqrt{3}}{2} \qquad\qquad \sin x = \frac{\sqrt{2}}{2}$$

$x = \dfrac{\pi}{6}, \dfrac{11\pi}{6} \qquad\qquad x = \dfrac{\pi}{4}, \dfrac{3\pi}{4}$

The solutions are $\dfrac{\pi}{6}, \dfrac{\pi}{4}, \dfrac{3\pi}{4}, \dfrac{11\pi}{6}$.

11.

$\csc x - \sqrt{2} = 0$

$\csc x = \sqrt{2}$

$x = \dfrac{\pi}{4}, \dfrac{3\pi}{4}$

13.

$2\sin^2 x + 1 = 3\sin x$

$2\sin^2 x - 3\sin x + 1 = 0$

$(2\sin x - 1)(\sin x - 1) = 0$

$2\sin x - 1 = 0 \qquad\qquad \sin x - 1 = 0$

$\sin x = \dfrac{1}{2} \qquad\qquad \sin x = 1$

$x = \dfrac{\pi}{6}, \dfrac{5\pi}{6} \qquad\qquad x = \dfrac{\pi}{2}$

The solutions are $\dfrac{\pi}{6}, \dfrac{\pi}{2}, \dfrac{5\pi}{6}$.

15.

$4\cos^2 x - 3 = 0$

$\cos^2 x = \dfrac{3}{4}$

$\cos x = \pm\dfrac{\sqrt{3}}{2}$

$x = \dfrac{\pi}{6}, \dfrac{5\pi}{6}, \dfrac{7\pi}{6}, \dfrac{11\pi}{6}$

17.

$2\sin^3 x = \sin x$

$2\sin^3 x - \sin x = 0$

$\sin x(2\sin^2 x - 1) = 0$

$\sin x = 0 \qquad\qquad 2\sin^2 x = 1$

$x = 0, \pi \qquad\qquad \sin x = \pm\dfrac{\sqrt{2}}{2}$

$\qquad\qquad\qquad x = \dfrac{\pi}{4}, \dfrac{3\pi}{4}, \dfrac{5\pi}{4}, \dfrac{7\pi}{4}$

The solutions are $0, \dfrac{\pi}{4}, \dfrac{3\pi}{4}, \pi, \dfrac{5\pi}{4}, \dfrac{7\pi}{4}$.

19.

$4\sin^2 x + 2\sqrt{3}\sin x - \sqrt{3} = 2\sin x$

$4\sin^2 x + 2\sqrt{3}\sin x - 2\sin x - \sqrt{3} = 0$

$2\sin x(2\sin x + \sqrt{3}) - (2\sin x + \sqrt{3}) = 0$

$(2\sin x + \sqrt{3})(2\sin x - 1) = 0$

$2\sin x + \sqrt{3} = 0 \qquad\qquad 2\sin x - 1 = 0$

$\sin x = -\dfrac{\sqrt{3}}{2} \qquad\qquad \sin x = \dfrac{1}{2}$

$x = \dfrac{4\pi}{3}, \dfrac{5\pi}{3} \qquad\qquad x = \dfrac{\pi}{6}, \dfrac{5\pi}{6}$

The solutions are $\dfrac{\pi}{6}, \dfrac{5\pi}{6}, \dfrac{4\pi}{3}, \dfrac{5\pi}{3}$.

21. $\sin^4 x = \sin^2 x$

$\sin^4 x - \sin^2 x = 0$

$\sin^2 x(\sin^2 x - 1) = 0$

$\sin^2 x = 0 \qquad\qquad \sin^2 x - 1 = 0$

$\quad \sin x = 0 \qquad\qquad\quad \sin x = \pm 1$

$\qquad x = 0,\ \pi \qquad\qquad\qquad x = \dfrac{\pi}{2},\ \dfrac{3\pi}{2}$

The solutions are $\ 0,\ \dfrac{\pi}{2},\ \pi,\ \dfrac{3\pi}{2}$.

23. $\cos x - 0.75 = 0$

$\quad \cos x = 0.75$

$\qquad x \approx 41.4°,\ 318.6°$

25. $3\sin x - 5 = 0$

$\quad 3\sin x = 5$

$\quad\ \sin x = \dfrac{5}{3}$

no solution

27. $3\sec x - 8 = 0$

$\quad 3\sec x = 8$

$\quad\ \sec x = \dfrac{8}{3}$

$\quad \dfrac{1}{\cos x} = \dfrac{8}{3}$

$\quad\ \cos x = \dfrac{3}{8}$

$\qquad x \approx 68.0°,\ 292.0°$

29. $\cos x + 3 = 0$

$\quad \cos x = -3$

no solution

31. $3 - 5\sin x = 4\sin x + 1$

$\quad -9\sin x = -2$

$\qquad \sin x = \dfrac{2}{9}$

$\qquad\quad x \approx 12.8°,\ 167.2°$

33. $\dfrac{1}{2}\sin x + \dfrac{2}{3} = \dfrac{3}{4}\sin x + \dfrac{3}{5}$

$\quad -\dfrac{1}{4}\sin x = \dfrac{3}{5} - \dfrac{2}{3}$

$\quad -\dfrac{1}{4}\sin x = -\dfrac{1}{15}$

$\qquad\quad \sin x = \dfrac{4}{15}$

$\qquad\qquad x \approx 15.5°,\ 164.5°$

35. $3\tan^2 x - 2\tan x = 0$

$\tan x(3\tan x - 2) = 0$

$\tan x = 0 \qquad\qquad 3\tan x - 2 = 0$

$\quad x = 0,\ 180° \qquad\qquad\ \tan x = \dfrac{2}{3}$

$\qquad\qquad\qquad\qquad\qquad x \approx 33.7°,\ 213.7°$

The solutions are 0°, 33.7°, 180°, 213.7°.

37. $3\cos x + \sec x = 0$

$3\cos x + \dfrac{1}{\cos x} = 0$

$3\cos^2 x + 1 = 0$

$\quad \cos^2 x = -\dfrac{1}{3}$

no solution

39. $\tan^2 x = 3\sec^2 x - 2$

$\tan^2 x = 3(1 + \tan^2 x) - 2$

$\tan^2 x = 3 + 3\tan^2 x - 2$

$-2\tan^2 x = 1$

$\quad \tan^2 x = -\dfrac{1}{2}$

no solution

41. $2\sin^2 x = 1 - \cos x$

$2(1 - \cos^2 x) = 1 - \cos x$

$2 - 2\cos^2 x = 1 - \cos x$

$0 = 2\cos^2 x - \cos x - 1$

$0 = (2\cos x + 1)(\cos x - 1)$

$2\cos x + 1 = 0 \qquad\qquad \cos x - 1 = 0$

$\quad \cos x = -\dfrac{1}{2} \qquad\qquad\quad \cos x = 1$

$\qquad\qquad\qquad\qquad\qquad\qquad\ x = 0°$

$\qquad x = 120°,\ 240°$

The solutions are 0°, 120°, 240°.

43. $3\cos^2 x + 5\cos x - 2 = 0$

$\cos x = \dfrac{-5 \pm \sqrt{5^2 - 4(3)(-2)}}{2 \cdot 3}$

$\quad = \dfrac{-5 \pm \sqrt{49}}{6} = \dfrac{-5 \pm 7}{6}$

$\cos x = \dfrac{1}{3} \qquad\qquad \cos x = -2$

$x \approx 70.5°,\ 289.5° \qquad$ no solution

The solutions are 70.5°, 289.5°.

45. $2\tan^2 x - \tan x - 10 = 0$

$(\tan x + 2)(2\tan x - 5) = 0$

$\tan x + 2 = 0 \qquad\qquad 2\tan x - 5 = 0$

$\tan x = -2 \qquad\qquad \tan x = \dfrac{5}{2}$

$x \approx 116.6°,\ 296.6° \quad x \approx 68.2°,\ 248.2°$

The solutions are 68.2°, 116.6°, 248.2°, 296.6°.

47. $3\sin x \cos x - \cos x = 0$

$\cos x(3\sin x - 1) = 0$

$\cos x = 0 \qquad\qquad 3\sin x - 1 = 0$

$\quad x = 90°,\ 270° \qquad\qquad \sin x = \dfrac{1}{3}$

$\qquad\qquad\qquad\qquad x \approx 19.5°,\ 160.5°$

The solutions are 19.5°, 90°, 160.5°, 270°.

49. $2\sin x \cos x - \sin x - 2\cos x + 1 = 0$

$\sin x(2\cos x - 1) - (2\cos x - 1) = 0$

$(2\cos x - 1)(\sin x - 1) = 0$

$2\cos x - 1 = 0 \qquad\qquad \sin x - 1 = 0$

$\cos x = \dfrac{1}{2} \qquad\qquad \sin x = 1$

$\qquad\qquad\qquad\qquad x = 90°$

$x = 60°,\ 300°$

The solutions are 60°, 90°, 300°.

51. $2\sin x - \cos x = 1$

$\qquad 2\sin x - 1 = \cos x$

$\qquad (2\sin x - 1)^2 = (\cos x)^2$

$4\sin^2 x - 4\sin x + 1 = \cos^2 x$

$4\sin^2 x - 4\sin x + 1 = 1 - \sin^2 x$

$\quad 5\sin^2 x - 4\sin x = 0$

$\quad \sin x(5\sin x - 4) = 0$

$\sin x = 0 \qquad\qquad 5\sin x - 4 = 0$

$\quad x = 180° \qquad\qquad \sin x = \dfrac{4}{5}$

$\qquad\qquad\qquad x \approx 53.1°$ or $126.9°$

126.9° does not check.

The solutions are 53.1°, 180°.

53. $2\sin x - 3\cos x = 1$

$2\sin x = 3\cos x + 1$

$(2\sin x)^2 = (3\cos x + 1)^2$

$4\sin^2 x = 9\cos^2 x + 6\cos x + 1$

$4(1 - \cos^2 x) = 9\cos^2 x + 6\cos x + 1$

$0 = 13\cos^2 x + 6\cos x - 3$

$\cos x = \dfrac{-6 \pm \sqrt{6^2 - 4(13)(-3)}}{2(13)}$

$\quad = \dfrac{-6 \pm \sqrt{192}}{26}$

$\cos x \approx 0.3022 \qquad\qquad \cos x \approx -0.7637$

$x \approx 72.4°$ or $287.6° \qquad x \approx 139.8°$ or $220.2°$

287.6° and 139.8° do not check.

The solutions are 72.4°, 220.2°.

55. $3\sin^2 x - \sin x - 1 = 0$

$\sin x = \dfrac{1 \pm \sqrt{(-1)^2 - 4(3)(-1)}}{2(3)}$

$\quad = \dfrac{1 \pm \sqrt{13}}{6}$

$\sin x = 0.7676 \qquad\qquad \sin x = -0.4343$

$x = 50.1°,\ 129.9° \qquad x = 205.7°,\ 334.3°$

The solutions are 50.1°, 129.9°, 205.7°, 334.3°.

57. $2\cos x - 1 + 3\sec x = 0$

$2\cos x - 1 + \dfrac{3}{\cos x} = 0$

$2\cos^2 x - \cos x + 3 = 0$

$\cos x = \dfrac{1 \pm \sqrt{(-1)^2 - 4(2)(3)}}{2(2)}$

$\quad = \dfrac{1 \pm \sqrt{-23}}{4}$

no solution

59. $\cos^2 x - 3\sin x + 2\sin^2 x = 0$

$1 - \sin^2 x - 3\sin x + 2\sin^2 x = 0$

$\sin^2 x - 3\sin x + 1 = 0$

$\sin x = \dfrac{3 \pm \sqrt{(-3)^2 - 4(1)(1)}}{2(1)}$

$\qquad = \dfrac{3 \pm \sqrt{5}}{2}$

$\sin x = 2.6180 \qquad\qquad \sin x = 0.3820$

no solution $\qquad\qquad\quad x = 22.5°, \ 157.5°$

The solutions are $22.5°, \ 157.5°$.

61. $\tan 2x - 1 = 0$

$\tan 2x = 1$

$2x = \dfrac{\pi}{4} + k\pi$

$x = \dfrac{\pi}{8} + \dfrac{k\pi}{2}$, where k is an integer

63. $\sin 5x = 1$

$5x = \dfrac{\pi}{2} + 2k\pi$

$x = \dfrac{\pi}{10} + \dfrac{2}{5}k\pi$, where k is an integer

65. $\sin 2x - \sin x = 0$

$2\sin x \cos x - \sin x = 0$

$\sin x(2\cos x - 1) = 0$

$\begin{array}{ll} \sin x = 0 & 2\cos x - 1 = 0 \\ \quad x = 0 + 2k\pi & \qquad \cos x = \dfrac{1}{2} \\ \qquad \text{or} & \\ \quad x = \pi + 2k\pi & \qquad x = \dfrac{\pi}{3} + 2k\pi \\ & \qquad\qquad \text{or} \\ & \qquad x = \dfrac{5\pi}{3} + 2k\pi \end{array}$

The solutions are $0 + 2k\pi, \ \dfrac{\pi}{3} + 2k\pi, \ \pi + 2k\pi, \ \dfrac{5\pi}{3} + 2k\pi$

where k is an integer.

67. $\sin\left(2x + \dfrac{\pi}{6}\right) = -\dfrac{1}{2}$

$2x + \dfrac{\pi}{6} = \dfrac{7\pi}{6} + 2k\pi \quad$ or $\quad 2x + \dfrac{\pi}{6} = \dfrac{11\pi}{6} + 2k\pi$

$2x = \pi + 2k\pi \qquad\qquad\qquad 2x = \dfrac{5\pi}{3} + 2k\pi$

$x = \dfrac{\pi}{2} + k\pi \qquad\qquad\qquad\ x = \dfrac{5\pi}{6} + k\pi$

The solutions are $\dfrac{\pi}{2} + k\pi, \ \dfrac{5\pi}{6} + k\pi$ where k is an integer.

69. $\sin^2 \dfrac{x}{2} + \cos x = 1$

$\left(\pm\sqrt{\dfrac{1 - \cos x}{2}}\right)^2 + \cos x = 1$

$\dfrac{1 - \cos x}{2} + \cos x = 1$

$1 - \cos x + 2\cos x = 2$

$\cos x = 1$

$x = 0 + 2k\pi$ where k is an integer

71.

$$\cos 2x = 1 - 3\sin x$$

$$1 - 2\sin^2 x = 1 - 3\sin x$$

$$0 = 2\sin^2 x - 3\sin x$$

$$0 = \sin x(2\sin x - 3)$$

$$\sin x = 0 \qquad 2\sin x - 3 = 0$$

$$x = 0,\ \pi \qquad \sin x = \frac{3}{2}$$

$$\text{no solution.}$$

The solutions are $0,\ \pi$.

73.

$$\sin 4x - \sin 2x = 0$$

$$2\sin 2x \cos 2x - \sin 2x = 0$$

$$\sin 2x(2\cos 2x - 1) = 0$$

$$\sin 2x = 0 \qquad\qquad 2\cos 2x - 1 = 0$$

$$2x = 0 + 2k\pi \qquad\qquad \cos 2x = \frac{1}{2}$$

$$\text{or}$$

$$2x = \pi + 2k\pi \qquad\qquad 2x = \frac{\pi}{3} + 2k\pi$$

$$\text{or}$$

$$2x = \frac{5\pi}{3} + 2k\pi$$

$$x = 0 + k\pi,\ \frac{\pi}{2} + k\pi,\ \frac{\pi}{6} + k\pi,\ \frac{5\pi}{6} + k\pi$$

The solutions are $0,\ \dfrac{\pi}{6},\ \dfrac{\pi}{2},\ \dfrac{5\pi}{6},\ \pi,\ \dfrac{7\pi}{6},\ \dfrac{3\pi}{2},\ \dfrac{11\pi}{6}$.

75.

$$\tan \frac{\pi}{2} = \sin x$$

$$\frac{1 - \cos x}{\sin x} = \sin x$$

$$1 - \cos x = \sin^2 x$$

$$1 - \cos x = 1 - \cos^2 x$$

$$\cos^2 x - \cos x = 0$$

$$\cos x(\cos x - 1) = 0$$

$$\cos x = 0 \qquad\qquad \cos x = 1$$

$$x = \frac{\pi}{2},\ \frac{3\pi}{2} \qquad\qquad x = 0$$

The solutions are $0,\ \dfrac{\pi}{2},\ \dfrac{3\pi}{2}$.

77.

$$\sin 2x \cos x + \cos 2x \sin x = 0$$

$$\sin(2x + x) = 0$$

$$\sin 3x = 0$$

$$3x = 0 + 2k\pi \quad \text{or} \quad 3x = \pi + 2k\pi$$

$$x = 0 + \frac{2}{3}k\pi \quad \text{or} \quad x = \frac{\pi}{3} + \frac{2}{3}k\pi$$

The solutions are $0,\ \dfrac{\pi}{3},\ \dfrac{2}{3}\pi,\ \pi,\ \dfrac{4}{3}\pi\ \dfrac{5}{3}\pi$.

79.

$$\sin x \cos 2x - \cos x \sin 2x = \frac{\sqrt{3}}{2}$$

$$\sin(x - 2x) = \frac{\sqrt{3}}{2}$$

$$\sin(-x) = \frac{\sqrt{3}}{2}$$

$$\sin x = -\frac{\sqrt{3}}{2}$$

$$x = \frac{4\pi}{3},\ \frac{5\pi}{3}$$

81.

$$\sin 3x - \sin x = 0$$

$$2\cos \frac{3x + x}{2} \sin \frac{3x - 2}{2} = 0$$

$$2\cos 2x \sin x = 0$$

$$2(1 - 2\sin^2 x)\sin x = 0$$

$$\sin x = 0 \qquad\qquad 1 - 2\sin^2 x = 0$$

$$x = 0,\ \pi \qquad\qquad \sin^2 x = \frac{1}{2}$$

$$\sin x = \pm\frac{\sqrt{2}}{2}$$

$$x = \frac{\pi}{4},\ \frac{3\pi}{4},\ \frac{5\pi}{4},\ \frac{7\pi}{4}$$

The solutions are $0,\ \dfrac{\pi}{4},\ \dfrac{3\pi}{4},\ \pi,\ \dfrac{5\pi}{4},\ \dfrac{7\pi}{4}$.

83. $2\sin x\cos x + 2\sin x - \cos x - 1 = 0$

$2\sin x(\cos x + 1) - (\cos x + 1) = 0$

$(\cos x + 1)(2\sin x - 1) = 0$

$\cos x + 1 = 0 \qquad\qquad 2\sin x - 1 = 0$

$\cos x = -1 \qquad\qquad\quad \sin x = \dfrac{1}{2}$

$x = \pi \qquad\qquad\qquad\quad x = \dfrac{\pi}{6}, \dfrac{5\pi}{6}$

The solutions are $\dfrac{\pi}{6}, \dfrac{5\pi}{6}, \pi$.

85.   0.7391

87.   −3.2957, 3.2957

89.   1.16

91.   Set your graphing utility to "degree" mode and graph $d = \dfrac{(288)^2}{16}\sin\theta\cos\theta$ and $d = 1295$ for $0° \le \theta \le 90°$, $-500 \le d \le 3000$.

Use the TRACE or INTERSECT feature of your graphing utility to determine the intersection of the two graphs.

Thus, $d = 1295$ for $\theta \approx 14.99°$ and $\theta \approx 75.01°$.

The sine regression functions in Exercises 93 –98 were obtained on a TI-83/TI-83 Plus/TI-84 Plus calculator by using an iteration value of 16. The use of a different iteration factor may produce a sine regression function that varies from the regression functions listed below.

93. a.

$f(x) = y \approx 1.1213\sin(0.01595x + 1.8362) + 6.6257$

b.   Set your graphing utility to "radian" mode.

$f(71) \approx 1.1213\sin(0.01595(71) + 1.8362) + 6.6257$

≈ 6.8186 hours

$\approx 6 + .8186(60) \rightarrow 6:49$

95. a.

$f(x) = y \approx 49.2125\sin(0.2130x - 1.4576) + 48.0550$

b.   Set your graphing utility to "radian" mode.

$f(31) \approx 49.2125\sin(0.2130(31) - 1.4576) + 48.0550$

$\approx 3\%$

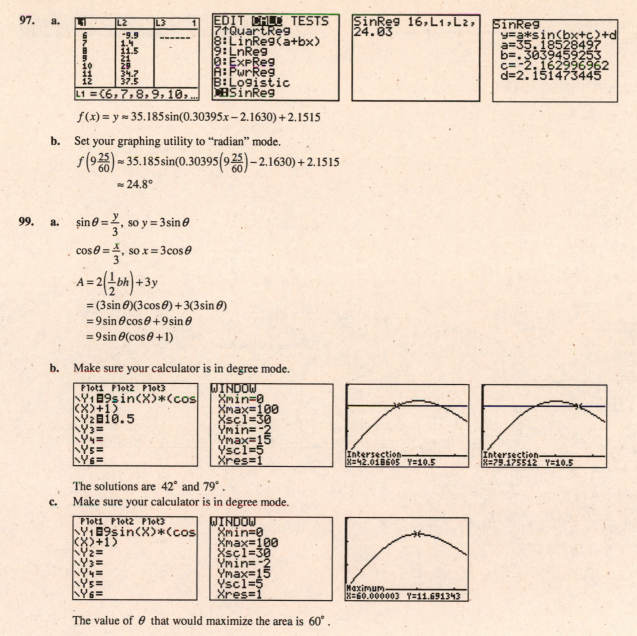

97. a.

$$f(x) = y \approx 35.185 \sin(0.30395x - 2.1630) + 2.1515$$

b. Set your graphing utility to "radian" mode.

$$f\left(9\frac{25}{60}\right) \approx 35.185 \sin\left(0.30395\left(9\frac{25}{60}\right) - 2.1630\right) + 2.1515$$

$$\approx 24.8°$$

99. a. $\sin\theta = \dfrac{y}{3}$, so $y = 3\sin\theta$

$\cos\theta = \dfrac{x}{3}$, so $x = 3\cos\theta$

$$A = 2\left(\frac{1}{2}bh\right) + 3y$$

$$= (3\sin\theta)(3\cos\theta) + 3(3\sin\theta)$$

$$= 9\sin\theta\cos\theta + 9\sin\theta$$

$$= 9\sin\theta(\cos\theta + 1)$$

b. Make sure your calculator is in degree mode.

The solutions are $42°$ and $79°$.

c. Make sure your calculator is in degree mode.

The value of θ that would maximize the area is $60°$.

101. $\theta = 20°$

$$x = \sqrt{(4 + 18\cot 20°)^2 + 100} - (4 + 18\cot 20°)$$

$x \approx 0.93$ ft.

$\theta = 30°$

$$x = \sqrt{(4 + 18\cot 30°)^2 + 100} - (4 + 18\cot 30°)$$

$x \approx 1.39$ ft.

103.　$\sqrt{3}\sin x + \cos x = \sqrt{3}$

$a = \sqrt{3},\ b = 1,\ k = \sqrt{(\sqrt{3})^2 + 1} = 2,\ \alpha$ is in first quadrant

$\tan \beta = \dfrac{1}{\sqrt{3}}$

$\beta = \dfrac{\pi}{6}$

$\alpha = \dfrac{\pi}{6}$

$2\sin\left(x + \dfrac{\pi}{6}\right) = \sqrt{3}$

$\sin\left(x + \dfrac{\pi}{6}\right) = \dfrac{\sqrt{3}}{2}$

$x + \dfrac{\pi}{6} = \dfrac{\pi}{3}$ 　　　　 $x + \dfrac{\pi}{6} = \dfrac{2\pi}{3}$

$x = \dfrac{\pi}{6}$ 　　　　　　 $x = \dfrac{\pi}{2}$

105.　$-\sin x + \sqrt{3}\cos x = \sqrt{3}$

$k = \sqrt{(-1)^2 + \left(\sqrt{3}\right)^2} = 2$ 　　　 $\tan \beta = \left|\dfrac{\sqrt{3}}{-1}\right| = \sqrt{3}$

$\beta = \dfrac{\pi}{3}$

$\alpha = \dfrac{2\pi}{3}$ second quadrant

$2\sin\left(x + \dfrac{2\pi}{3}\right) = \sqrt{3}$

$\sin\left(x + \dfrac{2\pi}{3}\right) = \dfrac{\sqrt{3}}{2}$

$x + \dfrac{2\pi}{3} = \dfrac{\pi}{3}$ 　　　　 $x + \dfrac{2\pi}{3} = \dfrac{2\pi}{3}$

$x = -\dfrac{\pi}{3}$ 　　　　　　 $x = 0$

$-\dfrac{\pi}{3} + 2\pi = \dfrac{5\pi}{3}$

The solutions are 0 and $\dfrac{5\pi}{3}$.

107.　　　　　$\cos 5x - \cos 3x = 0$

$-2\sin\dfrac{5x + 3x}{2}\sin\dfrac{5x - 3x}{2} = 0$

　　　　$-2\sin 4x \sin x = 0$

$\sin 4x = 0$ 　　　　　　　　　 $\sin x = 0$

$4x = 0 + 2k\pi$ 　　　 $4x = \pi + 2k\pi$ 　　　 $x = 0,\ \pi$

$x = 0 + \dfrac{1}{2}k\pi$ 　　　 $x = \dfrac{\pi}{4} + \dfrac{1}{2}k\pi$

$x = 0,\ \dfrac{\pi}{2},\ \pi,\ \dfrac{3\pi}{2}$ 　　　 $x = \dfrac{\pi}{4},\ \dfrac{3\pi}{4},\ \dfrac{5\pi}{4},\ \dfrac{7\pi}{4}$

The solutions are $0,\ \dfrac{\pi}{4},\ \dfrac{\pi}{2},\ \dfrac{3\pi}{4},\ \pi,\ \dfrac{5\pi}{4},\ \dfrac{3\pi}{2},\ \dfrac{7\pi}{4}$.

Approximate an Inverse Trigonometric Function with Polynomials

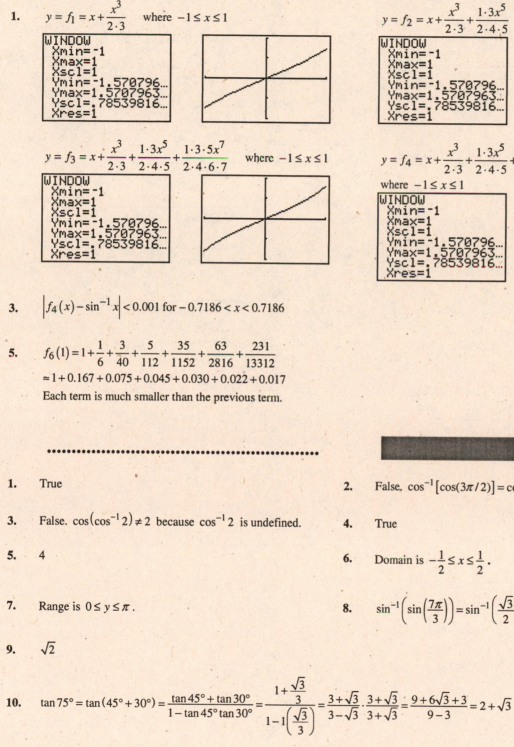

1. $y = f_1 = x + \dfrac{x^3}{2 \cdot 3}$ where $-1 \le x \le 1$

```
WINDOW
 Xmin=-1
 Xmax=1
 Xscl=1
 Ymin=-1.570796...
 Ymax=1.5707963...
 Yscl=.78539816...
 Xres=1
```

$y = f_2 = x + \dfrac{x^3}{2 \cdot 3} + \dfrac{1 \cdot 3 x^5}{2 \cdot 4 \cdot 5}$ where $-1 \le x \le 1$

```
WINDOW
 Xmin=-1
 Xmax=1
 Xscl=1
 Ymin=-1.570796...
 Ymax=1.5707963...
 Yscl=.78539816...
 Xres=1
```

$y = f_3 = x + \dfrac{x^3}{2 \cdot 3} + \dfrac{1 \cdot 3 x^5}{2 \cdot 4 \cdot 5} + \dfrac{1 \cdot 3 \cdot 5 x^7}{2 \cdot 4 \cdot 6 \cdot 7}$ where $-1 \le x \le 1$

```
WINDOW
 Xmin=-1
 Xmax=1
 Xscl=1
 Ymin=-1.570796...
 Ymax=1.5707963...
 Yscl=.78539816...
 Xres=1
```

$y = f_4 = x + \dfrac{x^3}{2 \cdot 3} + \dfrac{1 \cdot 3 x^5}{2 \cdot 4 \cdot 5} + \dfrac{1 \cdot 3 \cdot 5 x^7}{2 \cdot 4 \cdot 6 \cdot 7} + \dfrac{1 \cdot 3 \cdot 5 \cdot 7 x^9}{2 \cdot 4 \cdot 6 \cdot 8 \cdot 9}$

where $-1 \le x \le 1$

```
WINDOW
 Xmin=-1
 Xmax=1
 Xscl=1
 Ymin=-1.570796...
 Ymax=1.5707963...
 Yscl=.78539816...
 Xres=1
```

3. $\left| f_4(x) - \sin^{-1} x \right| < 0.001$ for $-0.7186 < x < 0.7186$

5. $f_6(1) = 1 + \dfrac{1}{6} + \dfrac{3}{40} + \dfrac{5}{112} + \dfrac{35}{1152} + \dfrac{63}{2816} + \dfrac{231}{13312}$

$\approx 1 + 0.167 + 0.075 + 0.045 + 0.030 + 0.022 + 0.017$

Each term is much smaller than the previous term.

1. True

2. False. $\cos^{-1}[\cos(3\pi/2)] = \cos^{-1}(0) = \pi/2 \ne 3\pi/2$.

3. False. $\cos(\cos^{-1} 2) \ne 2$ because $\cos^{-1} 2$ is undefined.

4. True

5. 4

6. Domain is $-\dfrac{1}{2} \le x \le \dfrac{1}{2}$.

7. Range is $0 \le y \le \pi$.

8. $\sin^{-1}\left(\sin\left(\dfrac{7\pi}{3}\right)\right) = \sin^{-1}\left(\dfrac{\sqrt{3}}{2}\right) = \dfrac{\pi}{3}$

9. $\sqrt{2}$

10. $\tan 75° = \tan(45° + 30°) = \dfrac{\tan 45° + \tan 30°}{1 - \tan 45° \tan 30°} = \dfrac{1 + \dfrac{\sqrt{3}}{3}}{1 - 1\left(\dfrac{\sqrt{3}}{3}\right)} = \dfrac{3 + \sqrt{3}}{3 - \sqrt{3}} \cdot \dfrac{3 + \sqrt{3}}{3 + \sqrt{3}} = \dfrac{9 + 6\sqrt{3} + 3}{9 - 3} = 2 + \sqrt{3}$

1. $\cos(45° + 30°) = \cos 45° \cos 30° - \sin 45° \sin 30°$　[3.2]

$$= \frac{\sqrt{2}}{2} \cdot \frac{\sqrt{3}}{2} - \frac{\sqrt{2}}{2} \cdot \frac{1}{2}$$

$$= \frac{\sqrt{6}}{4} - \frac{\sqrt{2}}{4}$$

$$= \frac{\sqrt{6} - \sqrt{2}}{4}$$

2. $\tan(210° - 45°) = \dfrac{\tan 210° - \tan 45°}{1 + \tan 210° \tan 45°}$　[3.2]

$$= \frac{\frac{1}{\sqrt{3}} - 1}{1 + \frac{1}{\sqrt{3}} \cdot 1} = \frac{\frac{1}{\sqrt{3}} - 1}{1 + \frac{1}{\sqrt{3}}} \cdot \frac{\sqrt{3}}{\sqrt{3}}$$

$$= \frac{1 - \sqrt{3}}{\sqrt{3} + 1} \cdot \frac{1 - \sqrt{3}}{1 - \sqrt{3}} = \frac{1 - 2\sqrt{3} + 3}{1 - 3}$$

$$= \frac{4 - 2\sqrt{3}}{-2} = \sqrt{3} - 2$$

3. $\sin\left(\dfrac{2\pi}{3} + \dfrac{\pi}{4}\right) = \sin\dfrac{2\pi}{3}\cos\dfrac{\pi}{4} + \cos\dfrac{2\pi}{3}\sin\dfrac{\pi}{4}$　[3.2]

$$= \frac{\sqrt{3}}{2} \cdot \frac{\sqrt{2}}{2} - \frac{1}{2} \cdot \frac{\sqrt{2}}{2}$$

$$= \frac{\sqrt{6}}{4} - \frac{\sqrt{2}}{4}$$

$$= \frac{\sqrt{6} - \sqrt{2}}{4}$$

4. $\sec\left(\dfrac{4\pi}{3} - \dfrac{\pi}{4}\right) = \dfrac{1}{\cos\left(\dfrac{4\pi}{3} - \dfrac{\pi}{4}\right)}$　[3.2]

$$= \frac{1}{\cos\frac{4\pi}{3}\cos\frac{\pi}{4} + \sin\frac{4\pi}{3}\sin\frac{\pi}{4}}$$

$$= \frac{1}{-\frac{1}{2} \cdot \frac{\sqrt{2}}{2} + \left(-\frac{\sqrt{3}}{2}\right) \cdot \frac{\sqrt{2}}{2}} \cdot \frac{-4}{-4}$$

$$= \frac{-4}{\sqrt{2} + \sqrt{6}} \cdot \frac{\sqrt{2} - \sqrt{6}}{\sqrt{2} - \sqrt{6}}$$

$$= \frac{-4\left(\sqrt{2} - \sqrt{6}\right)}{2 - 6} = \sqrt{2} - \sqrt{6}$$

5. $\sin(60° - 135°) = \sin 60° \cos 135° - \cos 60° \sin 135°$　[3.2]

$$= \frac{\sqrt{3}}{2}\left(-\frac{\sqrt{2}}{2}\right) - \frac{1}{2} \cdot \frac{\sqrt{2}}{2}$$

$$= -\frac{\sqrt{6}}{4} - \frac{\sqrt{2}}{4} = -\frac{\sqrt{6} + \sqrt{2}}{4}$$

6. $\cos\left(\dfrac{5\pi}{3} - \dfrac{7\pi}{4}\right) = \cos\dfrac{5\pi}{3}\cos\dfrac{7\pi}{4} + \sin\dfrac{5\pi}{3}\sin\dfrac{7\pi}{4}$　[3.2]

$$= \frac{1}{2} \cdot \frac{\sqrt{2}}{2} + \left(-\frac{\sqrt{3}}{2}\right)\left(-\frac{\sqrt{2}}{2}\right)$$

$$= \frac{\sqrt{2}}{4} + \frac{\sqrt{6}}{4} = \frac{\sqrt{2} + \sqrt{6}}{4}$$

7. $\sin 22.5° = \sin\dfrac{45°}{2}$　[3.3]

$$= \sqrt{\frac{1 - \cos 45°}{2}}$$

$$= \sqrt{\frac{1 - \frac{\sqrt{2}}{2}}{2}}$$

$$= \sqrt{\frac{2 - \sqrt{2}}{4}}$$

$$= \frac{\sqrt{2 - \sqrt{2}}}{2}$$

8. $\cos 105° = \cos\dfrac{210°}{2}$　[3.3]

$$= -\sqrt{\frac{1 + \cos 210°}{2}}$$

$$= -\sqrt{\frac{1 + \left(-\frac{\sqrt{3}}{2}\right)}{2}}$$

$$= -\sqrt{\frac{2 - \sqrt{3}}{4}}$$

$$= -\frac{\sqrt{2 - \sqrt{3}}}{2}$$

9. $\tan 67.5° = \tan\dfrac{135°}{2}$ [3.3]

$$= \dfrac{1-\cos 135°}{\sin 135°}$$

$$= \dfrac{1-\left(-\frac{1}{\sqrt{2}}\right)}{\frac{1}{\sqrt{2}}}$$

$$= \dfrac{1+\frac{1}{\sqrt{2}}}{\frac{1}{\sqrt{2}}}$$

$$= \sqrt{2}+1$$

10. $\sin 112.5° = \sin\dfrac{225°}{2}$ [3.3]

$$= \sqrt{\dfrac{1-\cos 225°}{2}}$$

$$= \sqrt{\dfrac{1-\left(-\frac{\sqrt{2}}{2}\right)}{2}}$$

$$= \sqrt{\dfrac{2+\sqrt{2}}{4}}$$

$$= \dfrac{\sqrt{2+\sqrt{2}}}{2}$$

11. $\sin\alpha = \dfrac{1}{2}, \cos\alpha = \dfrac{\sqrt{3}}{2}$, quadrant I, $\tan\alpha = \dfrac{\sqrt{3}}{3}$ [3.2/3.3]

$\cos\beta = \dfrac{1}{2}, \sin\beta = -\dfrac{\sqrt{3}}{2}$, quadrant IV, $\tan\beta = -\sqrt{3}$

a. $\cos(\alpha-\beta) = \cos\alpha\cos\beta + \sin\alpha\sin\beta$

$$= \dfrac{\sqrt{3}}{2}\cdot\dfrac{1}{2} + \dfrac{1}{2}\left(-\dfrac{\sqrt{3}}{2}\right)$$

$$= \dfrac{\sqrt{3}}{4} - \dfrac{\sqrt{3}}{4} = 0$$

b. $\tan 2\alpha = \dfrac{2\tan\alpha}{1-\tan^2\alpha}$

$$= \dfrac{2\left(\frac{\sqrt{3}}{3}\right)}{1-\left(\frac{\sqrt{3}}{3}\right)^2} = \dfrac{\frac{2\sqrt{3}}{3}}{1-\frac{1}{3}}\cdot\dfrac{3}{3}$$

$$= \dfrac{2\sqrt{3}}{3-1} = \sqrt{3}$$

c. $\sin\dfrac{\beta}{2} = \sqrt{\dfrac{1-\cos\beta}{2}}$

$$= \sqrt{\dfrac{1-\frac{1}{2}}{2}}$$

$$= \sqrt{\dfrac{1}{4}}$$

$$= \dfrac{1}{2}$$

12. $\sin\alpha = \dfrac{\sqrt{3}}{2}, \cos\alpha = -\dfrac{1}{2}$, quadrant II [3.2/3.3]

$\cos\beta = -\dfrac{1}{2}, \sin\beta = -\dfrac{\sqrt{3}}{2}$, quadrant III

a. $\sin(\alpha+\beta) = \sin\alpha\cos\beta + \cos\alpha\sin\beta$

$$= \dfrac{\sqrt{3}}{2}\cdot\left(-\dfrac{1}{2}\right) + \left(-\dfrac{1}{2}\right)\left(-\dfrac{\sqrt{3}}{2}\right)$$

$$= -\dfrac{\sqrt{3}}{4} + \dfrac{\sqrt{3}}{4} = 0$$

b. $\sec 2\beta = \dfrac{1}{\cos 2\beta} = \dfrac{1}{\cos^2\beta - \sin^2\beta}$

$$= \dfrac{1}{\left(-\frac{1}{2}\right)^2 - \left(-\frac{\sqrt{3}}{2}\right)^2}$$

$$= \dfrac{1}{\frac{1}{4}-\frac{3}{4}} = \dfrac{1}{-\frac{2}{4}} = -2$$

c. $\cos\dfrac{\alpha}{2} = \sqrt{\dfrac{1+\cos\alpha}{2}}$

$$= \sqrt{\dfrac{1+\left(-\frac{1}{2}\right)}{2}} = \sqrt{\dfrac{\frac{1}{2}}{2}}$$

$$= \sqrt{\dfrac{1}{4}}$$

$$= \dfrac{1}{2}$$

13. $\sin\alpha = -\dfrac{1}{2}$, $\cos\alpha = \dfrac{\sqrt{3}}{2}$, quadrant IV, $\tan\alpha = -\dfrac{\sqrt{3}}{3}$ [3.2/3.3]

$\cos\beta = -\dfrac{\sqrt{3}}{2}$, $\sin\beta = -\dfrac{1}{2}$, quadrant III

  a. $\sin(\alpha-\beta) = \sin\alpha\cos\beta - \cos\alpha\sin\beta$

$$= -\frac{1}{2}\left(-\frac{\sqrt{3}}{2}\right) - \frac{\sqrt{3}}{2}\left(-\frac{1}{2}\right)$$

$$= \frac{\sqrt{3}}{4} + \frac{\sqrt{3}}{4} = \frac{2\sqrt{3}}{4}$$

$$= \frac{\sqrt{3}}{2}$$

  b. $\tan 2\alpha = \dfrac{2\tan\alpha}{1-\tan^2\alpha}$

$$= \frac{2\left(-\frac{\sqrt{3}}{3}\right)}{1-\left(-\frac{\sqrt{3}}{3}\right)^2}$$

$$= \frac{-\frac{2\sqrt{3}}{3}}{1-\frac{1}{3}}\cdot\frac{3}{3}$$

$$= \frac{-2\sqrt{3}}{3-1}$$

$$= -\sqrt{3}$$

  c. $\cos\dfrac{\beta}{2} = -\sqrt{\dfrac{1+\cos\beta}{2}}$

$$= -\sqrt{\frac{1+\left(-\frac{\sqrt{3}}{2}\right)}{2}}$$

$$= -\sqrt{\frac{2-\sqrt{3}}{4}}$$

$$= -\frac{\sqrt{2-\sqrt{3}}}{2}$$

14. $\sin\alpha = \dfrac{\sqrt{2}}{2}$, $\cos\alpha = \dfrac{\sqrt{2}}{2}$, quadrant I [3.2/3.3]

$\cos\beta = \dfrac{\sqrt{3}}{2}$, $\sin\beta = -\dfrac{1}{2}$, quadrant IV, $\tan\beta = -\dfrac{\sqrt{3}}{3}$

  a. $\cos(\alpha-\beta) = \cos\alpha\cos\beta + \sin\alpha\sin\beta$

$$= \frac{\sqrt{2}}{2}\cdot\frac{\sqrt{3}}{2} + \frac{\sqrt{2}}{2}\left(-\frac{1}{2}\right)$$

$$= \frac{\sqrt{6}}{4} - \frac{\sqrt{2}}{4} = \frac{\sqrt{6}-\sqrt{2}}{4}$$

  b. $\tan 2\beta = \dfrac{2\tan\beta}{1-\tan^2\beta}$

$$= \frac{2\left(-\frac{\sqrt{3}}{3}\right)}{1-\left(-\frac{\sqrt{3}}{3}\right)^2}$$

$$= \frac{-2\left(\frac{\sqrt{3}}{3}\right)}{1-\frac{1}{3}}\cdot\frac{3}{3}$$

$$= \frac{2\sqrt{3}}{3-1}$$

$$= -\sqrt{3}$$

  c. $\sin 2\alpha = 2\sin\alpha\cos\alpha$

$$= 2\frac{\sqrt{2}}{2}\cdot\frac{\sqrt{2}}{2}$$

$$= 1$$

15. $2\sin 3x\cos 3x = \sin 2(3x)$ [3.3]

$\qquad\qquad\qquad = \sin 6x$

16. $\dfrac{\tan 2x + \tan x}{1-\tan 2x\tan x} = \tan(2x+x)$ [3.2]

$\qquad\qquad\qquad\qquad = \tan 3x$

17. $\sin 4x\cos x - \cos 4x\sin x = \sin(4x-x)$ [3.2]

$\qquad\qquad\qquad\qquad\quad = \sin 3x$

18. $\cos^2 2\theta - \sin^2 2\theta = \cos 2(2\theta)$ [3.3]

$\qquad\qquad\qquad\quad = \cos 4\theta$

19. $\dfrac{\sin 2\theta}{\cos 2\theta} = \tan 2\theta$ [3.1]

20. $\dfrac{1-\cos 2\theta}{\sin 2\theta} = \dfrac{1-(\cos^2\theta - \sin^2\theta)}{2\sin\theta\cos\theta}$ [3.3]

$$= \frac{\sin^2\theta + \cos^2\theta - \cos^2\theta + \sin^2\theta}{2\sin\theta\cos\theta}$$

$$= \frac{2\sin^2\theta}{2\sin\theta\cos\theta}$$

$$= \frac{\sin\theta}{\cos\theta}$$

$$= \tan\theta$$

21. $\cos 2\theta - \cos 4\theta = -2\sin\dfrac{2\theta + 4\theta}{2}\sin\dfrac{2\theta - 4\theta}{2}$ [3.4]

$\qquad\qquad = -2\sin 3\theta \sin(-\theta)$

$\qquad\qquad = 2\sin 3\theta \sin\theta$

22. $\sin 3\theta - \sin 5\theta = 2\cos\dfrac{3\theta + 5\theta}{2}\sin\dfrac{3\theta - 5\theta}{2}$ [3.4]

$\qquad\qquad = 2\cos 4\theta \sin(-\theta)$

$\qquad\qquad = -2\cos 4\theta \sin\theta$

23. $\sin 6\theta + \sin 2\theta = 2\sin\dfrac{6\theta + 2\theta}{2}\cos\dfrac{6\theta - 2\theta}{2}$ [3.4]

$\qquad\qquad = 2\sin 4\theta \cos 2\theta$

24. $\sin 5\theta - \sin\theta = 2\cos\dfrac{5\theta + \theta}{2}\sin\dfrac{5\theta - \theta}{2}$ [3.4]

$\qquad\qquad = 2\cos 3\theta \sin 2\theta$

25. $\dfrac{1}{\sin x - 1} + \dfrac{1}{\sin x + 1} = \dfrac{(\sin x + 1) + (\sin x - 1)}{(\sin x - 1)(\sin x + 1)}$

$\qquad\qquad = \dfrac{2\sin x}{\sin^2 x - 1}$

$\qquad\qquad = \dfrac{2\sin x}{-\cos^2 x}$

$\qquad\qquad = -2\tan x \sec x$

26. $\dfrac{\sin x}{1 - \cos x} = \dfrac{\sin x(1 + \cos x)}{(1 - \cos x)(1 + \cos x)}$

$\qquad\qquad = \dfrac{\sin x + \sin x \cos x}{1 - \cos^2 x}$

$\qquad\qquad = \dfrac{\sin x + \sin x \cos x}{\sin^2 x}$

$\qquad\qquad = \dfrac{\sin x}{\sin^2 x} + \dfrac{\sin x \cos x}{\sin^2 x}$

$\qquad\qquad = \csc x + \cot x,\ 0 < x < \dfrac{\pi}{2}$

27. $\dfrac{1 + \sin x}{\cos^2 x} = \dfrac{1}{\cos^2 x} + \dfrac{\sin x}{\cos^2 x}$

$\qquad\qquad = \sec^2 x + \tan x \sec x$

$\qquad\qquad = \tan^2 x + 1 + \tan x \sec x$

28. $\dfrac{\cos^2 2x - \sin^2 2x}{\cos 2x + \sin 2x} = \dfrac{(\cos 2x - \sin 2x)(\cos 2x + \sin 2x)}{\cos 2x + \sin 2x}$

$\qquad\qquad = \cos 2x - \sin 2x$

$\qquad\qquad = \cos^2 x - \sin^2 x - \sin 2x$

29. $\dfrac{1}{\cos x} - \cos x = \dfrac{1 - \cos^2 x}{\cos x}$

$\qquad\qquad = \dfrac{\sin^2 x}{\cos x}$

$\qquad\qquad = \tan x \sin x$

30. $\sin(270° - \theta) - \cos(270° - \theta) = \sin 270°\cos\theta - \cos 270°\sin\theta - \cos 270°\cos\theta - \sin 270°\sin\theta$

$\qquad\qquad = (-1)\cos\theta - 0 - 0 - (-1)\sin\theta$

$\qquad\qquad = -\cos\theta + \sin\theta$

$\qquad\qquad = \sin\theta - \cos\theta$

31. $\sin\left(\dfrac{\pi}{4} - \alpha\right) = \sin\dfrac{\pi}{4}\cos\alpha - \cos\dfrac{\pi}{4}\sin\alpha$

$\qquad\qquad = \dfrac{\sqrt{2}}{2}\cos\alpha - \dfrac{\sqrt{2}}{2}\sin\alpha$

$\qquad\qquad = \dfrac{\sqrt{2}}{2}(\cos\alpha - \sin\alpha)$

32. $\sin(180° - \alpha + \beta) = \sin[180° - (\alpha - \beta)]$

$\qquad\qquad = \sin 180°\cos(\alpha - \beta) - \cos 180°\sin(\alpha - \beta)$

$\qquad\qquad = 0[\cos(\alpha - \beta)] - (-1)[\sin\alpha\cos\beta - \cos\alpha\sin\beta]$

$\qquad\qquad = \sin\alpha\cos\beta - \cos\alpha\sin\beta$

33. $\dfrac{\sin 4x - \sin 2x}{\cos 4x - \cos 2x} = \dfrac{2\cos\dfrac{4x + 2x}{2}\sin\dfrac{4x - 2x}{2}}{-2\sin\dfrac{4x + 2x}{2}\sin\dfrac{4x - 2x}{2}}$

$\qquad\qquad = -\dfrac{\cos 3x \sin x}{\sin 3x \sin x}$

$\qquad\qquad = -\cot 3x$

34. $2\sin x \sin 3x = \cos(x - 3x) - \cos(x + 3x)$

$\qquad\qquad = \cos 2x - \cos 4x$

$\qquad\qquad = \cos 2x - (2\cos^2 2x - 1)$

$\qquad\qquad = 1 + \cos 2x - 2\cos^2 2x$

$\qquad\qquad = (1 - \cos 2x)(1 + 2\cos 2x)$

35. $\sin x - \cos 2x = \sin x - (1 - 2\sin^2 x)$

$$= 2\sin^2 x + \sin x - 1$$

$$= (2\sin x - 1)(\sin x + 1)$$

36. $\cos 4x = 1 - 2\sin^2 2x$

$$= 1 - 2(2\sin x \cos x)^2$$

$$= 1 - 2(4\sin^2 x \cos^2 x)$$

$$= 1 - 8\sin^2 x \cos^2 x$$

$$= 1 - 8\sin^2 x(1 - \sin^2 x)$$

$$= 1 - 8\sin^2 x + 8\sin^4 x$$

37. $\tan 4x = \dfrac{2\tan 2x}{1 - \tan^2 2x}$

$$= \dfrac{2\left(\dfrac{2\tan x}{1 - \tan^2 x}\right)}{1 - \left(\dfrac{2\tan x}{1 - \tan^2 x}\right)^2}$$

$$= \dfrac{\dfrac{4\tan x}{1 - \tan^2 x}}{\dfrac{(1 - \tan^2 x)^2 - (2\tan x)^2}{(1 - \tan^2 x)^2}} \cdot \dfrac{(1 - \tan^2 x)^2}{(1 - \tan^2 x)^2}$$

$$= \dfrac{4\tan x(1 - \tan^2 x)}{(1 - \tan^2 x)^2 - 4\tan^2 x}$$

$$= \dfrac{4\tan x - 4\tan^3 x}{1 - 2\tan^2 x + \tan^4 x - 4\tan^2 x}$$

$$= \dfrac{4\tan x - 4\tan^3 x}{1 - 6\tan^2 x + \tan^4 x}$$

38. $\dfrac{\sin 2x - \sin x}{\cos 2x + \cos x} = \dfrac{2\cos\dfrac{2x + x}{2}\sin\dfrac{2x - x}{2}}{2\cos\dfrac{2x + x}{2}\cos\dfrac{2x - x}{2}}$

$$= \dfrac{\sin\dfrac{\pi}{2}}{\cos\dfrac{\pi}{2}}$$

$$= \tan\dfrac{\pi}{2}$$

$$= \dfrac{1 - \cos x}{\sin x}$$

39. $2\sin 3x \cos 3x - 2\sin x \cos x = \sin 6x - \sin 2x$

$$= 2\cos\dfrac{6x + 2x}{2}\sin\dfrac{6x - 2x}{2}$$

$$= 2\cos 4x \sin 2x$$

40. $2\sin x \sin 2x = 2\sin x(2\sin x \cos x)$

$$= 4\cos x \sin^2 x$$

41. $\cos(x + y)\cos(x - y) = \dfrac{1}{2}[\cos(x + y + x - y) + \cos(x + y - x + y)]$

$$= \dfrac{1}{2}(\cos 2x + \cos 2y)$$

$$= \dfrac{1}{2}\left[2\cos^2 x - 1 + 2\cos^2 y - 1\right]$$

$$= \cos^2 x + \cos^2 y - 1$$

42. $\cos(x + y)\sin(x - y) = \dfrac{1}{2}[\sin(x + y + x - y) - \sin(x + y - x + y)]$

$$= \dfrac{1}{2}(\sin 2x - \sin 2y)$$

$$= \dfrac{1}{2}[2\sin x \cos x - 2\sin y \cos y]$$

$$= \sin x \cos x - \sin y \cos y$$

43. $y = \sec\left(\sin^{-1}\dfrac{12}{13}\right),\ \alpha = \sin^{-1}\dfrac{12}{13},\ \sin\alpha = \dfrac{12}{13},\ \cos\alpha = \dfrac{5}{13},\ \sec\alpha = \dfrac{13}{5}$ [3.5]

$$y = \sec\alpha = \dfrac{13}{5}$$

44. $y = \cos\left(\sin^{-1}\dfrac{3}{5}\right)$ [3.5]

$\alpha = \sin^{-1}\dfrac{3}{5}, \ \sin\alpha = \dfrac{3}{5}, \ y = \cos\alpha = \dfrac{4}{5}$

45. $\alpha = \sin^{-1}\left(-\dfrac{3}{5}\right)$ $\beta = \cos^{-1}\dfrac{5}{13}$ [3.5]

$\sin\alpha = -\dfrac{3}{5}$ $\cos\beta = \dfrac{5}{13}$

$\cos\alpha = \dfrac{4}{5}$ $\sin\beta = \dfrac{12}{13}$

$y = \cos\left[\sin^{-1}\left(-\dfrac{3}{5}\right) + \cos^{-1}\dfrac{5}{13}\right]$

$\quad = \cos(\alpha + \beta)$

$\quad = \cos\alpha\cos\beta - \sin\alpha\sin\beta$

$\quad = \dfrac{4}{5}\cdot\dfrac{5}{13} - \left(-\dfrac{3}{5}\right)\cdot\dfrac{12}{13}$

$\quad = \dfrac{20}{65} + \dfrac{36}{65}$

$\quad = \dfrac{56}{65}$

46. $y = \cos\left(2\sin^{-1}\dfrac{3}{5}\right)$ $\alpha = \sin^{-1}\dfrac{3}{5}$ [3.5]

$y = \cos 2\alpha$

$\quad = \cos^2\alpha - \sin^2\alpha$ $\sin\alpha = \dfrac{3}{5}$

$\quad = \left(\dfrac{4}{5}\right)^2 - \left(\dfrac{3}{5}\right)^2$ $\cos\alpha = \dfrac{4}{5}$

$\quad = \dfrac{16}{25} - \dfrac{9}{25}$

$\quad = \dfrac{7}{25}$

47. $2\sin^{-1}(x-1) = \dfrac{\pi}{3}$ [3.6]

$\quad \sin^{-1}(x-1) = \dfrac{\pi}{6}$

$\qquad x - 1 = \sin\dfrac{\pi}{6}$

$\qquad x - 1 = \dfrac{1}{2}$

$\qquad\quad x = \dfrac{3}{2}$

48. $\sin^{-1}x + \cos^{-1}\dfrac{4}{5} = \dfrac{\pi}{2}$ $\alpha = \cos^{-1}\dfrac{4}{5}$ [3.6]

$\quad \sin^{-1}x + \alpha = \dfrac{\pi}{2}$ $\cos\alpha = \dfrac{4}{5}$

$\qquad \sin^{-1}x = \dfrac{\pi}{2} - \alpha$ $\sin\alpha = \dfrac{3}{5}$

$x = \sin\left(\dfrac{\pi}{2} - \alpha\right)$

$\quad = \sin\dfrac{\pi}{2}\cos\alpha - \cos\dfrac{\pi}{2}\sin\alpha$

$\quad = 1\cdot\dfrac{4}{5} - 0\cdot\dfrac{3}{5}$

$\quad = \dfrac{4}{5}$

49. $4\sin^2 x + 2\sqrt{3}\sin x - 2\sin x - \sqrt{3} = 0$ [3.6]

$2\sin x\left(2\sin x + \sqrt{3}\right) - \left(2\sin x + \sqrt{3}\right) = 0$

$\qquad\quad \left(2\sin x + \sqrt{3}\right)\left(2\sin x - 1\right) = 0$

$2\sin x + \sqrt{3} = 0$ $2\sin x - 1 = 0$

$\sin x = -\dfrac{\sqrt{3}}{2}$ $\sin x = \dfrac{1}{2}$

$x = 240°, 300°$ $x = 30°, 150°$

The solutions are $30°, \ 150°, \ 240°, \ 300°$.

50. $2\sin x\cos x - \sqrt{2}\cos x - 2\sin x + \sqrt{2} = 0$ [3.6]

$\quad \cos x(2\sin x - \sqrt{2}) - (2\sin x - \sqrt{2}) = 0$

$\qquad\quad (2\sin x - \sqrt{2})(\cos x - 1) = 0$

$2\sin x - \sqrt{2} = 0$ $\cos x - 1 = 0$

$\sin x = \dfrac{\sqrt{2}}{2}$ $\cos x = 1$

$x = 45°, 135°$ $x = 0°$

The solutions are $0°, 45°, 135°$.

51. $3\cos^2 x + \sin x = 1$ [3.6]

$3(1 - \sin^2 x) + \sin x = 1$

$\qquad\quad 0 = 3\sin^2 x - \sin x - 2$

$\qquad\quad 0 = (3\sin x + 2)(\sin x - 1)$

$3\sin x + 2 = 0$ $\sin x - 1 = 0$

$\sin x = -\dfrac{2}{3}$ $\sin x = 1$

$x = 3.8713 \text{ or } 5.553$ $x = \dfrac{\pi}{2}$

The solutions are $\dfrac{\pi}{2} + 2k\pi, \ 3.8713 + 2k\pi, \ 5.553 + 2k\pi$

where k is an integer.

52. $\tan^2 x - 2\tan x - 3 = 0$ [3.6]

$(\tan x + 1)(\tan x - 3) = 0$

$\tan x = -1$ $\qquad\qquad$ $\tan x = 3$

$x = -\dfrac{\pi}{4} + k\pi$ $\qquad\qquad$ $x = 1.2490 + k\pi$

The solutions are $-\dfrac{\pi}{4} + k\pi$, $1.2490 + k\pi$ where k is an integer.

53. $\sin 3x \cos x - \cos 3x \sin x = \dfrac{1}{2}$ [3.6]

$\sin(3x - x) = \dfrac{1}{2}$

$\sin 2x = \dfrac{1}{2}$

$2x = \dfrac{\pi}{6} + 2k\pi$ $\qquad$ $2x = \dfrac{5\pi}{6} + 2k\pi$

$x = \dfrac{\pi}{12} + k\pi$ $\qquad$ $x = \dfrac{5\pi}{12} + k\pi$

The solutions are $\dfrac{\pi}{12}, \dfrac{5\pi}{12}, \dfrac{13\pi}{12}, \dfrac{17\pi}{12}$.

54. $\cos\left(2x - \dfrac{\pi}{3}\right) = -\dfrac{\sqrt{3}}{2}$ [3.6]

$2x - \dfrac{\pi}{3} = \dfrac{5\pi}{6} + 2k\pi$ $\qquad$ $2x - \dfrac{\pi}{3} = \dfrac{7\pi}{6} + 2k\pi$

$2x = \dfrac{7\pi}{6} + 2k\pi$ $\qquad\quad$ $2x = \dfrac{3\pi}{2} + 2k\pi$

$x = \dfrac{7\pi}{12} + k\pi$ $\qquad\quad$ $x = \dfrac{3\pi}{4} + k\pi$

The solutions are $\dfrac{7\pi}{12}, \dfrac{19\pi}{12}, \dfrac{3\pi}{4}, \dfrac{7\pi}{4}$.

55. $f(x) = \sqrt{3}\sin x + \cos x$ [3.4]

$f(x) = 2\sin\left(x + \dfrac{\pi}{6}\right)$

amplitude = 2

phase shift = $-\dfrac{\pi}{6}$

56. $f(x) = -2\sin x - 2\cos x$ [3.4]

$f(x) = 2\sqrt{2}\sin\left(x + \dfrac{5\pi}{4}\right)$

amplitude = $2\sqrt{2}$

phase shift = $-\dfrac{5\pi}{4}$

57. $f(x) = -\sin x - \sqrt{3}\cos x$ [3.4]

$f(x) = 2\sin\left(x + \dfrac{4\pi}{3}\right)$

amplitude = 2

phase shift = $-\dfrac{4\pi}{3}$

58. $f(x) = \dfrac{\sqrt{3}}{2}\sin x - \dfrac{1}{2}\cos x$ [3.4]

$f(x) = \sin\left(x + \dfrac{11\pi}{6}\right)$ or $\sin\left(x - \dfrac{\pi}{6}\right)$

amplitude = 1

phase shift = $\dfrac{\pi}{6}$

59. $f(x) = 2\cos^{-1} x$

60. $f(x) = \sin^{-1}(x - 1)$

61. $f(x) = \sin^{-1}\dfrac{x}{2}$

62. $f(x) = \sec^{-1} 2x$

63. a.

$$y \approx 1.1835\sin(0.01600x + 1.8497) + 6.4394$$

b. $y \approx 1.1835\sin(0.01600(104) + 1.8497) + 6.4394$ [3.6]

≈ 6.009

$= 6:01$

<hr />

Quantitative Reasoning

QR1. a. $s = \dfrac{(28)^2 \cos\alpha \sec\left(\dfrac{\pi}{18}\right)\sin\left(\alpha - \dfrac{\pi}{18}\right)}{16}$

$\alpha \approx 0.87266$

b. $\dfrac{\pi}{4} + \dfrac{\dfrac{\pi}{18}}{2} = \dfrac{\pi}{4} + \dfrac{\pi}{36} = \dfrac{9\pi + \pi}{36} = \dfrac{5\pi}{18} \approx 0.8727$

<hr />

Chapter Test

1.
$$\begin{aligned}1 + \sin^2 x \sec^2 x &= 1 + \sin^2 x \frac{1}{\cos^2 x}\\ &= 1 + \tan^2 x\\ &= \sec^2 x\end{aligned}$$

2.
$$\begin{aligned}\frac{1}{\sec x - \tan x} - \frac{1}{\sec x + \tan x} &= \frac{\sec x + \tan x - \sec x + \tan x}{\sec^2 x - \tan^2 x}\\ &= \frac{2\tan x}{1}\\ &= 2\tan x\end{aligned}$$

3.
$$\begin{aligned}\cos^3 x + \cos x \sin^2 x &= \cos x(\cos^2 x + \sin^2 x)\\ &= \cos x\end{aligned}$$

4.
$$\begin{aligned}\csc x - \cot x &= \frac{1}{\sin x} - \frac{\cos x}{\sin x}\\ &= \frac{1 - \cos x}{\sin x}\end{aligned}$$

5.
$$\begin{aligned}\sin 195° &= \sin(150° + 45°) \quad [3.2]\\ &= \sin 150°\cos 45° + \cos 150°\sin 45°\\ &= \frac{1}{2}\left(\frac{\sqrt{2}}{2}\right) + \left(-\frac{\sqrt{3}}{2}\right)\left(\frac{\sqrt{2}}{2}\right)\\ &= \frac{\sqrt{2} - \sqrt{6}}{4}\end{aligned}$$

6.
$$\sin\alpha = -\frac{3}{5}, \cos\alpha = -\frac{4}{5}, \cos\beta = -\frac{\sqrt{2}}{2}, \sin\beta\,\frac{\sqrt{2}}{2} \quad [3.2]$$
$$\sin(\alpha + \beta) = \sin\alpha\cos\beta + \cos\alpha\sin\beta$$
$$= \left(-\frac{3}{5}\right)\left(-\frac{\sqrt{2}}{2}\right) + \left(-\frac{4}{5}\right)\left(\frac{\sqrt{2}}{2}\right)$$
$$= \frac{3\sqrt{2} - 4\sqrt{2}}{10}$$
$$= -\frac{\sqrt{2}}{10}$$

<hr />

7. $\sin\left(\theta - \dfrac{3\pi}{2}\right) = \sin\theta\cos\dfrac{3\pi}{2} - \cos\theta\sin\dfrac{3\pi}{2}$

$\qquad\qquad = \sin\theta(0) - \cos\theta(-1)$

$\qquad\qquad = \cos\theta$

8. $\cos 6x\sin 3x + \sin 6x\cos 3x = \sin(6x + 3x)$ [3.2]

$\qquad\qquad\qquad\qquad\qquad\quad = \sin 9x$

9. $\sin\theta = \dfrac{4}{5}, \cos\theta = -\dfrac{3}{5}$ [3.3]

$\cos 2\theta = 2\cos^2\theta - 1$

$\qquad = 2\left(-\dfrac{3}{5}\right)^2 - 1$

$\qquad = \dfrac{18}{25} - 1 = -\dfrac{7}{25}$

10. $\tan\dfrac{\theta}{2} + \dfrac{\cos\theta}{\sin\theta} = \dfrac{1 - \cos\theta}{\sin\theta} + \dfrac{\cos\theta}{\sin\theta}$

$\qquad\qquad\quad = \dfrac{1 - \cos\theta + \cos\theta}{\sin\theta}$

$\qquad\qquad\quad = \csc\theta$

11. $\sin^2 2x + 4\cos^4 x = (2\sin x\cos x)^2 + 4\cos^4 x$

$\qquad\qquad\quad = 4\sin^2 x\cos^2 x + 4\cos^4 x$

$\qquad\qquad\quad = 4\cos^2 x(\sin^2 x + \cos^2 x)$

$\qquad\qquad\quad = 4\cos^2 x$

12. $\sin 15°\cos 75° = \dfrac{1}{2}\left[\sin(15° + 75°) + \sin(15° - 75°)\right]$ [3.4]

$\qquad\qquad\quad = \dfrac{1}{2}(\sin 90° - \sin 60°)$

$\qquad\qquad\quad = \dfrac{1}{2}\left(1 - \dfrac{\sqrt{3}}{2}\right)$

$\qquad\qquad\quad = \dfrac{1}{2} - \dfrac{\sqrt{3}}{4}$

$\qquad\qquad\quad = \dfrac{2 - \sqrt{3}}{4}$

13. $y = -\dfrac{\sqrt{3}}{2}\sin x + \dfrac{1}{2}\cos x$ [3.4]

$a = -\dfrac{\sqrt{3}}{2}, \; b = \dfrac{1}{2}$

$k = \sqrt{\left(-\dfrac{\sqrt{3}}{2}\right)^2 + \left(\dfrac{1}{2}\right)^2} = 1$

$\sin\beta = \left|\dfrac{\frac{1}{2}}{1}\right| = \dfrac{1}{2}$

$\beta = \dfrac{\pi}{6}$

$\alpha = \pi - \dfrac{\pi}{6}$

$\quad = \dfrac{5\pi}{6}$

$y = \sin\left(x + \dfrac{5\pi}{6}\right)$

14. $\theta = \cos^{-1}(0.7644)$ [3.5]

$\theta = 0.701$

15. $\sin\left(\cos^{-1}\dfrac{12}{13}\right)$ [3.5]

Let $\theta = \cos^{-1}\dfrac{12}{13}$ and find $\sin\theta$.

Then $\cos\theta = \dfrac{12}{13}$ and $0 \le \theta \le \pi$.

$\sin\theta = \dfrac{5}{13}$

$\sin\left(\cos^{-1}\dfrac{12}{13}\right) = \dfrac{5}{13}$

16. The graph of $y = \sin^{-1}(x + 2)$ is the graph of $y = \sin^{-1} x$ moved two units to the left.

17. $3\sin x - 2 = 0$ [3.6]

$\sin x = \dfrac{2}{3}$

$x = 41.8°, \; 138.2°$

18. $\sin x \cos x - \dfrac{\sqrt{3}}{2}\sin x = 0$ [3.6]

$\sin x\left(\cos x - \dfrac{\sqrt{3}}{2}\right) = 0$

$\sin x = 0$ $\qquad\qquad\qquad$ $\cos x - \dfrac{\sqrt{3}}{2} = 0$

$x = 0,\ \pi$ $\qquad\qquad\qquad\quad$ $\cos x = \dfrac{\sqrt{3}}{2}$

$\qquad\qquad\qquad\qquad\qquad\quad$ $x = \dfrac{\pi}{6},\ \dfrac{11\pi}{6}$

The solutions are $0,\ \dfrac{\pi}{6},\ \pi,\ \dfrac{11\pi}{6}$.

19. $\sin 2x + \sin x - 2\cos x - 1 = 0$

$2\sin x\cos x + \sin x - 2\cos x - 1 = 0$

$\sin x(2\cos x + 1) - (2\cos x + 1) = 0$

$(2\cos x + 1)(\sin x - 1) = 0$

$\cos x = -\dfrac{1}{2}$ $\qquad\qquad\qquad\qquad$ $\sin x = 1$

$x = \dfrac{2\pi}{3},\ \dfrac{4\pi}{3}$ $\qquad\qquad\qquad\qquad$ $x = \dfrac{\pi}{2}$

The solutions are $\dfrac{\pi}{2},\ \dfrac{2\pi}{3},\ \dfrac{4\pi}{3}$.

20. **a.**

L1	L2	L3	2
1	10.4	------	
32	10.883		
60	11.617		
91	12.467		
121	13.25		
152	13.8		
182	13.883		
L2(1)=10+24/60			

```
EDIT CALC TESTS
7↑QuartReg
8:LinReg(a+bx)
9:LnReg
0:ExpReg
A:PwrReg
B:Logistic
C:SinReg
```

```
SinReg 16,L1,L2,
365.25
```

```
SinReg
y=a*sin(bx+c)+d
a=1.756862376
b=.0167529869
c=-1.305554905
d=12.10996462
```

$f(x) = y \approx 1.7569\sin(0.0168x - 1.3056) + 12.1100$

b. $f(75) \approx 1.7569\sin(0.01675(75) - 1.3056) + 12.1100$ [3.6]

≈ 12.0233

≈ 12 hr 1 min

•••

1. $-2x + 1 < 7$

$-2x < 6$

$x > -3$

2. Shift the graph of $y = f(x)$ horizontally 1 unit to the left and up 2 units.

3. Reflect the graph of $y = f(x)$ across the x-axis.

4. $f(-x) = -x - \sin(-x)$

$= -x + \sin x$

$= -(x - \sin x)$

$= -f(x)$

odd function

5. $f(x) = \dfrac{5x}{x-1}$

$y = \dfrac{5x}{x-1}$

$x = \dfrac{5y}{y-1}$

$x(y-1) = 5y$

$xy - 5y = x$

$y(x-5) = x$

$y = \dfrac{x}{x-5}$

$f^{-1}(x) = \dfrac{x}{x-5}$

6. $240° = 240°\left(\dfrac{\pi}{180°}\right) = \dfrac{4\pi}{3}$ [2.1]

7. $\dfrac{5\pi}{3} = \dfrac{5\pi}{3}\left(\dfrac{180°}{\pi}\right) = 300°$ [2.1]

8.
$$\sin\frac{\pi}{3} = \frac{\sqrt{3}}{2}$$

9.
$$\csc 60° = \frac{1}{\sin 60°} = \frac{2}{\sqrt{3}} = \frac{2\sqrt{3}}{3}$$

10.
$$\sin\theta = \frac{\text{opp}}{\text{hyp}} = \frac{2}{3}$$
$$\text{adjacent side} = \sqrt{3^2 - 2^2}$$
$$= \sqrt{9-4}$$
$$= \sqrt{5}$$
$$\tan\theta = \frac{\text{opp}}{\text{adj}} = \frac{2}{\sqrt{5}} = \frac{2\sqrt{5}}{5}$$

11. $\cot\theta > 0$ in quadrant III
Positive [2.3]

12.
$$\theta = 310° \quad [2.3]$$
Since $270° < \theta < 360°$,
$$\theta = \theta' = 360°$$
$$\theta' = 50°$$

13.
$$\theta = \frac{5\pi}{3} \quad [2.3]$$
Since $\frac{3\pi}{2} < \theta < 2\pi$,
$$\theta = \theta' = 2\pi$$
$$\theta' = \frac{\pi}{3}$$

14.
$$t = \frac{\pi}{3} \quad [2.4]$$
$$y = \sin t = \sin\frac{\pi}{3} = \frac{\sqrt{3}}{2}$$
$$x = \cos t = \cos\frac{\pi}{3} = \frac{1}{2}$$
The point on the unit circle corresponding to $t = \frac{\pi}{3}$ is $\left(\frac{1}{2}, \frac{\sqrt{3}}{2}\right)$.

15.
$$y = 0.43\cos\left(2x - \frac{\pi}{6}\right) \quad [2.7]$$
amplitude: 0.43
$$0 \le 2x - \frac{\pi}{6} \le 2\pi$$
$$\frac{\pi}{6} \le 2x \le \frac{13\pi}{6}$$
$$\frac{\pi}{12} \le x \le \frac{13\pi}{12}$$
period $= \pi$, phase shift $= \frac{\pi}{12}$

16.
$$y = \sin^{-1}\frac{1}{2} \quad [3.5]$$
$$\sin y = \frac{1}{2} \quad -\frac{\pi}{2} \le y \le \frac{\pi}{2}$$
$$y = \frac{\pi}{6}$$

17. $\cos^{-1}(-0.8) = 2.498 \quad [3.5]$

18. Domain: [−1, 1]. [3.5]

19. Range: $\left(-\frac{\pi}{2}, \frac{\pi}{2}\right) \quad [3.5]$

20.
$$2\cos^2 x - 1 = -\sin x$$
$$1 - 2\sin^2 x = -\sin x$$
$$0 = 2\sin^2 x - \sin x - 1$$
$$0 = (2\sin x + 1)(\sin x - 1)$$

$$2\sin x + 1 = 0 \qquad \sin x - 1 = 0$$
$$\sin x = -\frac{1}{2} \qquad \sin x = 1$$
$$x = \frac{7\pi}{6}, \frac{11\pi}{6} \qquad x = \frac{\pi}{2}$$

The solutions are $\frac{\pi}{2}, \frac{7\pi}{6}, \frac{11\pi}{6}$.

Chapter 4
Applications of Trigonometry

Section 4.1

1.

$C = 180° - 42° - 61°$

$C = 77°$

$$\frac{b}{\sin B} = \frac{a}{\sin A}$$

$$\frac{b}{\sin 61°} = \frac{12}{\sin 42°}$$

$$b = \frac{12\sin 61°}{\sin 42°}$$

$$b \approx 16$$

$$\frac{c}{\sin C} = \frac{a}{\sin A}$$

$$\frac{c}{\sin 77°} = \frac{12}{\sin 42°}$$

$$c = \frac{12\sin 77°}{\sin 42°}$$

$$c \approx 17$$

3.

$B = 180° - 110° - 32°$

$B = 38°$

$$\frac{b}{\sin B} = \frac{a}{\sin A}$$

$$\frac{12}{\sin 38°} = \frac{a}{\sin 110°}$$

$$a = \frac{12\sin 110°}{\sin 38°}$$

$$a \approx 18$$

$$\frac{c}{\sin C} = \frac{b}{\sin B}$$

$$\frac{c}{\sin 32°} = \frac{12}{\sin 38°}$$

$$c = \frac{12\sin 32°}{\sin 38°}$$

$$c \approx 10$$

5.

$$\frac{a}{\sin A} = \frac{b}{\sin B}$$

$$\frac{22}{\sin 132°} = \frac{16}{\sin B}$$

$$\sin B = \frac{16\sin 132°}{22}$$

$$\sin B \approx 0.5405$$

$$B \approx 33°$$

$C \approx 180° - 33° - 132°$

$C \approx 15°$

$$\frac{a}{\sin A} = \frac{c}{\sin C}$$

$$\frac{22}{\sin 132°} \approx \frac{c}{\sin 15°}$$

$$c \approx \frac{22\sin 15°}{\sin 132°}$$

$$c \approx 7.7$$

7.

$C = 180° - 22.5° - 112.4°$

$C = 45.1°$

$$\frac{b}{\sin B} = \frac{a}{\sin A}$$

$$\frac{b}{\sin 112.4°} = \frac{16.3}{\sin 22.5°}$$

$$b = \frac{16.3\sin 112.4°}{\sin 22.5°}$$

$$b \approx 39.4$$

$$\frac{c}{\sin C} = \frac{a}{\sin A}$$

$$\frac{c}{\sin 45.1°} = \frac{16.3}{\sin 22.5°}$$

$$c = \frac{16.3\sin 45.1°}{\sin 22.5°}$$

$$c \approx 30.2$$

9.

$C = 180° - 65.4° - 82.0°$

$C = 32.6°$

$$\frac{c}{\sin C} = \frac{b}{\sin B}$$

$$\frac{c}{\sin 32.6°} = \frac{36.5}{\sin 65.4°}$$

$$c = \frac{36.5 \sin 32.6°}{\sin 65.4°}$$

$$c \approx 21.6$$

$$\frac{a}{\sin A} = \frac{b}{\sin B}$$

$$\frac{a}{\sin 82.0°} = \frac{36.5}{\sin 65.4°}$$

$$a = \frac{36.5 \sin 82.0°}{\sin 65.4°}$$

$$a \approx 39.8$$

11.

$B = 180° - 98.5° - 33.8°$

$B = 47.7°$

$$\frac{a}{\sin A} = \frac{c}{\sin C}$$

$$\frac{a}{\sin 33.8°} = \frac{102}{\sin 98.5°}$$

$$a = \frac{102 \sin 33.8°}{\sin 98.5°}$$

$$a \approx 57.4$$

$$\frac{b}{\sin B} = \frac{c}{\sin C}$$

$$\frac{b}{\sin 47.7°} = \frac{102}{\sin 98.5°}$$

$$b = \frac{102 \sin 47.7°}{\sin 98.5°}$$

$$b \approx 76.3$$

13.

$$\frac{c}{\sin C} = \frac{b}{\sin B}$$

$$\frac{87.2}{\sin 114.2°} = \frac{12.1}{\sin B}$$

$$\sin B = \frac{12.1 \sin 114.2°}{87.2} \approx 0.1266$$

$$B \approx 7.3°$$

$A \approx 180° - 114.2° - 7.3°$

$A \approx 58.5°$

$$\frac{c}{\sin C} = \frac{a}{\sin A}$$

$$\frac{87.2}{\sin 114.2°} \approx \frac{a}{\sin 58.5°}$$

$$a = \frac{87.2 \sin 58.5°}{\sin 114.2°} \approx 81.5$$

15.

$$\sin 37° = \frac{h}{40}$$

$$h = 40 \sin 37°$$

$$h \approx 24$$

Since $h < 28$, two triangles exist.

$$\frac{c}{\sin C} = \frac{a}{\sin A}$$

$$\frac{40}{\sin C} = \frac{28}{\sin 37°}$$

$$\sin C = \frac{40 \sin 37°}{28} = 0.8597$$

$$C \approx 59° \text{ or } 121°$$

$C = 59°$

$B = 180° - 37° - 59° = 84°$

$$\frac{b}{\sin 84°} = \frac{28}{\sin 37°}$$

$$b = \frac{28 \sin 84°}{\sin 37°} = 46$$

$C = 121°$

$B = 180° - 121° - 37° = 22°$

$$\frac{b}{\sin 22°} = \frac{28}{\sin 37°}$$

$$b = \frac{28 \sin 22°}{\sin 37°} \approx 17$$

17.

$$\sin 65° = \frac{h}{10}$$
$$h = 10 \sin 65°$$
$$h \approx 9.06$$
Since $h > 8$, no triangle is formed.

19.

$$\sin 30° = \frac{h}{2.4}$$
$$h = 2.4 \sin 30°$$
$$h \approx 1.2$$
Since $h > 1$, no triangle is formed.

21.

$$\sin 14.8° = \frac{h}{6.35}$$
$$h = 6.35 \sin 14.8°$$
$$h \approx 1.62$$

Since $h < 4.80$, two solutions exist.

$$\frac{c}{\sin C} = \frac{a}{\sin A}$$

$$\frac{6.35}{\sin C} = \frac{4.80}{\sin 14.8°}$$

$$\sin C = \frac{6.35 \sin 14.8°}{4.80}$$

$$\sin C = 0.3379$$

$$C \approx 19.8° \text{ or } 160.2°$$

$C = 19.8°$

$B = 180° - 19.8° - 14.8°$

$\qquad = 145.4°$

$$\frac{b}{\sin 145.4°} = \frac{4.80}{\sin 14.8°}$$

$$b = \frac{4.80 \sin 145.4°}{\sin 14.8°}$$

$$b \approx 10.7$$

$C = 160.2°$

$B = 180° - 160.2° - 14.8°$

$\qquad = 5.0°$

$$\frac{b}{\sin 5.0°} = \frac{4.80}{\sin 14.8°}$$

$$b = \frac{4.80 \sin 5.0°}{\sin 14.8°}$$

$$b \approx 1.64$$

23.

$$\sin 47.2° = \frac{h}{8.25}$$
$$h = 8.25 \sin 47.2°$$
$$h \approx 6.05$$

Since $h > 5.80$, no triangle is formed.

25.

C, 67.25, 15.05, $117.32°$, B, c, A

Since $b > a$, one triangle exists.

$$\frac{a}{\sin A} = \frac{b}{\sin B}$$

$$\frac{15.05}{\sin A} = \frac{67.25}{\sin 117.32°}$$

$$\sin A = \frac{15.04 \sin 117.32°}{67.25} \approx 0.1988$$

$$A = 11.47°$$

$$C = 180° - 11.47° - 117.32°$$
$$C = 51.21°$$

$$\frac{c}{\sin C} = \frac{b}{\sin B}$$

$$\frac{c}{\sin 51.21°} = \frac{67.25}{\sin 117.32°}$$

$$c = \frac{67.25 \sin 51.21°}{\sin 117.32°}$$

$$c \approx 59.00$$

27.

$$\sin 20.5° = \frac{h}{14.1}$$

$$h = 14.1 \sin 20.5°$$

$$h \approx 4.9$$

Since $h < 10.3$, two solutions exist.

$$\frac{a}{\sin A} = \frac{c}{\sin C}$$

$$\frac{10.3}{\sin 20.5°} = \frac{14.1}{\sin C}$$

$$\sin C = \frac{14.1 \sin 20.5°}{10.3}$$

$$\sin C = 0.4794$$

$$C \approx 28.6° \text{ or } 151.4°$$

$A = 20.5°$	$A = 20.5°$
$B = 180° - 28.6° - 20.5°$	$B = 180° - 151.4° - 20.5°$
$B = 130.9°$	$B = 8.1°$
$\dfrac{b}{\sin 130.9°} = \dfrac{10.3}{\sin 20.5°}$	$\dfrac{b}{\sin 8.1°} = \dfrac{10.3}{\sin 20.5°}$
$b = \dfrac{10.3 \sin 130.9°}{\sin 20.5°}$	$b = \dfrac{10.3 \sin 8.1°}{\sin 20.5°}$
$b \approx 22.2$	$b \approx 4.1$

29.

$$\angle B = 180° - (39.4° + 64.9°)$$

$$\angle B = 75.7°$$

$$\frac{a}{\sin A} = \frac{b}{\sin B}$$

$$\frac{a}{\sin 39.4°} = \frac{105}{\sin 75.7°}$$

$$a = \frac{105 \sin 39.4°}{\sin 75.7°}$$

$$a \approx 68.8 \text{ miles}$$

31. $a = 155$ yd, $c = 165$ yd, $A = 42.0°$

$$\frac{c}{\sin C} = \frac{a}{\sin A}$$

$$\frac{165}{\sin C} = \frac{155}{\sin 42.0°}$$

$$\sin C = \frac{155}{165 \sin 42.0°}$$

$$C = \sin^{-1}\left(\frac{155}{165 \sin 42.0°}\right) = 45.4°$$

$$B = 180° - 42.0° - 45.4° = 92.6°$$

$$\frac{b}{\sin B} = \frac{a}{\sin A}$$

$$\frac{b}{\sin 92.6°} = \frac{155}{\sin 42.0°}$$

$$b = \frac{155 \sin 92.6°}{165 \sin 42.0°} = 231 \text{ yd}$$

33.

$$B = 180° - 67° - 31°$$

$$B = 82°$$

$$\frac{c}{\sin C} = \frac{b}{\sin B}$$

$$\frac{c}{\sin 31°} = \frac{220}{\sin 82°}$$

$$c = \frac{220 \sin 31°}{\sin 82°}$$

$$c \approx 110 \text{ feet}$$

35.

$$ABC = 180° - 3°$$

$$= 177°$$

$$\frac{AC}{\sin ABC} = \frac{BC}{\sin CAB}$$

$$\frac{AC}{\sin 177°} = \frac{3550}{\sin 2.2°}$$

$$AC = \frac{3550 \sin 177°}{\sin 2.2°} \approx 4840 \text{ ft}$$

37.

$A = 32° - 11°$

$A = 21°$

$B = 180° - 90° - 32°$

$B = 58°$

$C = 180° - 58° - 21°$

$C = 101°$

$\dfrac{c}{\sin C} = \dfrac{a}{\sin A}$

$\dfrac{c}{\sin 101°} = \dfrac{35}{\sin 21°}$

$c = \dfrac{35\sin 101°}{\sin 21°}$

$c \approx 96$ feet

39.

$A = 5°$

$B = 180° - 90° - 75°$

$B = 15°$

$C = 180° - 15° - 5°$

$C = 160°$

$\dfrac{b}{\sin B} = \dfrac{a}{\sin A}$

$\dfrac{b}{\sin 15°} = \dfrac{12}{\sin 5°}$

$b = \dfrac{12\sin 15°}{\sin 5°}$

$\sin 70° = \dfrac{h}{b}$

$h = b\sin 70°$

$h = \left(\dfrac{12\sin 15°}{\sin 5°}\right)\sin 70°$

$h \approx 33$ feet

41.

$\phi = 360° - 332°$

$\phi = 28°$

$\alpha = 28°$

$C = 82° - 28°$

$C = 54°$

$A = 28° + 36°$

$A = 64°$

$B = 180° - 64° - 54°$

$B = 62°$

$\dfrac{a}{\sin A} = \dfrac{b}{\sin B}$

$\dfrac{a}{\sin 64°} = \dfrac{8.0}{\sin 62°}$

$a = \dfrac{8.0\sin 64°}{\sin 62°}$

$a \approx 8.1$ miles

43.

$A = 120° - 65°$

$A = 55°$

$\alpha = 65°$

$B = 38° + 65°$

$B = 103°$

$C = 180° - 103° - 55°$

$C = 22°$

$\dfrac{b}{\sin B} = \dfrac{c}{\sin C}$

$\dfrac{b}{\sin 103°} = \dfrac{450}{\sin 22°}$

$b = \dfrac{450\sin 103°}{\sin 22°}$

$b \approx 1200$ miles

Connecting Concepts

45.

$A = 180° - 67° - 68°$

$A = 45°$

$B = 67° + 11°$

$B = 78°$

$C = 180° - 45° - 78°$

$C = 57°$

$\dfrac{c}{\sin C} = \dfrac{b}{\sin B}$

$\dfrac{c}{\sin 57°} = \dfrac{300}{\sin 78°}$

$c = \dfrac{300\sin 57°}{\sin 78°}$

$c \approx 260$ meters

47.

$\dfrac{a}{\sin A} = \dfrac{b}{\sin B}$

$\dfrac{a}{b} = \dfrac{\sin A}{\sin B}$

$\dfrac{a}{b} + 1 = \dfrac{\sin A}{\sin B} + 1$

$\dfrac{a + b}{b} = \dfrac{\sin A + \sin B}{\sin B}$

49.

$$\frac{3}{d_1} = \cos\theta, \quad \frac{5}{d_2} = \sin\theta$$

$$d_1 = \frac{3}{\cos\theta}, \quad d_2 = \frac{5}{\sin\theta}$$

$$L = d_1 + d_2$$

$$L(\theta) = \frac{3}{\cos\theta} + \frac{5}{\sin\theta}$$

The graph of L is shown.
The minimum value of L is approximately 11.19 m.

PS1. $\sqrt{(10.0)^2 + (15.0)^2 - 2(10.0)(15.0)\cos 110.0°} \approx 20.7$

PS2. $A = \frac{1}{2}bh = \frac{1}{2}(6)(8.5) = 22.5 \text{ in.}^2$

PS3.
$$c^2 = a^2 + b^2 - 2ab\cos C$$
$$c^2 - a^2 - b^2 = -2ab\cos C$$
$$\cos C = \frac{c^2 - a^2 - b^2}{-2ab}$$
$$C = \cos^{-1}\left(\frac{c^2 - a^2 - b^2}{-2ab}\right) = \cos^{-1}\left(\frac{a^2 + b^2 - c^2}{2ab}\right)$$

PS4. $P = 6 + 9 + 10 = 25$
semiperimeter $= \frac{1}{2}(25) = 12.5 \text{ m}$

PS5. $s = \frac{a+b+c}{2} = \frac{3+4+5}{2} = 6$
$$\sqrt{6(6-3)(6-4)(6-5)} = \sqrt{6(3)(2)(1)} = 6$$

PS6. $c^2 = a^2 + b^2$

Section 4.2

1. $c^2 = a^2 + b^2 - 2ab\cos C$
$c^2 = 12^2 + 18^2 - 2(12)(18)\cos 44°$
$c^2 = 468 - 432\cos 44°$
$c = \sqrt{468 - 432\cos 44°}$
$c \approx 13$

3. $b^2 = a^2 + c^2 - 2ac\cos B$
$b^2 = 120^2 + 180^2 - 2(120)(180)\cos 56°$
$b^2 = 46,800 - 43,200\cos 56°$
$b = \sqrt{46,800 - 43,200\cos 56°}$
$b \approx 150$

5. $a^2 = b^2 + c^2 - 2bc\cos A$
$a^2 = 60^2 + 84^2 - 2(60)(84)\cos 13°$
$a^2 = 10,656 - 10,080\cos 13°$
$a = \sqrt{10,656 - 10,080\cos 13°}$
$a \approx 29$

7. $c^2 = a^2 + b^2 - 2ab\cos C$
$c^2 = 9.0^2 + 7.0^2 - 2(9.0)(7.0)\cos 72°$
$c^2 = 130 - 126\cos 72°$
$c = \sqrt{130 - 126\cos 72°}$
$c \approx 9.5$

9. $c^2 = a^2 + b^2 - 2ab\cos C$
$c^2 = 4.6^2 + 7.2^2 - 2(4.6)(7.2)\cos 124°$
$c^2 = 73 - 66.24\cos 124°$
$c = \sqrt{73 - 66.24\cos 124°}$
$c \approx 10$

11. $b^2 = a^2 + c^2 - 2ac\cos B$
$b^2 = 25.9^2 + 33.4^2 - 2(25.9)(33.4)\cos 84.0°$
$b^2 = 1786.37 - 1730.12\cos 84.0°$
$b = \sqrt{1786.37 - 1730\cos 84.0°}$
$b \approx 40.1$

13. $b^2 = a^2 + c^2 - 2ac\cos B$

$b^2 = 122^2 + 55.9^2 - 2(122)(55.9)\cos 44.2°$

$b^2 = 18,008.81 - 13,639.6\cos 44.2°$

$b = \sqrt{18,008.81 - 13,639.6\cos 44.2°}$

$b \approx 90.7$

15. $\cos A = \dfrac{b^2 + c^2 - a^2}{2bc}$

$\cos A = \dfrac{32^2 + 40^2 - 25^2}{2(32)(40)}$

$\cos A = \dfrac{1999}{2560}$

$A = \cos^{-1}\left(\dfrac{1999}{2560}\right) \approx 39°$

17. $\cos C = \dfrac{a^2 + b^2 - c^2}{2ab}$

$\cos C = \dfrac{8.0^2 + 9.0^2 - 12^2}{2(8.0)(9.0)}$

$\cos C = \dfrac{1}{144}$

$C = \cos^{-1}\left(\dfrac{1}{144}\right) \approx 90°$

19. $\cos B = \dfrac{a^2 + c^2 - b^2}{2ac}$

$\cos B = \dfrac{80^2 + 124^2 - 92^2}{2(80)(124)}$

$\cos B = \dfrac{13,312}{19,840}$

$B = \cos^{-1}\left(\dfrac{13312}{19840}\right) \approx 47.9°$

21. $\cos C = \dfrac{a^2 + b^2 - c^2}{2ab}$

$\cos C = \dfrac{1025^2 + 625^2 - 1420^2}{2(1025)(625)}$

$\cos C = \dfrac{-575,150}{1,281,250}$

$C = \cos^{-1}\left(\dfrac{-575,150}{1,281,250}\right) \approx 116.67°$

23. $\cos B = \dfrac{a^2 + c^2 - b^2}{2ac}$

$\cos B = \dfrac{32.5^2 + 29.6^2 - 40.1^2}{2(32.5)(29.6)}$

$\cos B = \dfrac{324.4}{1924}$

$B = \cos^{-1}\left(\dfrac{324.4}{1924}\right) \approx 80.3°$

25. $a^2 = b^2 + c^2 - 2bc\cos A$

$a = \sqrt{15.5^2 + 17.2^2 - 2(15.5)(17.2)\cos 39.4°}$

$a \approx 11.13860218$

$a \approx 11.1$

$\cos B = \dfrac{a^2 + c^2 - b^2}{2ac}$

$\cos B = \dfrac{11.13860218^2 + 17.2^2 - 15.5^2}{2(11.13860218)(17.2)}$

$B = \cos^{-1}\left(\dfrac{179.6584585}{383.167915}\right)$

$B \approx 62.0°$

$\cos C = \dfrac{a^2 + b^2 - c^2}{2ab}$

$\cos C = \dfrac{11.13860218^2 + 15.5^2 - 17.2^2}{2(11.13860218)(15.5)}$

$C = \cos^{-1}\left(\dfrac{68.47845852}{345.2966676}\right)$

$C \approx 78.6°$

27. $\cos A = \dfrac{b^2 + c^2 - a^2}{2bc}$

$\cos A = \dfrac{144^2 + 98.1^2 - 83.6^2}{2(144)(98.1)}$

$A = \cos^{-1}\left(\dfrac{23,370.65}{28,252.8}\right)$

$A \approx 34.2°$

$\cos B = \dfrac{a^2 + c^2 - b^2}{2ac}$

$\cos B = \dfrac{83.6^2 + 98.1^2 - 144^2}{2(83.6)(98.1)}$

$B = \cos^{-1}\left(\dfrac{-4123.43}{16,402.32}\right)$

$B \approx 104.6°$

$\cos C = \dfrac{a^2 + b^2 - c^2}{2ab}$

$\cos C = \dfrac{83.6^2 + 144^2 - 98.1^2}{2(83.6)(144)}$

$C = \cos^{-1}\left(\dfrac{18,101.35}{24,076.8}\right)$

$C \approx 41.3°$

29. $K = \dfrac{1}{2}bc\sin A$

$K = \dfrac{1}{2}(12)(24)\sin 105°$

$K \approx 140$ square units

31. $C = 180° - 42° - 76°$

$C = 62°$

$K = \dfrac{c^2 \sin A \sin B}{2\sin C}$

$K = \dfrac{12^2 \sin 42° \sin 76°}{2\sin 62°}$

$K \approx 53$ square units

33. $s = \dfrac{1}{2}(a+b+c)$

$s = \dfrac{1}{2}(16+12+14)$

$s = 21$

$K = \sqrt{s(s-a)(s-b)(s-c)}$

$K = \sqrt{21(21-16)(21-12)(21-14)}$

$K = \sqrt{21(5)(9)(7)}$

$K \approx 81$ square units

35. $\dfrac{a}{\sin A} = \dfrac{b}{\sin B}$

$\dfrac{22.4}{\sin A} = \dfrac{26.9}{\sin 54.3°}$

$\sin A = \dfrac{22.4 \sin 54.3°}{26.9}$

$\sin A \approx 0.6762$

$A \approx 42.5°$

$C = 180° - 42.5° - 54.3°$

$C = 83.2°$

$K = \dfrac{1}{2}ab\sin C$

$K = \dfrac{1}{2}(22.4)(26.9)\sin 83.2°$

$K \approx 299$ square units

37. $C = 180° - 116° - 34°$

$C = 30°$

$K = \dfrac{c^2 \sin A \sin B}{2 \sin C}$

$K = \dfrac{8.5^2 \sin 116° \sin 34°}{2 \sin 30°}$

$K \approx 36$ square units

39. $s = \dfrac{1}{2}(a+b+c)$

$s = \dfrac{1}{2}(3.6+4.2+4.8)$

$s = 6.3$

$K = \sqrt{s(s-a)(s-b)(s-c)}$

$K = \sqrt{6.3(6.3-3.6)(6.3-4.2)(6.3-4.8)}$

$K = \sqrt{6.3(2.7)(2.1)(1.5)}$

$K \approx 7.3$ square units

41.

$\alpha = 32°$

$\beta = 72°$

$B = 72° + 32°$

$B = 104°$

$b^2 = a^2 + c^2 - 2ac\cos 104°$

$b^2 = 320^2 + 560^2 - 2(320)(560)\cos 104°$

$b^2 = 416{,}000 - 358{,}400\cos 104°$

$b = \sqrt{416{,}000 - 358{,}400\cos 104°}$

$b \approx 710$ miles

43.

$a^2 = b^2 + c^2 - 2bc\cos A$

$a^2 = 26^2 + 90^2 - 2(26)(90)\cos 45°$

$a^2 = 8776 - 4680\cos 45°$

$a = \sqrt{8776 - 4680\cos 45°}$

$a \approx 74$ feet

45.

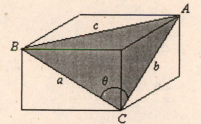

Let a = the length of the diagonal on the front of the box.
Let b = the length of the diagonal on the right side of the box.
Let c = the length of the diagonal on the top of the box.

$$a^2 = (4.75)^2 + (6.50)^2 = 64.8125$$

$$a = \sqrt{64.8125}$$

$$b^2 = (3.25)^2 + (4.75)^2 = 33.125$$

$$b = \sqrt{33.125}$$

$$c^2 = (6.50)^2 + (3.25)^2 = 52.8125$$

$$\theta = C$$

$$\cos C = \frac{a^2 + b^2 - c^2}{2ab}$$

$$\cos \theta = \frac{64.8125 + 33.125 - 52.8125}{2\sqrt{64.8125}\sqrt{33.125}}$$

$$\cos \theta = \frac{45.125}{2\sqrt{64.8125}\sqrt{33.125}}$$

$$\theta = \cos^{-1}\left(\frac{45.125}{2\sqrt{64.8125}\sqrt{33.125}}\right)$$

$$\theta \approx 60.9°$$

47.

$b = (18 \text{ mph})(10 \text{ hours}) = 180 \text{ miles}$

$c = (22 \text{ mph})(10 \text{ hours}) = 220 \text{ miles}$

$A = 318° - 198°$

$A = 120°$

$$a^2 = b^2 + c^2 - 2bc \cos A$$

$$a^2 = 180^2 + 220^2 - 2(180)(220)\cos 120°$$

$$a^2 = 120,400$$

$$a \approx 350 \text{ miles}$$

49.

$$A = \frac{360°}{6}$$

$$A = 60°$$

$$a^2 = 40^2 + 40^2 - 2(40)(40)\cos 60°$$

$$a^2 = 1600$$

$$a = 40 \text{ cm}$$

51.

$C = 90° + 14°$

$C = 104°$

$$a = \frac{180(5280)}{3600} \cdot 10$$

$a = 2640 \text{ feet}$

$$c^2 = 2640^2 + 400^2 - 2(2640)(400)(\cos 104°)$$

$$c^2 \approx 7,640,539$$

$$c \approx 2800 \text{ feet}$$

53.

$B = 108.5 + (180 - 124.6)$

$B = 163.9°$

$b^2 = a^2 + c^2 - 2ac\cos B$

$b^2 = 225^2 + 181^2 - 2(225)(181)\cos 163.9°$

$b^2 = 83,386 - 81,450\cos 163.9°$

$b = \sqrt{83,386 - 81,450\cos 163.9°}$

$b \approx 402.046592$

$b \approx 402$ mi

$\cos A = \dfrac{b^2 + c^2 - a^2}{2bc}$

$\cos A = \dfrac{402.046592^2 + 181^2 - 225^2}{2(402.046592)(181)}$

$A = \cos^{-1}\left(\dfrac{143,777.4621}{145,540.8663}\right)$

$A \approx 8.9°$

$\alpha = 180 - (108.5 + 8.9)$

$\alpha = 62.6°$

The distance is 402 mi and the bearing is S62.6°E.

55.

$s = \dfrac{1}{2}(a + b + c)$

$s = \dfrac{1}{2}(236 + 620 + 814)$

$s = 835$

$K = \sqrt{s(s-a)(s-b)(s-c)}$

$K = \sqrt{835(835 - 236)(835 - 620)(835 - 814)}$

$K = \sqrt{835(599)(215)(21)}$

$K \approx \sqrt{2,258,240,000}$

$K \approx 47,500$ square meters

57.

$\alpha = 90°$

$K = 4\left[\dfrac{1}{2}(9)(9)\sin 90°\right]$

$K = 162$ in^2

59.

$s = \dfrac{1}{2}(185 + 212 + 240)$

$s = 318.5$

$K = \sqrt{318.5(318.5 - 185)(318.5 - 212)(318.5 - 240)}$

$K \approx 18,854$ ft^2

cost $= 2.20(18,854)$

cost $\approx \$41,000$

61.

$s = \dfrac{1}{2}(680 + 800 + 1020)$

$s = 1250$

$K = \sqrt{1250(1250 - 680)(1250 - 800)(1250 - 1020)}$

$K \approx 271,558$ ft^2

Acres $= \dfrac{271,558}{43,560}$

Acres ≈ 6.23

63. For *ABC*,

$\dfrac{13.0 - 16.1}{10.4} \approx -0.2981$

$\dfrac{\sin\left(\dfrac{53.5 - 86.5}{2}\right)}{\cos\left(\dfrac{40.0}{2}\right)} \approx -0.3022$

Triangle *ABC* has correct dimensions.

For *DEF*,

$\dfrac{17.2 - 21.3}{22.8} \approx -0.1798$

$\dfrac{\sin\left(\dfrac{52.1 - 59.9}{2}\right)}{\cos\left(\dfrac{68.0}{2}\right)} \approx -0.0820$

Triangle *DEF* has an incorrect dimension.

65.

$$d(P_1, P_2) = \sqrt{[2-(-2)]^2 + (1-4)^2} = 5$$

$$d(P_1, P_3) = \sqrt{(-2-4)^2 + (4-(-3))^2} = \sqrt{85}$$

$$d(P_2, P_3) = \sqrt{(2-4)^2 + (1-(-3))^2} = 2\sqrt{5}$$

$$\cos\theta = \frac{5^2 + \left(\sqrt{85}\right)^2 - \left(2\sqrt{5}\right)^2}{2 \cdot 5 \cdot \sqrt{85}}$$

$$\cos\theta \approx 0.9762$$

$$\theta \approx 12.5°$$

67.

$$\cos A = \frac{b^2 + c^2 - a^2}{2bc} = \frac{b^2 + 2bc + c^2 - a^2 - 2bc}{2bc} = \frac{(b+c)^2 - a^2}{2bc} - \frac{2bc}{2bc} = \frac{(b+c-a)(b+c+a)}{2bc} - 1$$

69. $V = 18K$, where K = area of triangular base

$$V = \frac{1}{2}(4)(4)(\sin 72°)(18)$$

$$V \approx 140 \text{ in}^3$$

PS1. $\sqrt{\left(\frac{3}{5}\right)^2 + \left(-\frac{4}{5}\right)^2} = \sqrt{\frac{9}{25} + \frac{16}{25}} = \sqrt{\frac{25}{25}} = 1$

PS2. $10\cos 228° \approx -6.691$

PS3. $\tan\alpha = \left|\frac{-\sqrt{3}}{3}\right|$

$\tan\alpha = \frac{\sqrt{3}}{3}$

$\alpha = \tan^{-1}\left(\frac{\sqrt{3}}{3}\right) = 30°$

PS4. $\cos\alpha = \frac{-17}{\sqrt{338}}$

$\alpha = \cos^{-1}\left(\frac{-17}{\sqrt{338}}\right) \approx 157.6°$

PS5. $\frac{1}{\sqrt{5}} \cdot \frac{\sqrt{5}}{\sqrt{5}} = \frac{\sqrt{5}}{5}$

PS6. $\frac{28}{\sqrt{68}} = \frac{28}{2\sqrt{17}} = \frac{14}{\sqrt{17}} \cdot \frac{\sqrt{17}}{\sqrt{17}} = \frac{14\sqrt{17}}{17}$

Section 4.3

1. $a = 5 - 1 = 4$
$b = 4 - 2 = 2$

A vector equivalent to P_1P_2 is $\mathbf{v} = \langle 4,\ 2 \rangle$.

3. $a = -3 - 2 = -5$
$b = 5 - 1 = 4$

A vector equivalent to P_1P_2 is $\mathbf{v} = \langle -5,\ 4 \rangle$.

5. $a = 4 - (-3) = 7$
$b = -1 - 0 = -1$

A vector equivalent to P_1P_2 is $\mathbf{v} = \langle 7,\ -1 \rangle$.

7. $a = -3 - 4 = -7$
$b = -3 - 2 = -5$

A vector equivalent to P_1P_2 is $\mathbf{v} = \langle -7,\ -5 \rangle$.

9. $a = 2 - 2 = 0$
$b = 3 - (-5) = 8$

A vector equivalent to $\mathbf{P_1P_2}$ is $\mathbf{v} = \langle 0,\ 8 \rangle$.

11. $\|\mathbf{v}\| = \sqrt{(-3)^2 + 4^2}$ $\quad \alpha = \tan^{-1}\left|\dfrac{4}{-3}\right| = \tan^{-1}\dfrac{4}{3}$
$\|\mathbf{v}\| = \sqrt{9 + 16}$ $\quad \alpha \approx 53.1°$
$\|\mathbf{v}\| = 5$ $\quad \theta = 180° - \alpha$
$\quad\quad\quad\quad\quad\quad\quad \theta \approx 180° - 53.1°$
$\quad\quad\quad\quad\quad\quad\quad \theta \approx 126.9°$

$\mathbf{u} = \left\langle \dfrac{-3}{5}, \dfrac{4}{5} \right\rangle$

A unit vector in the direction of $\mathbf{v}$ is $\mathbf{u} = \left\langle -\dfrac{3}{5}, \dfrac{4}{5} \right\rangle$.

13. $\|\mathbf{v}\| = \sqrt{20^2 + (-40)^2}$ $\quad \alpha = \tan^{-1}\left|\dfrac{-40}{20}\right| = \tan^{-1}2$
$\|\mathbf{v}\| = \sqrt{400 + 1600}$ $\quad \alpha \approx 63.4°$
$\|\mathbf{v}\| = \sqrt{2000} = 20\sqrt{5}$ $\quad \theta = 360° - \alpha$
$\quad\quad \approx 44.7$ $\quad\quad\quad \theta \approx 360° - 63.4°$
$\quad\quad\quad\quad\quad\quad\quad \theta \approx 296.6°$

$\mathbf{u} = \left\langle \dfrac{20}{20\sqrt{5}}, \dfrac{-40}{20\sqrt{5}} \right\rangle = \left\langle \dfrac{\sqrt{5}}{5}, \dfrac{-2\sqrt{5}}{5} \right\rangle$

A unit vector in the direction of $\mathbf{v}$ is $\mathbf{u} = \left\langle \dfrac{\sqrt{5}}{5}, -\dfrac{2\sqrt{5}}{5} \right\rangle$.

15. $\|\mathbf{v}\| = \sqrt{2^2 + (-4)^2}$ $\quad \alpha = \tan^{-1}\left|\dfrac{-4}{2}\right| = \tan^{-1}2$
$\|\mathbf{v}\| = \sqrt{4 + 16}$ $\quad \alpha \approx 63.4°$
$\|\mathbf{v}\| = \sqrt{20} = 2\sqrt{5}$ $\quad \theta = 360° - \alpha$
$\quad\quad \approx 4.5$ $\quad\quad\quad \theta \approx 360° - 63.4°$
$\quad\quad\quad\quad\quad\quad\quad \approx 296.6°$

$\mathbf{u} = \left\langle \dfrac{2}{2\sqrt{5}}, \dfrac{-4}{2\sqrt{5}} \right\rangle = \left\langle \dfrac{\sqrt{5}}{5}, -\dfrac{2\sqrt{5}}{5} \right\rangle$

A unit vector in the direction of $\mathbf{v}$ is $\mathbf{u} = \left\langle \dfrac{\sqrt{5}}{5}, -\dfrac{2\sqrt{5}}{5} \right\rangle$.

17. $\|\mathbf{v}\| = \sqrt{42^2 + (-18)^2}$ $\quad \alpha = \tan^{-1}\left|\dfrac{-18}{42}\right| = \tan^{-1}\dfrac{3}{7}$
$\|\mathbf{v}\| = \sqrt{1764 + 324}$ $\quad \alpha \approx 23.2°$
$\|\mathbf{v}\| = \sqrt{2088}$ $\quad \theta = 360° - \alpha$
$\quad\quad = 6\sqrt{58}$ $\quad\quad\quad \theta \approx 360° - 23.2°$
$\quad\quad \approx 45.7$ $\quad\quad\quad \theta \approx 336.8°$

$\mathbf{u} = \left\langle \dfrac{42}{6\sqrt{58}}, \dfrac{-18}{6\sqrt{58}} \right\rangle = \left\langle \dfrac{7\sqrt{58}}{58}, -\dfrac{3\sqrt{58}}{58} \right\rangle$

A unit vector in the direction of $\mathbf{v}$ is $\mathbf{u} = \left\langle \dfrac{7\sqrt{58}}{58}, -\dfrac{3\sqrt{58}}{58} \right\rangle$.

19. $3\mathbf{u} = 3\langle -2,\ 4 \rangle = \langle -6,\ 12 \rangle$

21. $2\mathbf{u} - \mathbf{v} = 2\langle -2,\ 4 \rangle - \langle -3,\ -2 \rangle$
$\quad\quad\quad\quad = \langle -4,\ 8 \rangle - \langle -3,\ -2 \rangle$
$\quad\quad\quad\quad = \langle -1,\ 10 \rangle$

23. $\dfrac{2}{3}\mathbf{u} + \dfrac{1}{6}\mathbf{v} = \dfrac{2}{3}\langle -2,\ 4 \rangle + \dfrac{1}{6}\langle -3,\ -2 \rangle$
$\quad\quad\quad\quad = \left\langle -\dfrac{4}{3}, \dfrac{8}{3} \right\rangle + \left\langle -\dfrac{1}{2}, -\dfrac{1}{3} \right\rangle$
$\quad\quad\quad\quad = \left\langle -\dfrac{11}{6}, \dfrac{7}{3} \right\rangle$

25. $\|\mathbf{u}\| = \sqrt{(-2)^2 + 4^2} = \sqrt{20} = 2\sqrt{5}$

27. $3\mathbf{u} - 4\mathbf{v} = 3\langle -2,\ 4 \rangle - 4\langle -3,\ -2 \rangle$
$\quad\quad\quad\quad = \langle -6,\ 12 \rangle - \langle -12,\ -8 \rangle$
$\quad\quad\quad\quad = \langle 6,\ 20 \rangle$
$\|3\mathbf{u} - 4\mathbf{v}\| = \sqrt{6^2 + 20^2} = \sqrt{436} = 2\sqrt{109}$

29. $4\mathbf{v} = 4(-2\mathbf{i} + 3\mathbf{j})$
$\quad\quad = -8\mathbf{i} + 12\mathbf{j}$

31.
$$6\mathbf{u} + 2\mathbf{v} = 6(3\mathbf{i} - 2\mathbf{j}) + 2(-2\mathbf{i} + 3\mathbf{j})$$
$$= (18\mathbf{i} - 12\mathbf{j}) + (-4\mathbf{i} + 6\mathbf{j})$$
$$= (18 - 4)\mathbf{i} + (-12 + 6)\mathbf{j}$$
$$= 14\mathbf{i} - 6\mathbf{j}$$

33.
$$\frac{2}{3}\mathbf{v} + \frac{3}{4}\mathbf{u} = \frac{2}{3}(-2\mathbf{i} + 3\mathbf{j}) + \frac{3}{4}(3\mathbf{i} - 2\mathbf{j})$$
$$= \left(-\frac{4}{3}\mathbf{i} + 2\mathbf{j}\right) + \left(\frac{9}{4}\mathbf{i} - \frac{3}{2}\mathbf{j}\right)$$
$$= \left(-\frac{4}{3} + \frac{9}{4}\right)\mathbf{i} + \left(2 - \frac{3}{2}\right)\mathbf{j}$$
$$= \frac{11}{12}\mathbf{i} + \frac{1}{2}\mathbf{j}$$

35.
$$\mathbf{u} - 2\mathbf{v} = (3\mathbf{i} - 2\mathbf{j}) - 2(-2\mathbf{i} + 3\mathbf{j})$$
$$= (3\mathbf{i} - 2\mathbf{j}) - (-4\mathbf{i} + 6\mathbf{j})$$
$$= (3 + 4)\mathbf{i} + (-2 - 6)\mathbf{j}$$
$$= 7\mathbf{i} - 8\mathbf{j}$$

$$\|\mathbf{u} - 2\mathbf{v}\| = \sqrt{7^2 + (-8)^2} = \sqrt{113}$$

37.
$$a_1 = 5\cos 27° \approx 4.5$$
$$a_2 = 5\sin 27° \approx 2.3$$
$$\mathbf{v} = a_1\mathbf{i} + a_2\mathbf{j} \approx 4.5\mathbf{i} + 2.3\mathbf{j}$$

39.
$$a_1 = 4\cos\frac{\pi}{4} \approx 2.8$$
$$a_2 = 4\sin\frac{\pi}{4} \approx 2.8$$
$$\mathbf{v} = a_1\mathbf{i} + a_2\mathbf{j} \approx 2.8\mathbf{i} + 2.8\mathbf{j}$$

41. heading = 124° $\Rightarrow$ wind from the west $\Rightarrow$
direction angle = −34°   direction angle = 0°

$$\mathbf{AB} = 45\mathbf{i}$$
$$\mathbf{AD} = 340\cos(-34°)\mathbf{i} + 340\sin(-34°)\mathbf{j}$$
$$\mathbf{AD} \approx 281.9\mathbf{i} - 190.1\mathbf{j}$$
$$\mathbf{AC} = \mathbf{AB} + \mathbf{AD}$$
$$\mathbf{AC} = 45\mathbf{i} + 281.9\mathbf{i} - 190.1\mathbf{j}$$
$$\mathbf{AC} \approx 327\mathbf{i} - 190\mathbf{j}$$
$$\|\mathbf{AC}\| = \sqrt{327^2 + (-190)^2}$$
$$\|\mathbf{AC}\| \approx 380 \text{ mph}$$

The ground speed of the plane is approximately 380 mph.

43. heading = 96° $\Rightarrow$ heading = 37° $\Rightarrow$
direction angle = −6°   direction angle = 53°

$$\mathbf{AB} = 50\cos 53°\mathbf{i} + 50\sin 53°\mathbf{j}$$
$$\approx 30.1\mathbf{i} + 39.9\mathbf{j}$$
$$\mathbf{AD} = 225\cos(-6°)\mathbf{i} + 225\sin(-6°)\mathbf{j}$$
$$\approx 223.8\mathbf{i} - 23.5\mathbf{j}$$
$$\mathbf{AC} = \mathbf{AB} + \mathbf{AD}$$
$$\approx 30.1\mathbf{i} + 39.9\mathbf{j} + 223.8\mathbf{i} - 23.5\mathbf{j}$$
$$\approx 253.9\mathbf{i} + 16.4\mathbf{j}$$
$$\|\mathbf{AC}\| = \sqrt{(253.9)^2 + (16.4)^2} \approx 250$$

$$\alpha = \tan^{-1}\left|\frac{16.4}{253.9}\right|$$
$$= \tan^{-1}\frac{16.4}{253.9}$$
$$\alpha \approx 4°$$
$$\theta = 90° - \alpha$$
$$\theta \approx 90° - 4°$$
$$\theta \approx 86°$$

The ground speed of the plane is about 250 mph at a heading of approximately 86°.

45.

$$\sin 5.6^\circ = \frac{F}{3000}$$

$$F = 3000 \sin 5.6^\circ$$

$$F \approx 293 \text{ lb}$$

47. **a.** $\sin 22.4^\circ = \dfrac{\|\mathbf{F_1}\|}{345}$

$\|\mathbf{F_1}\| = 345 \sin 22.4^\circ$

$\|\mathbf{F_1}\| \approx 131 \text{ lb}$

b. $\cos 22.4^\circ = \dfrac{\|\mathbf{F_2}\|}{345}$

$\|\mathbf{F_2}\| = 345 \cos 22.4^\circ$

$\|\mathbf{F_2}\| \approx 319 \text{ lb}$

49. $\mathbf{F_1} + \mathbf{F_2} + \mathbf{F_3} = (18\mathbf{i} + 13.1\mathbf{j}) + (-12.4\mathbf{i} + 3.8\mathbf{j}) + (-5.8\mathbf{i} - 16.9\mathbf{j})$

$= (18 - 12.4 - 5.8)\mathbf{i} + (13.1 + 3.8 - 16.9)\mathbf{j}$

$= 0$

The forces are in equilibrium.

51. $\mathbf{F_1} + \mathbf{F_2} + \mathbf{F_3} = (155\mathbf{i} - 257\mathbf{j}) + (-124\mathbf{i} + 149\mathbf{j}) + (-31\mathbf{i} + 98\mathbf{j})$

$= (155 - 124 - 31)\mathbf{i} + (-257 + 149 + 98)\mathbf{j}$

$= 0\mathbf{i} - 10\mathbf{j}$

The forces are not in equilibrium. $\mathbf{F_4} = 0\mathbf{i} + 10\mathbf{j}$

53. $\mathbf{F_1} + \mathbf{F_2} + \mathbf{F_3} = (189.3\mathbf{i} + 235.7\mathbf{j}) + (45.8\mathbf{i} - 205.6\mathbf{j})$

$+ (-175.2\mathbf{i} - 37.7\mathbf{j}) + (-59.9\mathbf{i} + 7.6\mathbf{j})$

$= (189.3 + 45.8 - 175.2 - 59.9)\mathbf{i}$

$+ (235.7 - 205.6 - 37.7 + 7.6)\mathbf{j}$

$= 0$

The forces are in equilibrium.

55. $\mathbf{v} \cdot \mathbf{w} = \langle 3, -2 \rangle \cdot \langle 1, 3 \rangle$

$= 3(1) + (-2)3$

$= 3 - 6$

$= -3$

57. $\mathbf{v} \cdot \mathbf{w} = \langle 4, 1 \rangle \cdot \langle -1, 4 \rangle$

$= 4(-1) + 1(4)$

$= -4 + 4$

$= 0$

59. $\mathbf{v} \cdot \mathbf{w} = (\mathbf{i} + 2\mathbf{j}) \cdot (-\mathbf{i} + \mathbf{j})$

$= 1(-1) + 2(1)$

$= -1 + 2$

$= 1$

61. $\mathbf{v} \cdot \mathbf{w} = (6\mathbf{i} - 4\mathbf{j}) \cdot (-2\mathbf{i} - 3\mathbf{j})$

$= 6(-2) + (-4)(-3)$

$= -12 + 12$

$= 0$

63. $\cos\theta = \dfrac{\mathbf{v} \cdot \mathbf{w}}{\|\mathbf{v}\| \, \|\mathbf{w}\|}$

$\cos\theta = \dfrac{\langle 2, -1 \rangle \cdot \langle 3, 4 \rangle}{\sqrt{2^2 + (-1)^2} \sqrt{3^2 + 4^2}}$

$\cos\theta = \dfrac{2(3) + (-1)4}{\sqrt{5}\sqrt{25}}$

$\cos\theta = \dfrac{2}{5\sqrt{5}} \approx 0.1789$

$\theta \approx 79.7^\circ$

65. $\cos\theta = \dfrac{\mathbf{v} \cdot \mathbf{w}}{\|\mathbf{v}\| \, \|\mathbf{w}\|}$

$\cos\theta = \dfrac{\langle 0, 3 \rangle \cdot \langle 2, 2 \rangle}{\sqrt{0^2 + 3^2} \sqrt{2^2 + 2^2}}$

$\cos\theta = \dfrac{0(2) + 3(2)}{\sqrt{9}\sqrt{8}}$

$\cos\theta = \dfrac{6}{6\sqrt{2}} \approx 0.7071$

$\theta = 45^\circ$

67. $\cos\theta = \dfrac{\mathbf{v} \cdot \mathbf{w}}{\|\mathbf{v}\| \, \|\mathbf{w}\|}$

$\cos\theta = \dfrac{(5\mathbf{i} - 2\mathbf{j}) \cdot (2\mathbf{i} + 5\mathbf{j})}{\sqrt{5^2 + (-2)^2} \sqrt{2^2 + 5^2}}$

$\cos\theta = \dfrac{5(2) + (-2)(5)}{\sqrt{29}\sqrt{29}}$

$\cos\theta = \dfrac{0}{\sqrt{29}\sqrt{29}} = 0$

$\theta = 90^\circ$

Thus, the vectors are orthogonal.

69. $\cos\theta = \dfrac{\mathbf{v} \cdot \mathbf{w}}{\|\mathbf{v}\| \, \|\mathbf{w}\|}$

$\cos\theta = \dfrac{(5\mathbf{i} + 2\mathbf{j}) \cdot (-5\mathbf{i} - 2\mathbf{j})}{\sqrt{5^2 + 2^2} \sqrt{(-5)^2 + (-2)^2}}$

$\cos\theta = \dfrac{5(-5) + 2(-2)}{\sqrt{29}\sqrt{29}}$

$\cos\theta = \dfrac{-29}{\sqrt{29}\sqrt{29}} = -1$

$\theta = 180^\circ$

71. $\text{proj}_{\mathbf{w}}\mathbf{v} = \dfrac{\mathbf{v} \cdot \mathbf{w}}{\|\mathbf{w}\|}$

$\text{proj}_{\mathbf{w}}\mathbf{v} = \dfrac{\langle 6, 7 \rangle \cdot \langle 3, 4 \rangle}{\sqrt{3^2 + 4^2}} = \dfrac{18 + 28}{\sqrt{25}} = \dfrac{46}{5}$

73. $\text{proj}_{\mathbf{w}}\mathbf{v} = \dfrac{\mathbf{v} \cdot \mathbf{w}}{\|\mathbf{w}\|}$

$\text{proj}_{\mathbf{w}}\mathbf{v} = \dfrac{\langle -3,\ 4\rangle \cdot \langle 2,\ 5\rangle}{\sqrt{2^2 + 5^2}} = \dfrac{-6 + 20}{\sqrt{29}} = \dfrac{14}{\sqrt{29}} = \dfrac{14\sqrt{29}}{29} \approx 2.6$

75. $\text{proj}_{\mathbf{w}}\mathbf{v} = \dfrac{\mathbf{v} \cdot \mathbf{w}}{\|\mathbf{w}\|}$

$\text{proj}_{\mathbf{w}}\mathbf{v} = \dfrac{(2\mathbf{i} + \mathbf{j}) \cdot (6\mathbf{i} + 3\mathbf{j})}{\sqrt{6^2 + 3^2}} = \dfrac{12 + 3}{\sqrt{45}} = \dfrac{5}{\sqrt{5}} = \sqrt{5} \approx 2.2$

77. $\text{proj}_{\mathbf{w}}\mathbf{v} = \dfrac{\mathbf{v} \cdot \mathbf{w}}{\|\mathbf{w}\|}$

$\text{proj}_{\mathbf{w}}\mathbf{v} = \dfrac{(3\mathbf{i} - 4\mathbf{j}) \cdot (3\mathbf{i} - 4\mathbf{j})}{\sqrt{(-6)^2 + 12^2}} = \dfrac{-18 - 48}{\sqrt{180}} = -\dfrac{11}{\sqrt{5}}$

$= -\dfrac{11\sqrt{5}}{5} \approx -4.9$

79. $W = \mathbf{F} \cdot \mathbf{s}$

$W = \|\mathbf{F}\|\,\|\mathbf{s}\|\cos\alpha$

$W = (75)(15)(\cos 32°)$

$W \approx 954$ foot-pounds

81.

$W = \mathbf{F} \cdot \mathbf{s}$

$W = \|\mathbf{F}\|\,\|\mathbf{s}\|\cos\alpha$

$W = (75)(12)(\cos 30°)$

$W \approx 779$ foot-pounds

Connecting Concepts

83.

Thus, the sum is $\langle 6,\ 9\rangle$.

85.

The vector from $P_1(3,\ -1)$ to $P_2(5,\ -4)$ is equivalent to $2\mathbf{i} - 3\mathbf{j}$.

87.

$\mathbf{w} = 5\mathbf{i} + 2\mathbf{j}$

$\mathbf{v} = 2\mathbf{i} - 5\mathbf{j}$

$\mathbf{v} \cdot \mathbf{w} = (2\mathbf{i} - 5\mathbf{j}) \cdot (5\mathbf{i} + 2\mathbf{j})$

$= 10 - 10$

$= 0$

The two vectors are perpendicular.

89. $\mathbf{v} = \langle -2,\ 7\rangle$

$\langle -2,\ 7\rangle \cdot \langle a,\ b\rangle = 0$

$-2a + 7b = 0$

$a = \dfrac{7}{2}b$

Let $b = 2$

$a = 7$

Thus, $\mathbf{u} = \langle 7,\ 2\rangle$ is one example.

91. Let $\mathbf{u} = c\mathbf{i} + b\mathbf{j}$, $\mathbf{v} = c\mathbf{i} + d\mathbf{j}$, and $\mathbf{w} = e\mathbf{i} + f\mathbf{j}$.

$$(\mathbf{u} \cdot \mathbf{v}) \cdot \mathbf{w} = \left[(a\mathbf{i} + b\mathbf{j}) \cdot (c\mathbf{i} + d\mathbf{j})\right] \cdot (e\mathbf{i} + f\mathbf{j}).$$
$$= (ac + bd) \cdot (e\mathbf{i} + f\mathbf{j})$$

$ac + bd$ is a scalar quantity. The product of a scalar and a vector is not defined. Therefore, no, $(\mathbf{u} \cdot \mathbf{v}) \cdot \mathbf{w}$ does not equal $\mathbf{u} \cdot (\mathbf{v} \cdot \mathbf{w})$.

93. Let $\mathbf{v} = \langle a, b \rangle$ and $\mathbf{w} = \langle d, e \rangle$
$$c\mathbf{v} = \langle ca, ab \rangle$$

$$c(\mathbf{v} \cdot \mathbf{w}) = c\langle a, b \rangle \cdot \langle d, e \rangle = c(ad + be) = cad + cbe$$
$$(c\mathbf{v} \cdot \mathbf{w}) = \langle ca, ab \rangle \cdot \langle d, e \rangle = cad + cbe$$

Therefore, $c(\mathbf{v} \cdot \mathbf{w}) = (c\mathbf{v}) \cdot \mathbf{w}$.

95. Neither. If the force and the distance are the same, the work will be the same.

Exploring Concepts with Technology

Optimal Branching of Arteries

The following graph is the graph of R as given in Equation (2) with $a = 8$ cm, $b = 4$ cm, $r_1 = 0.4$ cm, and $r_2 = 0.3$ cm.

The following graph is a close-up of the graph of R, for $71° \leq \theta \leq 74°$.

According to this graph R is a minimum when $\theta = 72°$ (to the nearest degree).
Using Equation (3) yields

$$\cos\theta = \left(\frac{\frac{3}{4}r_1}{r_1}\right)^4$$
$$\cos\theta = \frac{81}{256}$$
$$\theta = \cos^{-1}\left(\frac{81}{256}\right) \approx 72°$$

Assessing Concepts

1. An oblique triangle that does not contain a right angle.

2. The Law of Cosines

3. SSA

4. The variable s represents the semiperimeter of the triangle.

5. A scalar

6. A scalar

7. True

8. False

9. True

10. True

1.

$B = 180° - 92° - 37°$
$B = 51°$

$\tan A = \dfrac{a}{4}$

$a = 14 \tan 37°$
$a \approx 11$

$\cos A = \dfrac{14}{c}$　　　[4.1]

$c = \dfrac{14}{\cos 37°}$
$c \approx 18$

2.

$A = 180° - 94.0° - 77.4°$
$A = 8.6°$

$\sin B = \dfrac{b}{11.8}$

$b = 11.8 \sin 77.4°$
$b \approx 11.5$

$\cos B = \dfrac{a}{11.8}$　　　[4.1]

$a = 11.8 \cos 77.4°$
$a \approx 2.57$

3.

$\cos B = \dfrac{12^2 + 20^2 - 15^2}{2(12)(20)}$

$\cos B \approx 0.6646$
$B \approx 48°$

$\cos C = \dfrac{12^2 + 15^2 - 20^2}{2(12)(15)}$

$\cos C \approx -0.0861$
$C \approx 95°$

$A = 180° - 48° - 95°$　[4.2]
$A = 37°$

4.

$\cos C = \dfrac{24^2 + 32^2 - 28^2}{2(24)(32)}$

$\cos C \approx 0.5313$
$C \approx 58°$

$\cos A = \dfrac{32^2 + 28^2 - 24^2}{2(32)(28)}$

$\cos A \approx 0.6875$
$A \approx 47°$

$B \approx 180° - 58° - 47°$　[4.2]
$B \approx 75°$

5.

$c^2 = 22^2 + 18^2 - 2(22)(18)\cos 35°$
$c^2 \approx 159$
$c = \sqrt{159}$
$c \approx 13$

$\dfrac{18}{\sin A} = \dfrac{\sqrt{159}}{\sin 35°}$

$\sin A = \dfrac{18 \sin 35°}{\sqrt{159}}$

$\sin A \approx 0.8188$
$A \approx 55°$

$B \approx 180 - 35° - 55°$　[4.2]
$B \approx 90°$

6.

$a^2 = 102^2 + 150^2 - 2(102)(150)\cos 82°$
$a^2 \approx 28645$
$a = \sqrt{28645}$
$a \approx 169$

$\dfrac{150}{\sin C} = \dfrac{\sqrt{28645}}{\sin 82°}$

$\sin C = \dfrac{150 \sin 82°}{\sqrt{28645}}$

$\sin C \approx 0.8776$
$C \approx 61°$

$B \approx 180° - 61° - 82°$　[4.2]
$B \approx 37°$

7.

$\dfrac{10}{\sin C} = \dfrac{8}{\sin 105°}$　[7.1]

$\sin C = \dfrac{10 \sin 105°}{8}$

$\sin C \approx 1.207$
No triangle is formed.

8.

$\dfrac{110}{\sin B} = \dfrac{80}{\sin 55°}$　[4.1]

$\sin B = \dfrac{110 \sin 55°}{80}$

$\sin B \approx 1.1263$
No triangle is formed.

9.

$C = 180° - 80° - 55°$

$C = 45°$

$$\frac{25}{\sin 45°} = \frac{a}{\sin 55°}$$

$$a = \frac{25\sin 55°}{\sin 45°}$$

$$a \approx 29$$

$$\frac{25}{\sin 45°} = \frac{b}{\sin 80°} \quad [4.1]$$

$$b = \frac{25\sin 80°}{\sin 45°}$$

$$b \approx 35$$

10.

$A = 180° - 40° - 25°$

$A = 115°$

$$\frac{a}{\sin 115°} = \frac{40}{\sin 40°}$$

$$a = \frac{40\sin 115°}{\sin 40°}$$

$$a \approx 56$$

$$\frac{b}{\sin 25°} = \frac{40}{\sin 40°} \quad [4.1]$$

$$b = \frac{40\sin 25°}{\sin 40°}$$

$$b \approx 26$$

11. $s = \dfrac{1}{2}(a+b+c)$ [4.2]

$s = \dfrac{1}{2}(24+30+36)$

$s = 45$

$K = \sqrt{s(s-a)(s-b)(s-c)}$

$K = \sqrt{45(45-24)(45-30)(45-36)}$

$K = \sqrt{127{,}575}$

$K \approx 360$ square units

12. $s = \dfrac{1}{2}(a+b+c)$ [4.2]

$s = \dfrac{1}{2}(9.0+7.0+12)$

$s = 14$

$K = \sqrt{s(s-a)(s-b)(s-c)}$

$K = \sqrt{14(14-9.0)(14-7.0)(14-12)}$

$K = \sqrt{980}$

$K \approx 31$ square units

13.

$K = \dfrac{1}{2}ab\sin C$

$K = \dfrac{1}{2}(60)(44)\sin 44°$

$K \approx 920$ square units

[4.2]

14.

$K = \dfrac{1}{2}bc\sin A$

$K = \dfrac{1}{2}(8.0)(12)\sin 75°$

$K \approx 46$ square units

[4.2]

15.

$$\frac{50}{\sin B} = \frac{75}{\sin 15°}$$

$$\sin B = \frac{50\sin 15°}{75}$$

$$\sin B \approx 0.1725$$

$$B \approx 10°$$

$A \approx 180° - 10° - 15°$

$A \approx 155°$

$K \approx \dfrac{1}{2}(50)(75)\sin 155°$ [4.2]

$K \approx 790$ square units

16.

$$\frac{18}{\sin B} = \frac{25}{\sin 68°}$$

$$\sin B = \frac{18\sin 68°}{25}$$

$$\sin B \approx 0.6676$$

$$B \approx 42°$$

$C \approx 180° - 42° - 68°$

$C \approx 70°$

$K \approx \dfrac{1}{2}(18)(25)\sin 70°$ [4.2]

$K \approx 210$ square units

17.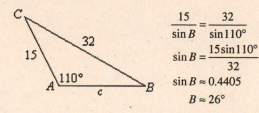

$$\frac{15}{\sin B} = \frac{32}{\sin 110°}$$

$$\sin B = \frac{15\sin 110°}{32}$$

$$\sin B \approx 0.4405$$

$$B \approx 26°$$

$C \approx 180° - 110° - 26°$

$C \approx 44°$

$K \approx \frac{1}{2}(15)(32)\sin 44°$ [4.2]

$K \approx 170$ square units

18.

$$\frac{18}{\sin B} = \frac{22}{\sin 45°}$$

$$\sin B \approx \frac{18\sin 45°}{22}$$

$$\sin B \approx 0.5785$$

$$B \approx 35°$$

$A \approx 180° - 45° - 35°$

$A \approx 100°$

$K \approx \frac{1}{2}(18)(22)\sin 100°$ [4.2]

$K \approx 190$ square units

19. Let $\mathbf{P_1P_2} = a_1\mathbf{i} + a_2\mathbf{j}$. [4.3]

$a_1 = 3 - (-2) = 5$

$a_2 = 7 - 4 = 3$

A vector equivalent to $\mathbf{P_1P_2}$ is $\mathbf{v} = \langle 5, \ 3\rangle$.

20. Let $\mathbf{P_1P_2} = a_1\mathbf{i} + a_2\mathbf{j}$. [4.3]

$a_1 = -3 - (-4) = 1$

$a_2 = 6 - 0 = 6$

A vector equivalent to $\mathbf{P_1P_2}$ is $\mathbf{v} = \langle 1, \ 6\rangle$.

21. $\|\mathbf{v}\| = \sqrt{(-4)^2 + 2^2}$

$\|\mathbf{v}\| = \sqrt{16 + 4}$

$\|\mathbf{v}\| \approx 4.5$

$\alpha \approx \tan^{-1}\left|\frac{2}{-4}\right| = \tan^{-1}\frac{1}{2}$ [4.3]

$\alpha \approx 26.6°$

$\theta \approx 180° - 26.6°$

$\theta \approx 153.4°$

22. $\|\mathbf{v}\| = \sqrt{6^2 + (-3)^2}$

$\|\mathbf{v}\| = \sqrt{36 + 9}$

$\|\mathbf{v}\| \approx 6.7$

$\alpha \approx \tan^{-1}\left|\frac{-3}{6}\right| = \tan^{-1}\frac{1}{2}$ [4.3]

$\alpha \approx 26.6°$

$\theta \approx 360° - 26.6°$

$\theta \approx 333.4°$

23. $\|\mathbf{u}\| = \sqrt{(-2)^2 + 3^2}$

$\|\mathbf{u}\| = \sqrt{4 + 9}$

$\|\mathbf{u}\| \approx 3.6$

$\alpha = \tan^{-1}\left|\frac{3}{-2}\right| = \tan^{-1}\frac{3}{2}$ [4.3]

$\alpha \approx 56.3°$

$\theta \approx 180° - 56.3°$

$\theta \approx 123.7°$

24. $\|\mathbf{u}\| = \sqrt{(-4)^2 + (-7)^2}$

$\|\mathbf{u}\| = \sqrt{16 + 49}$

$\|\mathbf{u}\| \approx 8.1$

$\alpha = \tan^{-1}\left|\frac{-7}{-4}\right| = \tan^{-1}\frac{7}{4}$ [4.3]

$\alpha \approx 60.3°$

$\theta \approx 180° + 60.3°$

$\theta \approx 240.3°$

25. $\|\mathbf{w}\| = \sqrt{(-8)^2 + 5^2}$

$\|\mathbf{w}\| = \sqrt{89}$

$\|\mathbf{u}\| = \left\langle \frac{-8}{\sqrt{89}}, \ \frac{5}{\sqrt{89}} \right\rangle = \left\langle -\frac{8\sqrt{89}}{89}, \ \frac{5\sqrt{89}}{89} \right\rangle$ [4.3]

A unit vector in the direction of $\|\mathbf{w}\|$ is $\|\mathbf{u}\| = \left\langle -\frac{8\sqrt{89}}{89}, \ \frac{5\sqrt{89}}{89} \right\rangle$.

26. $\|\mathbf{w}\| = \sqrt{7^2 + (-12)^2}$

$\|\mathbf{w}\| = \sqrt{193}$

$\mathbf{u} = \left\langle \frac{7}{\sqrt{193}}, \ \frac{-12}{\sqrt{193}} \right\rangle = \left\langle \frac{7\sqrt{193}}{193}, \ -\frac{12\sqrt{193}}{193} \right\rangle$ [4.3]

A unit vector in the direction of $\mathbf{w}$ is $\mathbf{u} = \left\langle \frac{7\sqrt{193}}{193}, \ -\frac{12\sqrt{193}}{193} \right\rangle$.

27. $\|\mathbf{v}\| = \sqrt{5^2 + 1^2}$

$\|\mathbf{v}\| = \sqrt{26}$

$\mathbf{u} = \frac{5}{\sqrt{26}}\mathbf{i} + \frac{1}{\sqrt{26}}\mathbf{j} = \frac{5\sqrt{26}}{26}\mathbf{i} + \frac{\sqrt{26}}{26}\mathbf{j}$ [4.3]

A unit vector in the direction of $\mathbf{v}$ is $\mathbf{u} = \frac{5\sqrt{26}}{26}\mathbf{i} + \frac{\sqrt{26}}{26}\mathbf{j}$.

28. $\|\mathbf{v}\| = \sqrt{3^2 + (-5)^2}$ $\qquad \mathbf{u} = \dfrac{3}{\sqrt{34}}\mathbf{i} - \dfrac{5}{\sqrt{34}}\mathbf{j} = \dfrac{3\sqrt{34}}{34}\mathbf{i} - \dfrac{5\sqrt{34}}{34}\mathbf{j}$ [4.3]

$\|\mathbf{v}\| = \sqrt{34}$

A unit vector in the direction of $\mathbf{v}$ is $\mathbf{u} = \dfrac{3\sqrt{34}}{34}\mathbf{i} - \dfrac{5\sqrt{34}}{34}\mathbf{j}$.

29. $\mathbf{v} - \mathbf{u} = \langle -4, -1 \rangle - \langle 3, 2 \rangle$ [4.3]

$\qquad = \langle -7, -3 \rangle$

30. $2\mathbf{u} - 3\mathbf{v} = 2\langle 3, 2 \rangle - 3\langle -4, -1 \rangle$ [4.3]

$\qquad = \langle 6, 4 \rangle - \langle -12, -3 \rangle$

$\qquad = \langle 18, 7 \rangle$

31. $-\mathbf{u} + \dfrac{1}{2}\mathbf{v} = -(10\mathbf{i} + 6\mathbf{j}) + \dfrac{1}{2}(8\mathbf{i} - 5\mathbf{j})$ [4.3]

$\qquad = (-10\mathbf{i} - 6\mathbf{j}) + \left(4\mathbf{i} - \dfrac{5}{2}\mathbf{j} \right)$

$\qquad = (-10 + 4)\mathbf{i} + \left(-6 - \dfrac{5}{2} \right)\mathbf{j}$

$\qquad = -6\mathbf{i} - \dfrac{17}{2}\mathbf{j}$

32. $\dfrac{2}{3}\mathbf{v} - \dfrac{3}{4}\mathbf{u} = \dfrac{2}{3}(8\mathbf{i} - 5\mathbf{j}) - \dfrac{3}{4}(10\mathbf{i} + 6\mathbf{j})$ [4.3]

$\qquad = \left(\dfrac{16}{3}\mathbf{i} - \dfrac{10}{3}\mathbf{j} \right) - \left(\dfrac{15}{2}\mathbf{i} + \dfrac{9}{2}\mathbf{j} \right)$

$\qquad = \left(\dfrac{16}{3} - \dfrac{15}{2} \right)\mathbf{i} + \left(-\dfrac{10}{3} - \dfrac{9}{2} \right)\mathbf{j}$

$\qquad = \dfrac{32 - 45}{6}\mathbf{i} + \dfrac{-20 = 27}{6}\mathbf{j}$

$\qquad = -\dfrac{13}{6}\mathbf{i} - \dfrac{47}{6}\mathbf{j}$

33. $\mathbf{v} = 400\sin 204°\mathbf{i} + 400\cos 204°\mathbf{j}$

$\mathbf{v} \approx -162.7\mathbf{i} - 365.4\mathbf{j}$

$\mathbf{w} = -45\mathbf{i}$

$\mathbf{R} = \mathbf{v} + \mathbf{w}$

$\mathbf{R} \approx -162.7\mathbf{i} - 365.4\mathbf{j} - 45\mathbf{i}$

$\mathbf{R} \approx -207.7\mathbf{i} - 365.4\mathbf{j}$

$\|\mathbf{R}\| \approx \sqrt{(-207.7)^2 + (-365.4)^2}$

$\|\mathbf{R}\| \approx 420 \text{ mph}$

$\alpha = \tan^{-1}\left| \dfrac{-365.4}{-207.7} \right| = \tan^{-1}\dfrac{365.4}{207.7}$

$\alpha \approx 60°$

$\theta \approx 180° + 60°$

$\theta \approx 240°$

The ground speed is approximately 420 mph at a heading of 240° [4.3]

34. $\theta = \sin^{-1}\dfrac{40}{320}$ [4.3]

$\theta \approx 7°$

35. $\mathbf{u} \cdot \mathbf{v} = \langle 3, 7 \rangle \cdot \langle -1, 3 \rangle$ [4.3]

$\qquad = (3)(-1) + (7)(3)$

$\qquad = 18$

36. $\mathbf{v} \cdot \mathbf{u} = \langle -8, 5 \rangle \cdot \langle 2, -1 \rangle$ [4.3]

$\qquad = (-8)(2) + (5)(-1)$

$\qquad = -21$

37. $\mathbf{v} \cdot \mathbf{u} = (-4\mathbf{i} - \mathbf{j}) \cdot (2\mathbf{i} + \mathbf{j})$ [4.3]

$\qquad = (-4)(2) + (-1)(1)$

$\qquad = -9$

38. $\mathbf{u} \cdot \mathbf{v} = (-3\mathbf{i} + 7\mathbf{j}) \cdot (-2\mathbf{i} + 2\mathbf{j})$ [4.3]

$\qquad = (-3)(-2) + (7)(2)$

$\qquad = 20$

39. $\cos\alpha = \dfrac{\langle 7, -4 \rangle \cdot \langle 2, 3 \rangle}{\sqrt{7^2 + (-4)^2}\sqrt{2^2 + 3^2}}$ [4.3]

$\cos\alpha = \dfrac{14 + (-12)}{\sqrt{65}\sqrt{13}}$

$\cos\alpha \approx 0.0688$

$\alpha \approx 86°$

40. $\cos\alpha = \dfrac{\langle -5, 2 \rangle \cdot \langle 2, -4 \rangle}{\sqrt{(-5)^2 + 2^2}\sqrt{2^2 + (-4)^2}}$

$\cos\alpha = \dfrac{-10 - 8}{\sqrt{29}\sqrt{20}}$

$\cos\alpha \approx -0.7474$ [4.3]

$\alpha = 138°$

41. $\cos\alpha = \dfrac{(6\mathbf{i} - 11\mathbf{j}) \cdot (2\mathbf{i} - 4\mathbf{j})}{\sqrt{6^2 + (-11)^2}\sqrt{2^2 + 4^2}}$

$\cos\alpha = \dfrac{12 - 44}{\sqrt{157}\sqrt{20}}$

$\cos\alpha \approx -0.5711$ [4.3]

$\cos\alpha \approx 125°$

42. $\cos\alpha = \dfrac{(\mathbf{i} - 5\mathbf{j}) \cdot (\mathbf{i} + 5\mathbf{j})}{\sqrt{1^2 + (-5)^2}\sqrt{1^2 + 5^2}}$

$\cos\alpha = \dfrac{1 - 25}{\sqrt{26}\sqrt{26}}$

$\cos\alpha \approx -0.9231$ [4.3]

$\alpha \approx 157°$

43. $\text{proj}_{\mathbf{w}}\mathbf{v} = \dfrac{\mathbf{v} \cdot \mathbf{w}}{\|\mathbf{w}\|}$ [4.3]

$\text{proj}_{\mathbf{w}}\mathbf{v} = \dfrac{\langle -2,5 \rangle \cdot \langle 5,4 \rangle}{\sqrt{5^2 + 4^2}}$

$= \dfrac{-10+20}{\sqrt{41}}$

$= \dfrac{10}{\sqrt{41}}$

$= \dfrac{10\sqrt{41}}{41}$

44. $\text{proj}_{\mathbf{w}}\mathbf{v} = \dfrac{\mathbf{v} \cdot \mathbf{w}}{\|\mathbf{w}\|}$ [4.3]

$\text{proj}_{\mathbf{w}}\mathbf{v} = \dfrac{(4\mathbf{i}-7\mathbf{j}) \cdot (-2\mathbf{i}-5\mathbf{j})}{\sqrt{(-2)^2 + (-5)^2}}$

$= \dfrac{-8+35}{\sqrt{29}}$

$= \dfrac{27}{\sqrt{29}}$

$= \dfrac{27\sqrt{29}}{29}$

45. $w = \|\mathbf{F}\|\,\|\mathbf{S}\|\cos\theta$ [4.3]

$w = 60 \cdot 14\cos 38°$

$w \approx 662$ foot-pounds

Quantitative Reasoning

QR1. The distance between MCO and LAX is

$d = \cos^{-1}[\cos(\text{lat1})\cos(\text{lat2})\cos(\text{lon1} - \text{lon2}) + \sin(\text{lat1})\sin(\text{lat2})]$

$d = \cos^{-1}[\cos(0.496187)\cos(0.592409)\cos(-1.419110 - (-2.066611)) + \sin(0.496187)\sin(0.592409)]$

$d \approx 0.559146$ radian

The great circle distance between MCO and LAX is

$d \approx 0.559146 \times 3960$

$d \approx 2210$ mi

QR3. The distance between JFK and DEN is

$d = \cos^{-1}[\cos(\text{lat1})\cos(\text{lat2})\cos(\text{lon1} - \text{lon2}) + \sin(\text{lat1})\sin(\text{lat2})]$

$d = \cos^{-1}[\cos(0.709476)\cos(0.695717)\cos(-1.287756 - (-1.826892)) + \sin(0.709476)\sin(0.695717)]$

$d \approx 0.409558$ radian

The great circle distance between JFK and DEN is

$d \approx 0.409558 \times 3960$

$d \approx 1620$ mi

Chapter Test

1.

$B = 180° - 70° - 16°$

$B = 94°$

$\dfrac{c}{\sin C} = \dfrac{a}{\sin A}$

$a = \dfrac{14\sin 70°}{\sin 16°}$

$a \approx 48$

$\dfrac{c}{\sin C} = \dfrac{b}{\sin B}$ [4.1]

$b = \dfrac{14\sin 94°}{\sin 16°}$

$b \approx 51$

2.

$\dfrac{b}{\sin B} = \dfrac{a}{\sin A}$

$B = \sin^{-1}\left(\dfrac{b\sin A}{a}\right)$ [4.1]

$B = \sin^{-1}\left(\dfrac{13\sin 140°}{45}\right)$

$B \approx 11°$

3.

$$c^2 = a^2 + b^2 - 2ab\cos C$$
$$c^2 = 20^2 + 12^2 - 2(20)(12)\cos 42° \quad [4.2]$$
$$c \approx 14$$

4.

$$\cos B = \frac{a^2 + c^2 - b^2}{2ac}$$
$$B = \cos^{-1}\left(\frac{32^2 + 18^2 - 24^2}{2(32)(18)}\right) \quad [4.2]$$
$$B \approx 48°$$

5.

$$K = \frac{1}{2}ab\sin C$$
$$K = \frac{1}{2}(7)(12)(\sin 110°)$$
$$K \approx 39 \text{ square units}$$
$$[4.2]$$

6.

$$A = 180° - 42° - 75°$$
$$A = 63°$$
$$K = \frac{b^2 \sin A \, \sin C}{2\sin B}$$
$$K = \frac{12^2 \sin 63° \sin 75°}{2\sin 42°}$$
$$K \approx 93 \text{ square units}$$
$$[4.2]$$

7.

$$s = \frac{1}{2}(a + b + c)$$
$$s = \frac{1}{2}(17 + 55 + 42) = 57$$
$$K = \sqrt{s(s-a)(s-b)(s-c)}$$
$$K = \sqrt{57(57 - 17)(57 - 55)(57 - 42)}$$
$$K \approx 260 \text{ square units} \qquad [4.2]$$

8. $\quad |\mathbf{v}| = \sqrt{(-2)^2 + (3)^2} = \sqrt{13} \quad [4.3]$

9.
$$a_1 = 12\cos 220° \approx -9.2 \quad [4.3]$$
$$a_2 = 12\sin 220° \approx -7.7$$
$$\mathbf{v} = a_1\mathbf{i} + a_2\mathbf{j}$$
$$\mathbf{v} = -9.2\mathbf{i} - 7.7\mathbf{j}$$

10.
$$3\mathbf{u} - 5\mathbf{v} = 3(2\mathbf{i} - 3\mathbf{j}) - 5(5\mathbf{i} + 4\mathbf{j}) \quad [4.3]$$
$$= (6\mathbf{i} - 9\mathbf{j}) - (25\mathbf{i} + 20\mathbf{j})$$
$$= (6 - 25)\mathbf{i} + (-9 - 20)\mathbf{j}$$
$$= -19\mathbf{i} - 29\mathbf{j}$$

11.
$$\mathbf{u} \cdot \mathbf{v} = (-2\mathbf{i} + 3\mathbf{j}) \cdot (5\mathbf{i} + 3\mathbf{j}) \quad [4.3]$$
$$= (-2 \cdot 5) + (3 \cdot 3)$$
$$= -10 + 9$$
$$= -1$$

12.
$$\cos\theta = \frac{\mathbf{u} \cdot \mathbf{v}}{\|\mathbf{u}\| \, \|\mathbf{v}\|} = \frac{\langle 3, 5\rangle \cdot \langle -6, 2\rangle}{\sqrt{3^2 + 5^2}\sqrt{(-6)^2 + 2^2}} \quad [4.3]$$
$$\cos\theta = \frac{-18 + 10}{\sqrt{34}\sqrt{40}} = \frac{-8}{\sqrt{34}\sqrt{40}}$$
$$\theta \approx 103°$$

13.

$A = 142° - 65° = 77°$

$R^2 = 24^2 + 18^2 - 2(24)(18)\cos 77°$

$R \approx 27$ miles   [4.3]

14.

$A = 82° - 32° = 50°$

$C = 180° - 50° - 32° - 72° = 26°$

$\dfrac{a}{\sin 50°} = \dfrac{12}{\sin 26°}$

$a = \dfrac{12\sin 50°}{\sin 26°}$

$a \approx 21$ miles   [4.3]

15.

$S = \dfrac{1}{2}(112 + 165 + 140) = 208.5$

$K = \sqrt{208.5(208.5 - 112)(208.5 - 165)(208.5 - 140)}$

$K \approx 7743$

$\text{cost} \approx 8.50(7743)$

$\text{cost} \approx \$66,000$ [4.2]

Cumulative Review

1. $d = \sqrt{(-3-4)^2 + (4-(-1))^2} = \sqrt{49+25} = \sqrt{74}$ [1.2]

2. $f(x) + g(x) = \sin x + \cos x$ [1.5]

3. $(f \circ g)(x) = f[g(x)]$ [1.5]
$= f[\cos x]$
$= \sec(\cos x))$

4. $f(x) = \dfrac{1}{2}x - 3$ [1.6]

$y = \dfrac{1}{2}x - 3$

$x = \dfrac{1}{2}y - 3$

$2(x+3) = y$

$f^{-1}(x) = 2x + 6$

5. Shifted 2 units to the right and 3 units up. [1.4]

6. $\sin 27° = \dfrac{15}{a}$ [2.2]

$a = \dfrac{15}{\sin 27°} \approx 33$ cm

7. $y = 3\sin(2\pi x)$

[2.5]

8. $y = \dfrac{1}{4}\tan(2x)$

[2.6]

9. $y = 2\sin\left(\dfrac{\pi x}{2}\right) + 1$

[2.7]

10. $y = 3\sin\left(\dfrac{1}{3}x - \dfrac{\pi}{2}\right)$ [2.7]

$0 \le \dfrac{1}{3}x - \dfrac{\pi}{2} \le 2\pi$

$\dfrac{\pi}{2} \le \dfrac{1}{3}x \le \dfrac{5\pi}{2}$

$\dfrac{3\pi}{2} \le x \le \dfrac{15\pi}{2}$

amplitude = 3, period $= 6\pi$,   phase shift $= \dfrac{3\pi}{2}$

11. $y = \sin x + \cos x$ [2.7]

Amplitude: $\sqrt{2}$, period: 2π , phase shift: $\dfrac{\pi}{4}$

12. $c^2 = a^2 + b^2 - 2ab\cos C$ [4.2]

$c^2 = (10)^2 + (12)^2 - 2(10)(12)\cos 50°$

$\quad = 244 - 240\cos 50°$

$\quad \approx 9.473$

$c \approx 9.5$ cm

13. $\dfrac{1}{\cos x} - \cos x = \dfrac{1 - \cos^2 x}{\cos x}$ [3.1]

$\qquad\qquad\qquad = \dfrac{\sin^2 x}{\cos x}$

$\qquad\qquad\qquad = \sin x \tan x$

14. $\sin^{-1}\sin\left(\dfrac{2\pi}{3}\right) = \sin^{-1}\left(\dfrac{\sqrt{3}}{2}\right) = \dfrac{\pi}{3}$ [3.5]

15. $\tan\left(\cos^{-1}\left(\dfrac{12}{13}\right)\right)$ [3.5]

Let $\theta = \cos^{-1}\dfrac{12}{13}$ and find $y = \tan\theta$.

Then $\cos\theta = \dfrac{12}{13}$, and $0 \le \theta \le \pi$.

$\sqrt{13^2 - 12^2} = \sqrt{25} = 5$

Thus $\tan\theta = \dfrac{5}{12}$.

$y = \dfrac{5}{12}$

16. $\sin x \tan x - \dfrac{1}{2}\tan x = 0$ [3.6]

$\tan x\left(\sin x - \dfrac{1}{2}\right) = 0$

$\tan x = 0 \qquad\qquad \sin x - \dfrac{1}{2} = 0$

$\quad x = 0, \pi \qquad\qquad\quad \sin x = \dfrac{1}{2}$

$\qquad\qquad\qquad\qquad\quad x = \dfrac{\pi}{6}, \dfrac{5\pi}{6}$

17. $\|\mathbf{v}\| = \sqrt{4^2 + (-3)^2}$ $\alpha = \tan^{-1}\left|\dfrac{-3}{4}\right| = \tan^{-1}\dfrac{3}{4}$ [4.3]

$\|\mathbf{v}\| = \sqrt{16 + 9}$ $\quad \alpha \approx 36.9°$

$\|\mathbf{v}\| = 5$ $\qquad\quad \theta = 360° - \alpha$

$\qquad\qquad\qquad\quad \theta \approx 360° - 36.9°$

$\qquad\qquad\qquad\quad \theta \approx 323.1°$

18. $\cos\theta = \dfrac{\mathbf{v} \cdot \mathbf{w}}{\|\mathbf{v}\|\,\|\mathbf{w}\|}$ [4.3]

$\cos\theta = \dfrac{\langle 1,\, 2\rangle \cdot \langle -2,\, 3\rangle}{\sqrt{1^2 + (2)^2}\,\sqrt{(-2)^2 + 3^2}}$

$\cos\theta = \dfrac{1(-2) + (2)(3)}{\sqrt{5}\sqrt{13}}$

$\cos\theta = \dfrac{4}{\sqrt{5}\sqrt{13}} \approx 0.49613$

$\theta = 60.3°$

19.

$$\alpha = \sin^{-1}\left|\frac{1}{3}\right| = \sin^{-1}\frac{1}{3}$$

$$\alpha \approx 19.5°$$

heading $= \theta = 270° + \alpha$ [4.3]

$$\theta \approx 270° + 19.5°$$

$$\theta \approx 289.5°$$

20. $\mathbf{AB} = 515(\cos 36\mathbf{i} + \sin 36\mathbf{j}) \approx 416.6\mathbf{i} + 302.7\mathbf{j}$ [4.3]

$$\mathbf{AD} = 150[\cos(-30°)\mathbf{i} + \sin(-30°)\mathbf{j}] \approx 129.9\mathbf{i} - 75\mathbf{j}$$

$$\mathbf{AC} = \mathbf{AB} + \mathbf{AD}$$

$$\mathbf{AC} = 416.6\mathbf{i} + 302.7\mathbf{j} + 129.9\mathbf{i} - 75\mathbf{j}$$

$$\mathbf{AC} \approx 546.5\mathbf{i} + 227.7\mathbf{j}$$

$$\|\mathbf{AC}\| = \sqrt{546.5^2 + (227.7)^2}$$

$$\|\mathbf{AC}\| \approx 592 \text{ mph}$$

$$\alpha = 90° - \theta = 90° - \tan^{-1}\left(\frac{227.7}{546.5}\right) \approx 67.4°$$

Chapter 5
Complex Numbers

Section 5.1

1. $\sqrt{-81} = i\sqrt{81} = 9i$

3. $\sqrt{-98} = i\sqrt{98} = 7i\sqrt{2}$

5. $\sqrt{16} + \sqrt{-81} = 4 + i\sqrt{81}$
$= 4 + 9i$

7. $5 + \sqrt{-49} = 5 + i\sqrt{49} = 5 + 7i$

9. $8 - \sqrt{-18} = 8 - i\sqrt{18} = 8 - 3i\sqrt{2}$

11. $(5 + 2i) + (6 - 7i) = 5 + 2i + 6 - 7i$
$= (5 + 6) + (2i - 7i)$
$= 11 - 5i$

13. $(-2 - 4i) - (5 - 8i) = -2 - 4i - 5 + 8i$
$= (-2 - 5) + (-4i + 8i)$
$= -7 + 4i$

15. $(1 - 3i) + (7 - 2i) = 1 - 3i + 7 - 2i$
$= (1 + 7) + (-3i - 2i)$
$= 8 - 5i$

17. $(-3 - 5i) - (7 - 5i) = -3 - 5i - 7 + 5i$
$= (-3 - 7) + (-5i + 5i)$
$= -10$

19. $8i - (2 - 8i) = 8i - 2 + 8i$
$= -2 + (8i + 8i)$
$= -2 + 16i$

21. $5i \cdot 8i = 40i^2$
$= 40(-1)$
$= -40$

23. $\sqrt{-50} \cdot \sqrt{-2} = i\sqrt{50} \cdot i\sqrt{2} = 5i\sqrt{2} \cdot i\sqrt{2}$
$= 5i^2(\sqrt{2})^2 = 5(-1)(2)$
$= -10$

25. $3(2 + 5i) - 2(3 - 2i) = 6 + 15i - 6 + 4i$
$= (6 - 6) + (15i + 4i)$
$= 19i$

27. $(4 + 2i)(3 - 4i) = 4(3 - 4i) + (2i)(3 - 4i)$
$= 12 - 16i + 6i - 8i^2$
$= 12 - 16i + 6i - 8(-1)$
$= 12 - 16i + 6i + 8$
$= (12 + 8) + (-16i + 6i)$
$= 20 - 10i$

29. $(-3 - 4i)(2 + 7i) = -3(2 + 7i) - 4i(2 + 7i)$
$= -6 - 21i - 8i - 28i^2$
$= -6 - 21i - 8i - 28(-1)$
$= -6 - 21i - 8i + 28$
$= (-6 + 28) + (-21i - 8i)$
$= 22 - 29i$

31. $(4 - 5i)(4 + 5i) = 4(4 + 5i) - 5i(4 + 5i)$
$= 16 + 20i - 20i - 25i^2$
$= 16 + 20i - 20i - 25(-1)$
$= 16 + 20i - 20i + 25$
$= (16 + 25) + (20i - 20i)$
$= 41$

33. $(3 + \sqrt{-4})(2 - \sqrt{-9}) = (3 + i\sqrt{4})(2 - i\sqrt{9})$
$= (3 + 2i)(2 - 3i)$
$= 3(2 - 3i) + 2i(2 - 3i)$
$= 6 - 9i + 4i - 6i^2$
$= 6 - 9i + 4i - 6(-1)$
$= 6 - 9i + 4i + 6$
$= (6 + 6) + (-9i + 4i)$
$= 12 - 5i$

35. $(3 + 2\sqrt{-18})(2 + 2\sqrt{-50}) = (3 + 2i\sqrt{18})(2 + 2i\sqrt{50}) = [3 + 2i(3\sqrt{2})][2 + 2i(5\sqrt{2})] = (3 + 6i\sqrt{2})(2 + 10i\sqrt{2})$
$= 3(2 + 10i\sqrt{2}) + 6i\sqrt{2}(2 + 10i\sqrt{2}) = 6 + 30i\sqrt{2} + 12i\sqrt{2} + 60i^2(\sqrt{2})^2$
$= 6 + 30i\sqrt{2} + 12i\sqrt{2} + 60(-1)(2) = 6 + 30i\sqrt{2} + 12i\sqrt{2} - 120$
$= (6 - 120) + (30i\sqrt{2} + 12i\sqrt{2}) = -114 + 42i\sqrt{2}$

37. $\dfrac{6}{i} = \dfrac{6}{i} \cdot \dfrac{i}{i} = \dfrac{6i}{i^2} = \dfrac{6i}{-1} = -6i$

39. $\dfrac{6+3i}{i} = \dfrac{6+3i}{i} \cdot \dfrac{i}{i} = \dfrac{6i+3i^2}{i^2} = \dfrac{6i+3(-1)}{-1} = \dfrac{6i-3}{-1} = 3 - 6i$

41. $\dfrac{1}{7+2i} = \dfrac{1}{7+2i} \cdot \dfrac{7-2i}{7-2i} = \dfrac{1(7-2i)}{(7+2i)(7-2i)} = \dfrac{7-2i}{49-4i^2} = \dfrac{7-2i}{49-4(-1)} = \dfrac{7-2i}{49+4} = \dfrac{7-2i}{53} = \dfrac{7}{53} - \dfrac{2}{53}i$

43. $\dfrac{2i}{1+i} = \dfrac{2i}{1+i} \cdot \dfrac{1-i}{1-i} = \dfrac{2i(1-i)}{(1+i)(1-i)} = \dfrac{2i-2i^2}{1-i^2} = \dfrac{2i-2(-1)}{1-(-1)} = \dfrac{2i+2}{1+1} = \dfrac{2+2i}{2} = \dfrac{2}{2} + \dfrac{2}{2}i = 1 + i$

45. $\dfrac{5-i}{4+5i} = \dfrac{5-i}{4+5i} \cdot \dfrac{4-5i}{4-5i} = \dfrac{(5-i)(4-5i)}{(4+5i)(4-5i)} = \dfrac{5(4-5i)-i(4-5i)}{4(4-5i)+5i(4-5i)} = \dfrac{20-25i-4i+5i^2}{16-20i+20i-25i^2}$

$= \dfrac{20-25i-4i+5(-1)}{16-25(-1)} = \dfrac{20-25i-4i-5}{16+25} = \dfrac{(20-5)+(-25i-4i)}{16+25} = \dfrac{15-29i}{41} = \dfrac{15}{41} - \dfrac{29}{41}i$

47. $\dfrac{3+2i}{3-2i} = \dfrac{3+2i}{3-2i} \cdot \dfrac{3+2i}{3+2i} = \dfrac{(3+2i)^2}{(3-2i)(3+2i)} = \dfrac{3^2+2(3)(2i)+(2i)^2}{3^2-(2i)^2} = \dfrac{9+12i+4i^2}{9-4i^2} = \dfrac{9+12i+4(-1)}{9-4(-1)}$

$= \dfrac{9+12i-4}{9+4} = \dfrac{5+12i}{13} = \dfrac{5}{13} + \dfrac{12}{13}i$

49. $\dfrac{-7+26i}{4+3i} = \dfrac{-7+26i}{4+3i} \cdot \dfrac{4-3i}{4-3i} = \dfrac{(-7+26i)(4-3i)}{(4+3i)(4-3i)} = \dfrac{-7(4-3i)+26i(4-3i)}{4^2-(3i)^2} = \dfrac{-28+21i+104i-78i^2}{16-9i^2}$

$= \dfrac{-28+21i+104i-78(-1)}{16-9(-1)} = \dfrac{-28+21i+104i+78}{16+9} = \dfrac{50+125i}{25} = \dfrac{50}{25} + \dfrac{125}{25}i = 2 + 5i$

51. $\begin{aligned}(3-5i)^2 &= 3^3 + 2(3)(-5i) + (-5i)^2 \\ &= 9 - 30i + 25i^2 \\ &= 9 - 30i + 25(-1) \\ &= 9 - 30i - 25 \\ &= -16 - 30i\end{aligned}$

53. $\begin{aligned}(1+2i)^3 &= (1+2i)(1+2i)^2 \\ &= (1+2i)[1^2 + 2(1)(2i) + (2i)^2] \\ &= (1+2i)[1 + 4i + 4i^2] \\ &= (1+2i)[1 + 4i + 4(-1)] \\ &= (1+2i)[1 + 4i - 4] \\ &= (1+2i)(-3+4i) \\ &= 1(-3+4i) + 2i(-3+4i) \\ &= -3 + 4i - 6i + 8i^2 \\ &= -3 + 4i - 6i - 8 \\ &= -11 - 2i\end{aligned}$

55. Use the Powers of i Theorem.
The remainder of $15 \div 4$ is 3.
$i^{15} = i^3 = -i$

57. Use the Powers of i Theorem.
The remainder of $40 \div 4$ is 0.
$-i^{40} = -(i^0) = -1$

59. Use the Powers of i Theorem.
The remainder of $25 \div 4$ is 1.
$\dfrac{1}{i^{25}} = \dfrac{1}{i} = \dfrac{1}{i} \cdot \dfrac{i}{i} = \dfrac{i}{i^2} = \dfrac{i}{-1} = -i$

61. Use the Powers of i Theorem.
The remainder of $34 \div 4$ is 2.
$i^{-34} = \dfrac{1}{i^{34}} = \dfrac{1}{i^2} = \dfrac{1}{-1} = -1$

63.
$z_1 = (-2+i)+4+3i = 2+4i$
$z_2 = (2+4i)+4+3i = 6+7i$
$z_3 = (6+7i)+4+3i = 10+10i$
$z_4 = (10+10i)+4+3i = 14+13i$
$z_5 = (14+13i)+4+3i = 18+16i$

65.
$z_1 = i(1-i) = i-i^2 = 1+i$
$z_2 = i(1+i) = i+i^2 = -1+i$
$z_3 = i(-1+i) = -i+i^2 = -1-i$
$z_4 = i(-1-i) = -i-i^2 = 1-i$
$z_5 = z_1$
$z_6 = z_2$
$z_7 = z_3$
$z_8 = z_4$

67. Use $a = 3, b = -3, c = 3$.

$$\frac{-b+\sqrt{b^2-4ac}}{2a} = \frac{-(-3)+\sqrt{(-3)^2-4(3)(3)}}{2(3)}$$

$$= \frac{3+\sqrt{9-36}}{6} = \frac{3+\sqrt{-27}}{6}$$

$$= \frac{3+i\sqrt{27}}{6} = \frac{3+3i\sqrt{3}}{6}$$

$$= \frac{3}{6}+\frac{3\sqrt{3}}{6}i = \frac{1}{2}+\frac{\sqrt{3}}{2}i$$

69. Use $a = 2, b = 6, c = 6$.

$$\frac{-b+\sqrt{b^2-4ac}}{2a} = \frac{-(6)+\sqrt{(6)^2-4(2)(6)}}{2(2)}$$

$$= \frac{-6+\sqrt{36-48}}{4} = \frac{-6+\sqrt{-12}}{4}$$

$$= \frac{-6+i\sqrt{12}}{4} = \frac{-6+2i\sqrt{3}}{4}$$

$$= \frac{-6}{4}+\frac{2i\sqrt{3}}{4} = -\frac{3}{2}+\frac{\sqrt{3}}{2}i$$

71. Use $a = 4, b = -4, c = 2$.

$$\frac{-b+\sqrt{b^2-4ac}}{2a} = \frac{-(-4)+\sqrt{(-4)^2-4(4)(2)}}{2(4)}$$

$$= \frac{4+\sqrt{16-32}}{8} = \frac{4+\sqrt{-16}}{8}$$

$$= \frac{4+i\sqrt{16}}{8} = \frac{4+4i}{8}$$

$$= \frac{4}{8}+\frac{4i}{8} = \frac{1}{2}+\frac{1}{2}i$$

Connecting Concepts

73. $x^2+16 = x^2+4^2 = (x+4i)(x-4i)$

75. $z^2+25 = z^2+5^2 = (z+5i)(z-5i)$

77. $4x^2+81 = (2x)^2+9^2 = (2x+9i)(2x-9i)$

79. If $x = 1+2i$, then $x^2-2x+5 = (1+2i)^2-2(1+2i)+5 = 1+4i+4i^2-2-4i+5 = 1+4i+4(-1)-2-4i+5$
$= 1+4i-4-2-4i+5 = (1-4-2+5)+(4i-4i) = 0$

81. Verify that $(-1+i\sqrt{3})^3 = 8$.

$(-1+i\sqrt{3})^3 = (-1+i\sqrt{3})(-1+i\sqrt{3})^2 = (-1+i\sqrt{3})[(-1)^2 + 2(-1)(i\sqrt{3}) + (i\sqrt{3})^2]$

$= (-1+i\sqrt{3})[1-2i\sqrt{3}+3i^2] = (-1+i\sqrt{3})[1-2i\sqrt{3}+3(-1)] = (-1+i\sqrt{3})[1-2i\sqrt{3}-3]$

$= (-1+i\sqrt{3})(-2-2i\sqrt{3}) = -1(-2-2i\sqrt{3}) + i\sqrt{3}(-2-2i\sqrt{3}) = 2+2i\sqrt{3}-2i\sqrt{3}-2i^2(\sqrt{3})^2$

$= 2+2i\sqrt{3}-2i\sqrt{3}-2(-1)(3) = 2+2i\sqrt{3}-2i\sqrt{3}+6 = (2+6)+(2i\sqrt{3}-2i\sqrt{3})$

$= 8$

Verify that $(-1-i\sqrt{3})^3 = 8$.

$(-1-i\sqrt{3})^3 = (-1-i\sqrt{3})(-1-i\sqrt{3})^2 = (-1-i\sqrt{3})[(-1)^2 + 2(-1)(-i\sqrt{3}) + (-i\sqrt{3})^2]$

$= (-1-i\sqrt{3})[1+2i\sqrt{3}+3i^2] = (-1-i\sqrt{3})[1+2i\sqrt{3}+3(-1)] = (-1-i\sqrt{3})[1+2i\sqrt{3}-3]$

$= (-1-i\sqrt{3})(-2+2i\sqrt{3}) = -1(-2+2i\sqrt{3}) - i\sqrt{3}(-2+2i\sqrt{3}) = 2-2i\sqrt{3}+2i\sqrt{3}-2i^2(\sqrt{3})^2$

$= 2-2i\sqrt{3}+2i\sqrt{3}-2(-1)(3) = 2-2i\sqrt{3}+2i\sqrt{3}+6 = (2+6)+(-2i\sqrt{3}+2i\sqrt{3})$

$= 8$

83. $i+i^2+i^3+i^4+...+i^{28} = 7(i+i^2+i^3+i^4) = 7(i+(-1)+(-i)+1) = 7(0) = 0$

···   **P**repare for **S**ection **5.2**

PS1. $(1+i)(2+i) = 2+3i+i^2 = 1+3i$

PS2. $\dfrac{2+i}{3-i} \cdot \dfrac{3+i}{3+i} = \dfrac{6+5i+i^2}{9-i^2} = \dfrac{5+5i}{10} = \dfrac{1}{2}+\dfrac{1}{2}i$

PS3. $2-3i$

PS4. $3+5i$

PS5. $x = \dfrac{-1\pm\sqrt{1^2-4(1)(1)}}{2(1)} = \dfrac{-1\pm\sqrt{-3}}{2} = -\dfrac{1}{2}\pm\dfrac{\sqrt{3}}{2}i$

PS6. $x^2+9=0$
$x^2=-9$
$x=\pm 3i$

Section 5.2

1. $|z| = \sqrt{(-2)^2 + (-2)^2}$
$= \sqrt{8} = 2\sqrt{2}$

3. $|z| = \sqrt{(\sqrt{3})^2 + (-1)^2}$
$= \sqrt{3+1} = \sqrt{4}$
$= 2$

5. $|z| = \sqrt{0^2 + (-2)^2} = 2$

7. $|z| = \sqrt{3^2 + (-5)^2}$
$= \sqrt{34}$

9. $r = \sqrt{1^2 + (-1)^2}$
$r = \sqrt{2}$

$\alpha = \tan^{-1}\left|\dfrac{-1}{1}\right|$
$= \tan^{-1}1$
$= 45°$
$\theta = 360° - 45° = 315°$

$z = \sqrt{2}\text{ cis }315°$

11. $r = \sqrt{(\sqrt{3})^2 + (-1)^2}$
$r = 2$

$\alpha = \tan^{-1}\left|\dfrac{-1}{\sqrt{3}}\right|$
$= \tan^{-1}\dfrac{1}{\sqrt{3}} = 30°$
$\alpha = 360° - 30° = 330°$

$z = 2\text{ cis }330°$

13. $r = \sqrt{0^2 + 3^2}$

$r = 3$

$\theta = 90°$

$z = 3 \text{ cis } 90°$

15. $r = \sqrt{(-5)^2 + 0^2}$

$r = 5$

$\theta = 180°$

$z = 5 \text{ cis } 180°$

17. $r = \sqrt{(-8)^2 + (8\sqrt{3})^2}$

$r = 16$

$\alpha = \tan^{-1} \left| \dfrac{8\sqrt{3}}{-8} \right|$

$= \tan^{-1} \sqrt{3} = 60°$

$\theta = 180° - 60° = 120°$

$z = 16 \text{ cis } 120°$

19. $r = \sqrt{(-2)^2 + (-2\sqrt{3})^2}$

$r = 4$

$\alpha = \tan^{-1} \left| \dfrac{-2\sqrt{3}}{-2} \right|$

$= \tan^{-1} \sqrt{3} = 60°$

$\theta = 180° + 60° = 240°$

$z = 4 \text{ cis } 240°$

21. $z = 2(\cos 45° + i \sin 45°)$

$z = 2\left(\dfrac{\sqrt{2}}{2} + \dfrac{\sqrt{2}}{2}i \right)$

$z = \sqrt{2} + i\sqrt{2}$

23. $z = \cos 315° + i \sin 315°$

$z = \dfrac{\sqrt{2}}{2} - \dfrac{\sqrt{2}}{2}i$

25. $z = 6 \text{ cis } 135°$

$z = 6(\cos 135° + i \sin 135°)$

$z = 6\left(-\dfrac{\sqrt{2}}{2} + \dfrac{\sqrt{2}}{2}i \right)$

$z = -3\sqrt{2} + 3i\sqrt{2}$

27. $z = 8 \text{ cis } 0°$

$z = 8(\cos 0° + i \sin 0°)$

$z = 8(1 + 0i)$

$z = 8$

29. $z = 2\left(\cos \dfrac{5\pi}{6} + i \sin \dfrac{5\pi}{6} \right)$

$z = 2\left(-\dfrac{\sqrt{3}}{2} + \dfrac{1}{2}i \right)$

$z = -\sqrt{3} + i$

31. $z = 3\left(\cos \dfrac{3\pi}{2} + i \sin \dfrac{3\pi}{2} \right)$

$z = 3(0 - i)$

$z = -3i$

33. $z = 8 \text{ cis } \dfrac{3\pi}{4}$

$= 8\left(\cos \dfrac{3\pi}{4} + i \sin \dfrac{3\pi}{4} \right)$

$z = 8\left(-\dfrac{\sqrt{2}}{2} + \dfrac{i\sqrt{2}}{2} \right)$

$z = -4\sqrt{2} + 4i\sqrt{2}$

35. $z = 9 \text{ cis } \dfrac{11\pi}{6}$

$z = 9\left(\cos \dfrac{11\pi}{6} + i \sin \dfrac{11\pi}{6} \right)$

$z = 9\left(\dfrac{\sqrt{3}}{2} - \dfrac{1}{2}i \right)$

$z = \dfrac{9\sqrt{3}}{2} - \dfrac{9}{2}i$

37. $z = 2 \text{ cis } 2$

$z = 2(\cos 2 + i \sin 2)$

$z \approx 2(-0.4161 + 0.9093i)$

$z \approx -0.832 + 1.819i$

39. $z_1 z_2 = 2 \text{ cis } 30° \cdot 3 \text{ cis } 225°$

$z_1 z_2 = 6 \text{ cis}(30° + 225°)$

$z_1 z_2 = 6 \text{ cis } 255°$

41. $z_1 z_2 = 3(\cos 122° + i \sin 122°) \cdot 4(\cos 213° + i \sin 213°)$

$z_1 z_2 = 12[\cos(122° + 213°) + i \sin(122° + 213°)]$

$z_1 z_2 = 12(\cos 335° + i \sin 335°)$

$z_1 z_2 = 12 \text{ cis } 335°$

43. $z_1 z_2 = 5\left(\cos \dfrac{2\pi}{3} + i \sin \dfrac{2\pi}{3} \right) \cdot 2\left(\cos \dfrac{2\pi}{5} + i \sin \dfrac{2\pi}{5} \right)$

$z_1 z_2 = 10\left[\cos\left(\dfrac{2\pi}{3} + \dfrac{2\pi}{5} \right) + i \sin\left(\dfrac{2\pi}{3} + \dfrac{2\pi}{5} \right) \right]$

$z_1 z_2 = 10\left(\cos \dfrac{16\pi}{15} + i \sin \dfrac{16\pi}{15} \right)$

$z_1 z_2 = 10 \text{ cis } \dfrac{16\pi}{15}$

45.

$z_1 z_2 = 4 \text{ cis } 2.4 \cdot 6 \text{ cis } 4.1$

$z_1 z_2 = 24 \text{ cis } (2.4 + 4.1)$

$z_1 z_2 = 24 \text{ cis } 6.5$

47.

$\dfrac{z_1}{z_2} = \dfrac{32 \text{ cis } 30°}{4 \text{ cis } 150°}$

$\dfrac{z_1}{z_2} = 8 \text{ cis}(30° - 150°)$

$\dfrac{z_1}{z_2} = 8 \text{ cis}(-120°)$

$\dfrac{z_1}{z_2} = 8 \, (\cos 120° - i \sin 120°)$

$\dfrac{z_1}{z_2} = 8 \left(-\dfrac{1}{2} - \dfrac{i\sqrt{3}}{2} \right) = -4 - 4i\sqrt{3}$

49.

$\dfrac{z_1}{z_2} = \dfrac{27(\cos 315° + i \sin 315°)}{9(\cos 225° + i \sin 225°)}$

$\dfrac{z_1}{z_2} = 3\,[\cos(315° - 225°) + i \sin(315° - 225°)]$

$\dfrac{z_1}{z_2} = 3(\cos 90° + i \sin 90°) = 3(0 + i) = 3i$

51.

$\dfrac{z_1}{z_2} = \dfrac{12\left(\cos \frac{2\pi}{3} + i \sin \frac{2\pi}{3}\right)}{4\left(\cos \frac{11\pi}{6} + i \sin \frac{11\pi}{6}\right)}$

$\dfrac{z_1}{z_2} = 3\left[\cos\left(\dfrac{2\pi}{3} - \dfrac{11\pi}{6}\right) + i \sin\left(\dfrac{2\pi}{3} - \dfrac{11\pi}{6}\right) \right]$

$\dfrac{z_1}{z_2} = 3\left(\cos \dfrac{7\pi}{6} - i \sin \dfrac{7\pi}{6} \right)$

$\dfrac{z_1}{z_2} = 3\left[-\dfrac{\sqrt{3}}{2} - \left(-\dfrac{1}{2}i\right) \right]$

$\dfrac{z_1}{z_2} = -\dfrac{3\sqrt{3}}{2} + \dfrac{3}{2}i$

53.

$\dfrac{z_1}{z_2} = \dfrac{25 \text{ cis } 3.5}{5 \text{ cis } 1.5}$

$\dfrac{z_1}{z_2} = 5 \text{ cis } (3.5 - 1.5)$

$\dfrac{z_1}{z_2} = 5 \text{ cis } 2$

$\dfrac{z_1}{z_2} = 5 \, (\cos 2 + i \sin 2)$

$\dfrac{z_1}{z_2} \approx 5 \, (-0.4161 + 0.9093i)$

$\dfrac{z_1}{z_2} \approx -2.081 + 4.546i$

55.

$z_1 = 1 - i\sqrt{3}$

$r_1 = \sqrt{1^2 + \left(\sqrt{3}\right)^2}$ $\qquad \alpha = \tan^{-1}\left|\dfrac{-\sqrt{3}}{1}\right| = 60°$

$r_1 = 2$

$\qquad\qquad\qquad \theta_1 = 300°$

$z_1 = 2(\cos 300° + i \sin 300°)$

$z_2 = 1 + i$

$r_2 = \sqrt{1^2 + 1^2}$ $\qquad \alpha = \tan^{-1}\left|\dfrac{1}{1}\right| = 45°$

$r_2 = \sqrt{2}$

$\qquad\qquad\qquad \theta_2 = 45°$

$z_2 = \sqrt{2}(\cos 45° + i \sin 45°)$

$z_1 z_2 = 2(\cos 300° + i \sin 300°) \cdot \sqrt{2}(\cos 45° + i \sin 45°)$

$z_1 z_2 = 2\sqrt{2}[\cos(300° + 45°) + i \sin(300° + 45°)]$

$z_1 z_2 = 2\sqrt{2}(\cos 345° + i \sin 345°)$

$z_1 z_2 \approx 2.732 - 0.732i$

57. $z_1 = 3 - 3i$

$r_1 = \sqrt{3^2 + (-3)^2}$ $\alpha = \tan^{-1}\left|\frac{-3}{3}\right| = 45°$

$r_1 = 3\sqrt{2}$ $\theta_1 = 315°$

$z_1 = 3\sqrt{2}(\cos 315° + i \sin 315°)$

$z_2 = 1 + i$

$r_2 = \sqrt{1^2 + 1^2}$ $\alpha = \tan^{-1}\left|\frac{1}{1}\right| = 45°$

$r_2 = \sqrt{2}$ $\theta_2 = 45°$

$z_2 = \sqrt{2}(\cos 45° + i \sin 45°)$

$z_1 z_2 = 3\sqrt{2}(\cos 315° + i \sin 315°) \cdot \sqrt{2}(\cos 45° + i \sin 45°)$

$z_1 z_2 = 6[\cos(315° + 45°) + i \sin(315° + 45°)]$

$z_1 z_2 = 6(\cos 360° + i \sin 360°)$

$z_1 z_2 = 6 + 0i$

$z_1 z_2 = 6$

59. $z_1 = 1 + i\sqrt{3}$

$r_1 = \sqrt{1^2 + (\sqrt{3})^2}$ $\alpha_1 = \tan^{-1}\left|\frac{\sqrt{3}}{1}\right| = 60°$

$r_1 = 2$ $\theta_1 = 60°$

$z_1 = 2(\cos 60° + i \sin 60°)$

$z_2 = 1 - i\sqrt{3}$

$r_2 = \sqrt{1^2 + (\sqrt{3})^2}$ $\alpha = \tan^{-1}\left|\frac{-\sqrt{3}}{1}\right| = 60°$

$r_2 = 2$ $\theta_2 = 300°$

$z_2 = 2(\cos 300° + i \sin 300°)$

$\frac{z_1}{z_2} = \frac{2(\cos 60° + i \sin 60°)}{2(\cos 300° + i \sin 300°)}$

$\frac{z_1}{z_2} = \cos(60° - 300°) + i \sin(60° - 300°)$

$\frac{z_1}{z_2} = \cos 240° - i \sin 240° = -\frac{1}{2} + \frac{\sqrt{3}}{2}i$

61. $z_1 = \sqrt{2} - i\sqrt{2}$

$r_1 = \sqrt{(\sqrt{2})^2 + (\sqrt{2})^2}$ $\alpha_1 = \tan^{-1}\left|\frac{-\sqrt{2}}{2}\right| = 45°$

$r_1 = 2$ $\theta_1 = 315°$

$z_1 = 2(\cos 315° + i \sin 315°)$

$z_2 = 1 + i$

$r_2 = \sqrt{1^2 + 1^2}$ $\alpha_2 = \tan^{-1}\left|\frac{1}{1}\right| = 45°$

$r_2 = \sqrt{2}$ $\theta_2 = 45°$

$z_2 = \sqrt{2}(\cos 45° + i \sin 45°)$

$\frac{z_1}{z_2} = \frac{2(\cos 315° + i \sin 315°)}{\sqrt{2}(\cos 45° + i \sin 45°)}$

$\frac{z_1}{z_2} = \sqrt{2}[\cos(315° - 45°) + i \sin(315° - 45°)]$

$\frac{z_1}{z_2} = \sqrt{2}(\cos 270° + i \sin 270°)$

$\frac{z_1}{z_2} = \sqrt{2}[0 + i(-1)] = \sqrt{2}(0 - 1i) = 0 - \sqrt{2}i = -\sqrt{2}i$ or $-i\sqrt{2}$

63.

$z_1 = \sqrt{3} - 1$

$r_1 = \sqrt{(\sqrt{3})^2 + (-1)^2}$

$r_1 = 2$

$\alpha_1 = \tan^{-1}\left|\dfrac{-1}{\sqrt{3}}\right| = 30°$

$\theta_1 = 330°$

$z_1 = 2(\cos 330° + i \sin 330°)$

$z_2 = 2 + 2i$

$r_2 = \sqrt{2^2 + 2^2}$

$r_2 = 2\sqrt{2}$

$\alpha_2 = \tan^{-1}\left|\dfrac{2}{2}\right| = 45°$

$\theta_2 = 45°$

$z_2 = 2\sqrt{2}(\cos 45° + i \sin 45°)$

$z_3 = 2 - 2i\sqrt{3}$

$r_3 = \sqrt{2^2 + (-2\sqrt{3})^2}$

$r_3 = 4$

$\alpha_3 = \tan^{-1}\left|\dfrac{-2\sqrt{3}}{2}\right| = 60°$

$\theta_3 = 300°$

$z_3 = 4(\cos 300° + i \sin 300°)$

$z_1 z_2 z_3 = 2(\cos 330° + i \sin 330°) \cdot 2\sqrt{2}(\cos 45° + i \sin 45°) \cdot 4(\cos 300° + i \sin 300°)$

$z_1 z_2 z_3 = 16\sqrt{2}[\cos(330° + 45° + 300°) + i \sin(330° + 45° + 300°)]$

$z_1 z_2 z_3 = 16\sqrt{2}(\cos 675° + i \sin 675°)$

$z_1 z_2 z_3 = 16\sqrt{2}(\cos 315° + i \sin 315°)$

$z_1 z_2 z_3 = 16\sqrt{2}\left(\dfrac{1}{\sqrt{2}} - \dfrac{1}{\sqrt{2}}i\right) = 16 - 16i$

65.

$z_1 = \sqrt{3} + i\sqrt{3}$

$r_1 = \sqrt{(\sqrt{3})^2 + (\sqrt{3})^2}$

$r_1 = \sqrt{6}$

$\alpha_1 = \tan^{-1}\left|\dfrac{\sqrt{3}}{\sqrt{3}}\right| = 45°$

$\theta_1 = 45°$

$z_1 = \sqrt{6}(\cos 45° + i \sin 45°)$

$z_2 = 1 - i\sqrt{3}$

$r_2 = \sqrt{1^2 + (-\sqrt{3})^2}$

$r_2 = 2$

$\alpha_2 = \tan^{-1}\left|\dfrac{-\sqrt{3}}{1}\right| = 60°$

$\theta_2 = 300°$

$z_2 = 2(\cos 300° + i \sin 300°)$

$z_3 = 2 - 2i$

$r_3 = \sqrt{2^2 + (-2)^2}$

$r_3 = 2\sqrt{2}$

$\alpha_3 = \tan^{-1}\left|\dfrac{-2}{2}\right| = 45°$

$\theta_3 = 315°$

$z_3 = 2\sqrt{2}(\cos 315° + i \sin 315°)$

$\dfrac{z_1}{z_2 z_3} = \dfrac{\sqrt{6}(\cos 45° + i \sin 45°)}{2(\cos 300° + i \sin 300°) \cdot 2\sqrt{2}(\cos 315° + i \sin 315°)}$

$\dfrac{z_1}{z_2 z_3} = \dfrac{\sqrt{6}(\cos 45° + i \sin 45°)}{4\sqrt{2}[\cos(300° + 315°) + i \sin(300° + 315°)]}$

$\dfrac{z_1}{z_2 z_3} = \dfrac{\sqrt{6}(\cos 45° + i \sin 45°)}{4\sqrt{2}(\cos 255° + i \sin 255°)}$

$\dfrac{z_1}{z_2 z_3} = \dfrac{\sqrt{3}}{4}[\cos(45° - 255°) + i \sin(45° - 255°)]$

$\dfrac{z_1}{z_2 z_3} = \dfrac{\sqrt{3}}{4}(\cos 210° - i \sin 210°) = \dfrac{\sqrt{3}}{4}\left(-\dfrac{\sqrt{3}}{2} + \dfrac{i}{2}\right) = -\dfrac{3}{8} + \dfrac{\sqrt{3}}{8}i$

67.

$z_1 = 1 - 3i$	$z_2 = 2 + 3i$	$z_3 = 4 + 5i$
$r_1 = \sqrt{1^2 + (-3)^2}$	$r_2 = \sqrt{2^2 + 3^2}$	$r_3 = \sqrt{4^2 + 5^2}$
$r_1 = \sqrt{10}$	$r_2 = \sqrt{13}$	$r_3 = \sqrt{41}$
$\alpha_1 = \tan^{-1}\left|\dfrac{-3}{1}\right| \approx 71.57°$	$\alpha_2 = \tan^{-1}\left|\dfrac{3}{2}\right| \approx 56.31°$	$\alpha_3 = \tan^{1}\left|\dfrac{5}{4}\right| \approx 51.34°$
$\theta_1 = 288.43°$	$\theta_2 = 56.31°$	$\theta_3 = 51.34°$
$z_1 = \sqrt{10}(\cos 288.4° + i\sin 288.4°)$	$z_2 = \sqrt{13}(\cos 56.3° + i\sin 56.3°)$	$z_3 = \sqrt{41}(\cos 51.3° + i\sin 51.3°)$

$z_1 z_2 z_3 = \sqrt{10}(\cos 288.4° + i\sin 288.4°) \cdot \sqrt{13}(\cos 56.3° + i\sin 56.3°) \cdot \sqrt{41}(\cos 51.3° + i\sin 51.3°)$

$z_1 z_2 z_3 = \sqrt{10} \cdot \sqrt{13} \cdot \sqrt{41}[\cos(288.43° + 56.31° + 51.34°) + i\sin(288.43° + 56.31° + 51.34°)]$

$z_1 z_2 z_3 \approx 73.0(\cos 396.08° + i\sin 396.08°)$

$z_1 z_2 z_3 = 73.0(\cos 36.08° + i\sin 36.08°)$

$z_1 z_2 z_3 \approx 59.0 + 43.0i$

69.

$z = r(\cos\theta + i\sin\theta) \qquad \overline{z} = r(\cos\theta - i\sin\theta)$

$z \cdot \overline{z} = r(\cos\theta + i\sin\theta) \cdot r(\cos\theta - i\sin\theta)$

$z \cdot \overline{z} = r(\cos\theta + i\sin\theta) \cdot r[\cos(-\theta) + i\sin(-\theta)]$

$z \cdot \overline{z} = r^2[\cos(\theta - \theta) + i\sin(\theta - \theta)]$

$z \cdot \overline{z} = r^2(\cos 0 + i\sin 0)$

$z \cdot \overline{z} = r^2$ or $a^2 + b^2$

•••

Prepare for Section 5.3

PS1. $\left(\dfrac{\sqrt{2}}{2} + \dfrac{\sqrt{2}}{2}i\right)^2 = \dfrac{2}{4} + 2\dfrac{2}{4}i + \dfrac{2}{4}i^2 = i$

PS2. $x^3 - 8 = (x - 2)(x^2 + 2x + 4)$

$(x^2 + 2x + 4)$ yields 2 complex solutions

$(x - 2)$ yields 1 real solution

The real root is 2.

PS3. $x^5 - 243 = (x - 3)(3x^4 + 3x^3 + 9x^2 + 27x + 81)$

$(3x^4 + 3x^3 + 9x^2 + 27x + 81)$ yields 4 complex solutions

$(x - 3)$ yields 1 real solution

The real root is 3.

PS4. $r = \sqrt{2^2 + 2^2} = 2\sqrt{2}$

$\alpha = \tan^{-1}\left|\dfrac{2}{2}\right|$

$\quad = \tan^{-1} 1 = 45°$

$\theta = 45°$

$z = 2\sqrt{2} \text{ cis } 45°$ or $2\sqrt{2} \text{ cis } \dfrac{\pi}{4}$

PS5. $2(\cos 150° + i\sin 150°) = 2\left(-\dfrac{\sqrt{3}}{2} + \dfrac{1}{2}i\right) = -\sqrt{3} + i$

PS6. $|z| = \sqrt{\left(\dfrac{\sqrt{2}}{2}\right)^2 + \left(-\dfrac{\sqrt{2}}{2}\right)^2}$

$\quad = \sqrt{\dfrac{2}{4} + \dfrac{2}{4}}$

$\quad = 1$

Section 5.3

1. $[2(\cos 30° + i \sin 30°)]^8 = 2^8[\cos(8 \cdot 30°) + i \sin(8 \cdot 30°)]$
$$= 256(\cos 240° + i \sin 240°)$$
$$= -128 - 128i\sqrt{3}$$

3. $[2(\cos 240° + i \sin 240°)]^5 = 2^5[\cos(5 \cdot 240°) + i \sin(5 \cdot 240°)]$
$$= 32[\cos 1200° + i \sin 1200°]$$
$$= 32(\cos 120° + i \sin 120°)$$
$$= -16 + 16i\sqrt{3}$$

5. $[2\operatorname{cis}(225°)]^5 = 2^5 \operatorname{cis}(5 \cdot 225°)$
$$= 32(\cos 1125° + i \sin 1125°)$$
$$= 32(\cos 45° + i \sin 45°)$$
$$= 16\sqrt{2} + 16i\sqrt{2}$$

7. $[2\operatorname{cis}(120°)]^6 = 2^6 \operatorname{cis}(6 \cdot 2\pi/3)$
$$= 64(\cos 720° + i \sin 720°)$$
$$= 64(\cos 0° + i \sin 0°)$$
$$= 64$$

9. $z = 1 - i$
$r = \sqrt{1^2 + (-1)^2} \qquad \alpha = \tan^{-1}\left|\dfrac{-1}{1}\right| = 45°$
$r = \sqrt{2}$
$$\theta = 315°$$
$$z = \sqrt{2}(\cos 315° + i \sin 315°)$$
$(1 - i)^{10} = [\sqrt{2}(\cos 315° + i \sin 315°)]^{10}$
$$= (\sqrt{2})^{10}[\cos(10 \cdot 315°) + i \sin(10 \cdot 315°)]$$
$$= 32(\cos 3150° + i \sin 3150°)$$
$$= 32(\cos 270° + i \sin 270°)$$
$$= 0 - 32i = -32i$$

11. $z = 1 + i$
$r = \sqrt{1^2 + 1^2} \qquad \alpha = \tan^{-1}\left|\dfrac{1}{1}\right| = 45°$
$r = \sqrt{2}$
$$\theta = 45°$$
$$z = \sqrt{2}(\cos 45° + i \sin 45°)$$
$(1 + i)^4 = [\sqrt{2}(\cos 45° + i \sin 45°)]^4$
$$= (\sqrt{2})^4[\cos(4 \cdot 45°) + i \sin(4 \cdot 45°)]$$
$$= 4(\cos 180° + i \sin 180°)$$
$$= -4 + 0i = -4$$

13. $z = 2 + 2i$
$r = \sqrt{2^2 + 2^2} \qquad\qquad \alpha = \tan^{-1}\left|\dfrac{2}{2}\right| = 45°$
$r = 2\sqrt{2}$
$$\theta = 45°$$
$$z = 2\sqrt{2}(\cos 45° + i \sin 45°)$$
$(2 + 2i)^7 = [2\sqrt{2}(\cos 45° + i \sin 45°)]^7$
$$= 1024\sqrt{2}[\cos(7 \cdot 45°) + i \sin(7 \cdot 45°)]$$
$$= 1024\sqrt{2}(\cos 315° + i \sin 315°)$$
$$= 1024 - 1024i$$

15. $z = \dfrac{\sqrt{2}}{2} + i\dfrac{\sqrt{2}}{2}$
$r = \sqrt{(\sqrt{2}/2)^2 + (\sqrt{2}/2)^2} \qquad \alpha = \tan^{-1}\left|\dfrac{\sqrt{2}/2}{\sqrt{2}/2}\right| = 45°$
$r = 1$
$$\theta = 45°$$
$$z = \cos 45° + i \sin 45°$$
$\left(\dfrac{\sqrt{2}}{2} - i\dfrac{\sqrt{2}}{2}\right)^6 = (\cos 45° + i \sin 45°)^6$
$$= \cos(6 \cdot 45°) + i \sin(6 \cdot 45°)$$
$$= \cos 270° + i \sin 270°$$
$$= 0 - 1i = -i$$

17. $9 = 9(\cos 0° + i \sin 0°)$
$$w_k = 9^{1/2}\left(\cos \dfrac{0° + 360°k}{2} + i \sin \dfrac{0° + 360°k}{2}\right) \qquad k = 0, 1$$
$w_0 = 3(\cos 0° + i \sin 0°)$
$w_0 = 3 + 0i = 3$
$$w_1 = 3\left(\cos \dfrac{0° + 360°}{2} + i \sin \dfrac{0° + 360°}{2}\right)$$
$w_1 = 3(\cos 180° + i \sin 180°)$
$w_1 = -3 + 0i = -3$

19. $64 = 64(\cos 0° + i \sin 0°)$

$$w_k = 64^{1/6}\left(\cos\frac{0° + 360°k}{6} + i \sin\frac{0° + 360°k}{6}\right) \quad k = 0, 1, 2, 3, 4, 5$$

$w_0 = 2(\cos 0° + i \sin 0°)$

$w_0 = 2 + 0i = 2$

$w_1 = 2\left(\cos\frac{0 + 360°}{6} + i \sin\frac{0 + 360°}{6}\right)$

$w_1 = 2(\cos 60° + i \sin 60°)$

$w_1 = 2\left(\frac{1}{2} + i\frac{\sqrt{3}}{2}\right)$

$w_1 = 1 + i\sqrt{3}$

$w_2 = 2\left(\cos\frac{0° + 360° \cdot 2}{6} + i \sin\frac{0° + 360° \cdot 2}{6}\right)$

$w_2 = 2(\cos 120° + i \sin 120°)$

$w_2 = 2\left(-\frac{1}{2} + i\frac{\sqrt{3}}{2}\right)$

$w_2 = -1 + i\sqrt{3}$

$w_3 = 2\left(\cos\frac{0° + 360° \cdot 3}{6} + i \sin\frac{0° + 360° \cdot 3}{6}\right)$

$w_3 = 2(\cos 180° + i \sin 180°)$

$w_3 = 2(-1 + 0i)$

$w_3 = -2 + 0i = -2$

$w_4 = 2\left(\cos\frac{0° + 360° \cdot 4}{6} + i \sin\frac{0° + 360° \cdot 4}{6}\right)$

$w_4 = 2(\cos 240° + i \sin 240°)$

$w_4 = 2\left(-\frac{1}{2} - i\frac{\sqrt{3}}{2}\right)$

$w_4 = -1 - i\sqrt{3}$

$w_5 = 2\left(\cos\frac{0° + 360° \cdot 5}{6} + i \sin\frac{0° + 360° \cdot 5}{6}\right)$

$w_5 = 2(\cos 300° + i \sin 300°)$

$w_5 = 2\left(\frac{1}{2} - i\frac{\sqrt{3}}{2}\right)$

$w_5 = 1 - i\sqrt{3}$

21. $-1 = 1(\cos 180° + i \sin 180°)$

$$w_k = 1^{1/5}\left(\cos\frac{180° + 360°k}{5} + i \sin\frac{180° + 360°k}{5}\right) \quad k = 0, 1, 2, 3, 4$$

$w_0 = 1(\cos 36° + i \sin 36°)$

$w_0 \approx 0.809 + 0.588i$

$w_1 = \cos\frac{180° + 360°}{5} + i \sin\frac{180° + 360°}{5}$

$w_1 = \cos 108° + i \sin 108°$

$w_1 \approx -0.309 + 0.951i$

$w_2 = \cos\frac{180° + 360° \cdot 2}{5} + i \sin\frac{180° + 360° \cdot 2}{5}$

$w_2 = \cos 180° + i \sin 180°$

$w_2 = -1 + 0i = -1$

$w_3 = \cos\frac{180° + 360° \cdot 3}{5} + i \sin\frac{180° + 360° \cdot 3}{5}$

$w_3 = \cos 252° + i \sin 252°$

$w_3 \approx -0.309 - 0.951i$

$w_4 = \cos\frac{180° + 360° \cdot 4}{5} + i \sin\frac{180° + 360° \cdot 4}{5}$

$w_4 = \cos 324° + i \sin 324°$

$w_4 \approx 0.809 - 0.588i$

23. $1 = \cos 0° + i \sin 0°$

$$w_k = \cos\frac{0° + 360°k}{3} + i \sin\frac{0° + 360°k}{3} \quad k = 0, 1, 2$$

$w_0 = \cos\frac{0°}{3} + i \sin\frac{0°}{3}$

$w_0 = \cos 0° + i \sin 0°$

$w_0 = 1 + 0i = 1$

$w_1 = \cos\frac{0° + 360°}{3} + i \sin\frac{0° + 360°}{3}$

$w_1 = \cos 120° + i \sin 120°$

$w_1 = -\frac{1}{2} + \frac{\sqrt{3}}{2}i$

$w_2 = \cos\frac{0° + 360° \cdot 2}{3} + i \sin\frac{0° + 360° \cdot 2}{3}$

$w_2 = \cos 240° + i \sin 240°$

$w_2 = -\frac{1}{2} - \frac{\sqrt{3}}{2}i$

25. $1+i = \sqrt{2}(\cos 45° + i\sin 45°)$

$$w_k = \left(\sqrt{2}\right)^{1/4}\left(\cos\frac{45° + 360°k}{4} + i\sin\frac{45° + 360°k}{4}\right) \quad k = 0, 1, 2, 3$$

$w_0 = 2^{1/8}\left(\cos\dfrac{45°}{4} + i\sin\dfrac{45°}{4}\right)$ $\qquad$ $w_1 = 2^{1/8}\left(\cos\dfrac{45° + 360°}{4} + i\sin\dfrac{45° + 360°}{4}\right)$

$w_0 = 2^{1/8}(\cos 11.25° + i\sin 11.25°)$ $\qquad$ $w_1 = 2^{1/8}(\cos 101.25° = i\sin 101.25°)$

$w_0 \approx 1.070 + 0.213i$ $\qquad$ $w_1 \approx -0.213 - 1.070i$

$w_2 = 2^{1/8}\left(\cos\dfrac{45° + 360°\cdot 2}{4} + i\sin\dfrac{45° + 360°\cdot 2}{4}\right)$ $\qquad$ $w_3 = 2^{1/8}\left(\cos\dfrac{45° + 360°\cdot 3}{4} + i\sin\dfrac{45° + 360°\cdot 3}{4}\right)$

$w_2 = 2^{1/8}(\cos 191.25° + i\sin 191.25°)$ $\qquad$ $w_3 = 2^{1/8}(\cos 281.25° + i\sin 281.25°)$

$w_2 \approx -1.070 - 0.213i$ $\qquad$ $w_3 \approx 0.213 - 1.070i$

27. $2 - 2i\sqrt{3} = 4(\cos 300° + i\sin 300°) \quad k = 0, 1, 2$

$$w_k = 4^{1/3}\left(\cos\frac{300° + 360°k}{3} + i\sin\frac{300° + 360°k}{3}\right)$$

$w_0 = 4^{1/3}\left(\cos\dfrac{300°}{3} + i\sin\dfrac{300°}{3}\right)$ $\qquad$ $w_1 = 4^{1/3}\left(\cos\dfrac{300° + 360°}{3} + i\sin\dfrac{300° + 360°}{3}\right)$

$w_0 = 4^{1/3}(\cos 100° + i\sin 100°)$ $\qquad$ $w_1 = 4^{1/3}(\cos 220° + i\sin 220°)$

$w_0 \approx -0.276 + 1.563i$ $\qquad$ $w_1 \approx -1.216 - 1.020i$

$w_2 = 4^{1/3}\left(\cos\dfrac{300° + 360°\cdot 2}{3} + i\sin\dfrac{300° + 360°\cdot 2}{3}\right)$

$w_2 = 4^{1/3}(\cos 340° + i\sin 340°)$

$w_2 \approx 1.492 - 0.543i$

29. $-16 + 16i\sqrt{3} = 32(\cos 120° + i\sin 120°)$

$$w_k = 32^{1/2}\left(\cos\frac{120° + 360°k}{2} + i\sin\frac{120° + 360°k}{2}\right) \quad k = 0, 1$$

$w_0 = 4\sqrt{2}\left(\cos\dfrac{120°}{2} + i\sin\dfrac{120°}{2}\right)$ $\qquad$ $w_1 = 4\sqrt{2}\left(\cos\dfrac{120° + 360°}{2} + i\sin\dfrac{120° + 360°}{2}\right)$

$w_0 = 4\sqrt{2}(\cos 60° + i\sin 60°)$ $\qquad$ $w_1 = 4\sqrt{2}(\cos 240° + i\sin 240°)$

$w_0 \approx 2\sqrt{2} + 2i\sqrt{6}$ $\qquad$ $w_1 \approx -2\sqrt{2} - 2i\sqrt{6}$

31. $x^3 + 8 = 0$

$x^3 = -8$

Find the three cube roots of -8.

$-8 = 8(\cos 180° + i \sin 180°)$

$x_k = 8^{1/3}\left(\cos\dfrac{180° + 360°k}{3} + i\sin\dfrac{180° + 360°k}{3}\right) \quad k = 0, 1, 2$

$w_0 = 2\left(\cos\dfrac{180°}{3} + i\sin\dfrac{180°}{3}\right)$ $\qquad\qquad w_1 = 2\left(\cos\dfrac{180° + 360°}{3} + i\sin\dfrac{180° + 360°}{3}\right)$

$w_0 = 2(\cos 60° + i \sin 60°)$ $\qquad\qquad\qquad\qquad w_1 = 2(\cos 180° + i \sin 180°)$

$w_0 = 2 \text{ cis } 60°$ $\qquad\qquad\qquad\qquad\qquad\qquad w_1 = 2 \text{ cis } 180°$

$w_2 = 2\left(\cos\dfrac{180° + 360° \cdot 2}{3} + i\sin\dfrac{180° + 360° \cdot 2}{3}\right)$

$w_2 = 2(\cos 300° + i \sin 300°)$

$w_2 = 2 \text{ cis } 300°$

33. $x^4 + i = 0$

$x^4 = -i$

Find the four fourth roots of $-i$.

$-i = (\cos 270° + i \sin 270°)$

$w_k = \cos\dfrac{270° + 360°k}{4} + i\sin\dfrac{270° + 360°k}{4} \quad k = 0, 1, 2, 3$

$w_0 = \text{cis } \dfrac{270°}{4}$ $\qquad w_1 = \text{cis } \dfrac{270° + 360°}{4}$ $\qquad w_2 = \text{cis } \dfrac{270° + 360° \cdot 2}{4}$ $\qquad w_3 = \text{cis } \dfrac{270° + 360° \cdot 3}{4}$

$w_0 = \text{cis } 67.5°$ $\qquad\quad w_1 = \text{cis } 157.5°$ $\qquad\qquad w_2 = \text{cis } 247.5°$ $\qquad\qquad w_3 = \text{cis } 337.5°$

35. $x^3 - 27 = 0$

$x^3 = 27$

Find the three cube roots of 27.

$27 = 27(\cos 0° + i \sin 0°)$

$w_k = 3\left(\cos\dfrac{0° + 360°k}{3} + i\sin\dfrac{0° + 360°k}{3}\right) \quad k = 0, 1, 2$

$w_0 = 3 \text{ cis } \dfrac{0°}{3}$ $\qquad\qquad\qquad w_1 = 3 \text{ cis } \dfrac{0° + 360°}{3}$ $\qquad\qquad\qquad w_2 = 3 \text{ cis } \dfrac{0° + 360° \cdot 2}{3}$

$w_0 = 3 \text{ cis } 0°$ $\qquad\qquad\qquad\quad w_1 = 3 \text{ cis } 120°$ $\qquad\qquad\qquad\qquad w_2 = 3 \text{ cis } 240°$

37. $x^4 + 81 = 0$

$x^4 = -81$

Find the four fourth roots of -81.

$-81 = 81(\cos 180° + i \sin 180°)$

$w_k = 81^{1/4}\left(\cos\dfrac{180° + 360°k}{4} + i\sin\dfrac{180° + 360°k}{4}\right) \quad k = 0, 1, 2, 3$

$w_0 = 3 \text{ cis } \dfrac{180°}{4}$ $\qquad w_1 = 3 \text{ cis } \dfrac{180° + 360°}{4}$ $\qquad w_2 = 2 \text{ cis } \dfrac{0° + 360° \cdot 2}{5}$ $\qquad w_3 = 3 \text{ cis } \dfrac{180° + 360° \cdot 3}{4}$

$w_0 = 3 \text{ cis } 45°$ $\qquad\quad w_1 = 3 \text{ cis } 135°$ $\qquad\qquad w_2 = 2 \text{ cis } 225°$ $\qquad\qquad w_3 = 3 \text{ cis } 315°$

39. $x^4 - (1 - i\sqrt{3}) = 0$

$$x^4 = 1 - i\sqrt{3}$$

Find the four fourth roots of $1 - i\sqrt{3}$.

$$1 - i\sqrt{3} = 2(\cos 300° + i \sin 300°)$$

$$w_k = 2^{1/4}\left(\cos\frac{300° + 360°k}{4} + i\sin\frac{300° + 360°k}{4}\right) \quad k = 0, 1, 2, 3$$

$w_0 = \sqrt[4]{2}\ \text{cis}\ \dfrac{300°}{4}$
$w_1 = \sqrt[4]{2}\ \text{cis}\ \dfrac{300° + 360°}{4}$
$w_2 = \sqrt[4]{2}\ \text{cis}\ \dfrac{300° + 360° \cdot 2}{4}$
$w_3 = \sqrt[4]{2}\ \text{cis}\ \dfrac{300° + 360° \cdot 3}{4}$

$w_0 = \sqrt[4]{2}\ \text{cis}\ 75°$
$w_1 = \sqrt[4]{2}\ \text{cis}\ 165°$
$w_2 = \sqrt[4]{2}\ \text{cis}\ 255°$
$w_3 = \sqrt[4]{2}\ \text{cis}\ 345°$

41. $x^3 + (1 + i\sqrt{3}) = 0$

$$x^3 = -1 - i\sqrt{3}$$

Find the three cube roots of $-1 - i\sqrt{3}$.

$$-1 - i\sqrt{3} = 2(\cos 240° + i \sin 240°)$$

$$w_k = 2^{1/3}\left(\cos\frac{240° + 360°k}{3} + i\sin\frac{240° + 360°k}{3}\right) \quad k = 0, 1, 2$$

$w_0 = \sqrt[3]{2}\ \text{cis}\ \dfrac{240°}{3}$
$w_1 = \sqrt[3]{2}\ \text{cis}\ \dfrac{240° + 360°}{3}$
$w_2 = \sqrt[3]{2}\ \text{cis}\ \dfrac{240° + 360° \cdot 2}{3}$

$w_0 = \sqrt[3]{2}\ \text{cis}\ 80°$
$w_1 = \sqrt[3]{2}\ \text{cis}\ 200°$
$w_2 = \sqrt[3]{2}\ \text{cis}\ 320°$

•   **Connecting Concepts**

43. Let $z = a + bi$. Then $\bar{z} = a - bi$ by definition.

Substitute $a = r\cos\theta$ and $b = r\sin\theta$.

Thus $\bar{z} = r\cos\theta - ri\sin\theta = r(\cos\theta - i\sin\theta)$

45.
$$z = r(\cos\theta + i\sin\theta)$$

$$z^2 = r^2(\cos 2\theta + i\sin 2\theta)$$

$$\frac{1}{z^2} = \frac{1}{r^2(\cos 2\theta + i\sin 2\theta)}$$

$$= \frac{\cos 2\theta - i\sin 2\theta}{r^2(\cos 2\theta + i\sin 2\theta)(\cos 2\theta - i\sin 2\theta)}$$

$$= \frac{\cos 2\theta - i\sin 2\theta}{r^2(\cos^2 2\theta - i^2 \sin^2 2\theta)} = \frac{\cos 2\theta - i\sin 2\theta}{r^2(\cos^2 2\theta + \sin^2 2\theta)}$$

$$z^{-2} = r^{-2}(\cos 2\theta - i\sin 2\theta)$$

47. For $n = 2$, the two square roots of 1 are
1 and -1.
The sum of these roots is
$1 + (-1) = 0$.

For $n = 3$, the three cube roots of 1 are (from exercise 23)
$1, \ -\frac{1}{2} + \frac{\sqrt{3}}{2}i$ and $-\frac{1}{2} - \frac{\sqrt{3}}{2}i$.
The sum of these roots is
$1 - \frac{1}{2} + \frac{\sqrt{3}}{2}i - \frac{1}{2} - \frac{\sqrt{3}}{2}i = 0$.

For $n = 4$, the four fourth roots of 1 are
$1, -1, i,$ and $-i$.
The sum of these roots is
$1 - 1 + i - i = 0$

For $n = 5$, the five fifth roots of 1 are
$1, \ \text{cis } 72°, \ \text{cis } 144°, \ \text{cis } 216°, \ \text{cis } 288°$
The sum of these roots is
$1 + \text{cis } 72° + \text{cis } 144° + \text{cis } 216° + \text{cis } 288° = 0$

For $n = 6$, the six sixth roots of 1 are
$1, -1, \ -\frac{1}{2} + \frac{\sqrt{3}}{2}i, \ -\frac{1}{2} - \frac{\sqrt{3}}{2}i, \ \frac{1}{2} + \frac{\sqrt{3}}{2}i, \ \frac{1}{2} - \frac{\sqrt{3}}{2}i$
The sum of these roots is
$1 - 1 - \frac{1}{2} + \frac{\sqrt{3}}{2}i - \frac{1}{2} - \frac{\sqrt{3}}{2}i + \frac{1}{2} + \frac{\sqrt{3}}{2}i + \frac{1}{2} - \frac{\sqrt{3}}{2}i = 0$

For $n \geq 2$, the sum of the nth roots of 1 is 0.

•   E**xploring** C**oncepts with** T**echnology**

The Mandelbrot Iteration Procedure

1.

$z_5 \approx 0.4001878333$ $z_{10} \approx 0.4353725328$ $z_{100} \approx 0.4906925008$ $z_{200} \approx 0.4951947016$

Answers will vary.

•   A**ssessing** C**oncepts**

1. True   **2.** True   **3.** True   **4.** False

5. The four roots are equally spaced around a circle with center   **6.** 1
(0, 0) and radius 1.

7. 5   **8.** No   **9.** $-3 + 5i$ **10.** 2

1. $3 - \sqrt{-64} = 3 - 8i$ [5.1]
   $3 + 8i$

2. $\sqrt{-4} + 6 = 6 + 2i$ [5.1]
   $6 - 2i$

3. $-2 + \sqrt{-5} = -2 + i\sqrt{5}$ [5.1]
   $-2 - i\sqrt{5}$

4. $-5 + \sqrt{-27} = -5 + 3i\sqrt{3}$ [5.1]
   $-5 - 3i\sqrt{3}$

5. $\left(\sqrt{-4}\right)\left(\sqrt{-4}\right) = (2i)(2i) = 4i^2 = -4$
   [5.1]

6. $\left(-\sqrt{-27}\right)\left(\sqrt{-3}\right) = \left(-3i\sqrt{3}\right)\left(i\sqrt{3}\right)$ [5.1]
   $= -9i^2 = 9$

7. $(3 + 7i) + (2 - 5i) = 5 + 2i$ [5.1]

8. $(3 - 4i) + (-6 + 8i) = -3 + 4i$ [5.1]

9. $(6 - 8i) - (9 - 11i) = -3 + 3i$ [5.1]

10. $(-3 - 5i) - (2 + 10i) = -5 - 15i$ [5.1]

11. $(5 + 3i)(2 - 5i) = 10 - 19i - 15i^2$ [5.1]
    $= 25 - 19i$

12. $(-2 - 3i)(-4 + 7i) = 8 - 2i - 21i^2$ [5.1]
    $= 29 - 2i$

13. $\dfrac{-2i}{3 - 4i} \cdot \dfrac{3 + 4i}{3 + 4i} = \dfrac{-6i - 8i^2}{9 + 16} = \dfrac{8}{25} - \dfrac{6}{25}i$ [5.1]

14. $\dfrac{4 + i}{7 - 2i} \cdot \dfrac{7 + 2i}{7 + 2i} = \dfrac{28 + 15i + 2i^2}{49 + 4} = \dfrac{26}{53} + \dfrac{15}{53}i$ [5.1]

15. $i(2i) - (1 + i)^2 = 2i^2 - 1 - 2i - i^2$ [5.1]
    $= -2 - 2i$

16. $(2 - i)^3 = (4 - 4i + i^2)(2 - i)$ [5.1]
    $= (3 - 4i)(2 - i)$
    $= 6 - 11i + 4i^2$
    $= 2 - 11i$

17. $(3 + \sqrt{-4}) - (-3 - \sqrt{-16}) = 3 + 2i + 3 + 4i = 6 + 6i$ [5.1]

18. $(-2 + \sqrt{-9}) + (-3 - \sqrt{-81}) = -2 + 3i - 3 - 9i = -5 - 6i$ [5.1]

19. $(2 - \sqrt{-3})(2 + \sqrt{-3}) = 4 - (-3) = 7$ [5.1]

20. $(3 - \sqrt{-5})(2 + \sqrt{-5}) = 6 + i\sqrt{5} - (-5)$ [5.1]
    $= 11 + i\sqrt{5}$

21. $i^{27} = i^3 = -i$ [5.1]

22. $i^{105} = i$ [5.1]

23. $\dfrac{i}{i^{17}} = \dfrac{1}{i^{16}} = \dfrac{1}{1} = 1$ [5.1]

24. $i^{62} = i^2 = -1$ [5.1]

25. $|-8i| = \sqrt{(0)^2 + (-8)^2}$ [5.2]
    $= 8$

26. $|2 - 3i| = \sqrt{(2)^2 + (-3)^2}$ [5.2]
    $= \sqrt{4 + 9} = \sqrt{13}$

27. $|-4 + 5i| = \sqrt{(-4)^2 + (5)^2}$ [5.2]
    $= \sqrt{16 + 25} = \sqrt{41}$

28. $|-1 - i| = \sqrt{(-1)^2 + (-1)^2}$ [5.2]
    $= \sqrt{1 + 1} = \sqrt{2}$

29. $r = \sqrt{(2)^2 + (-2)^2}$ [5.2]
    $r = 2\sqrt{2}$

    $\alpha = \tan^{-1}\left|\dfrac{-2}{2}\right|$

    $= \tan^{-1} 1 = 45°$

    $\alpha = 360° - 45° = 315°$

    $z = 2\sqrt{2} \text{ cis } 315°$

30. $r = \sqrt{\left(-\sqrt{3}\right)^2 + (1)^2}$ [5.2]
    $r = 2$

    $\alpha = \tan^{-1}\left|\dfrac{1}{\sqrt{3}}\right|$

    $= \tan^{-1} \dfrac{1}{\sqrt{3}} = 30°$

    $\alpha = 180° - 30° = 150°$

    $z = 2 \text{ cis } 150°$

31.
$$r = \sqrt{(-3)^2 + (2)^2} \quad [5.2]$$
$$r = \sqrt{13}$$

$$\alpha = \tan^{-1}\left|\frac{2}{-3}\right|$$

$$= \tan^{-1}\frac{2}{3} = 33.7°$$

$$\alpha = 180° - 33.7° = 146.3°$$

$$z = \sqrt{13} \text{ cis } 146.3°$$

32.
$$r = \sqrt{(4)^2 + (-1)^2} \quad [5.2]$$
$$r = \sqrt{17}$$

$$\alpha = \tan^{-1}\left|\frac{-1}{4}\right|$$

$$= \tan^{-1}\frac{1}{4} = 14.04°$$

$$\alpha = 360° - 14.04° = 345.96°$$

$$z = \sqrt{17} \text{ cis } 345.96°$$

33.
$$z = 5(\cos 315° + i \sin 315°) \quad [5.2]$$
$$z = 5\left(\frac{\sqrt{2}}{2} - \frac{\sqrt{2}}{2}i\right)$$
$$z = \frac{5\sqrt{2}}{2} - \frac{5\sqrt{2}}{2}i$$
$$\approx 3.536 - 3.536i$$

34.
$$z = 6\left(\cos\frac{4\pi}{3} + i \sin\frac{4\pi}{3}\right) \quad [5.2]$$
$$z = 6\left(-\frac{1}{2} - \frac{\sqrt{3}}{2}i\right)$$
$$z = -3 - 3\sqrt{3}i$$

35.
$$z = 2(\cos 2 + i \sin 2) \quad [5.2]$$
$$z = 2(-0.4161 + 0.9093i)$$
$$z \approx -0.832 + 1.819i$$

36.
$$z = 3(\cos 115° + i \sin 115°) \quad [5.2]$$
$$z \approx 3(-0.4226 + 0.9063i)$$
$$z \approx -1.27 + 2.72i$$

37.
$$z_1 z_2 = 3(\cos 225° + i \sin 225°) \cdot 10(\cos 45° + i \sin 45°) \quad [5.2]$$
$$z_1 z_2 = 30[\cos(225° + 45°) + i \sin(225° + 45°)]$$
$$z_1 z_2 = 30(\cos 270° + i \sin 270°)$$
$$z_1 z_2 = 30(0 - 1i)$$
$$z_1 z_2 = -30i$$

38.
$$z_1 z_2 = 5(\cos 162° + i \sin 162°) \cdot 2(\cos 63° + i \sin 63°) \quad [5.2]$$
$$z_1 z_2 = 10[\cos(162° + 63°) + i \sin(162° + 63°)]$$
$$z_1 z_2 = 10(\cos 225° + i \sin 225°)$$
$$z_1 z_2 = 10\left(-\frac{\sqrt{2}}{2} - \frac{\sqrt{2}}{2}i\right)$$
$$z_1 z_2 = -5\sqrt{2} - 5i\sqrt{2}$$

39.
$$z_1 z_2 = 3(\cos 12° + i \sin 12°) \cdot 4(\cos 126° + i \sin 126°) \quad [5.2]$$
$$z_1 z_2 = 12[\cos(12° + 126°) + i \sin(12° + 126°)]$$
$$z_1 z_2 = 12(\cos 138° + i \sin 138°)$$
$$z_1 z_2 \approx 12(-0.74314 + 0.66913i)$$
$$z_1 z_2 \approx -8.918 + 8.030i$$

40.
$$z_1 z_2 = (\cos 23° + i \sin 23°) \cdot 4(\cos 233° + i \sin 233°) \quad [5.2]$$
$$z_1 z_2 = 4[\cos(23° + 233°) + i \sin(23° + 233°)]$$
$$z_1 z_2 = 4(\cos 256° + i \sin 256°)$$
$$z_1 z_2 \approx 4(-0.24192 - 0.97030i)$$
$$z_1 z_2 \approx -0.968 - 3.881i$$

41.
$$z_1 z_2 = 3(\cos 1.8 + i \sin 1.8) \cdot 5(\cos 2.5 + i \sin 2.5) \quad [5.2]$$
$$z_1 z_2 = 15[\cos(1.8 + 2.5) + i \sin(1.8 + 2.5)]$$
$$z_1 z_2 = 15(\cos 4.3 + i \sin 4.3)$$
$$z_1 z_2 = 15(-0.4008 - 0.9162i)$$
$$z_1 z_2 \approx -6.012 - 13.743i$$

42.
$$z_1 z_2 = 6(\cos 3.1 + i \sin 1.8) \cdot 5(\cos 4.3 + i \sin 4.3) \quad [5.2]$$
$$z_1 z_2 = 30[\cos(3.1 + 4.3) + i \sin(3.1 + 4.3)]$$
$$z_1 z_2 = 30(\cos 7.4 + i \sin 7.4)$$
$$z_1 z_2 \approx 30(0.439 + 0.899i)$$
$$z_1 z_2 \approx 13.2 + 27.0i$$

43.
$$\frac{z_1}{z_2} = \frac{6(\cos 50° + i \sin 50°)}{2(\cos 150° + i \sin 150°)} \quad [5.2]$$

$$\frac{z_1}{z_2} = 3[\cos(50° - 150°) + i \sin(50° - 150°)]$$

$$\frac{z_1}{z_2} = 3(\cos -100° + i \sin -100°) = 3\text{cis}(-100°)$$

44.
$$\frac{z_1}{z_2} = \frac{30(\cos 165° + i \sin 165°)}{10(\cos 55° + i \sin 55°)} \quad [5.2]$$

$$\frac{z_1}{z_2} = 3[\cos(165° - 55°) + i \sin(165° - 55°)]$$

$$\frac{z_1}{z_2} = 3(\cos 110° + i \sin 110°) = 3\text{cis}(110°)$$

45.
$$\frac{z_1}{z_2} = \frac{40(\cos 66° + i \sin 66°)}{8(\cos 125° + i \sin 125°)} \quad [5.2]$$

$$\frac{z_1}{z_2} = 5 \left[\cos(66° - 125°) + i \sin(66° - 125°) \right]$$

$$\frac{z_1}{z_2} = 5(\cos -59° + i \sin -59°) = 5 \text{cis} (-59°)$$

46.
$$\frac{z_1}{z_2} = \frac{2(\cos 150° + i \sin 150°)}{\sqrt{2}(\cos 200° + i \sin 200°)} \quad [5.2]$$

$$\frac{z_1}{z_2} = \sqrt{2} \left[\cos(150° - 200°) + i \sin(150° - 200°) \right]$$

$$\frac{z_1}{z_2} = \sqrt{2}(\cos -50° + i \sin -50°) = \sqrt{2} \text{cis} (-50°)$$

47.
$$\frac{z_1}{z_2} = \frac{10(\cos 3.7 + i \sin 3.7)}{6(\cos 1.8 + i \sin 1.8)} \quad [5.2]$$

$$\frac{z_1}{z_2} = \frac{5}{3} \left[\cos(3.7 - 1.8) + i \sin(3.7 - 1.8) \right]$$

$$\frac{z_1}{z_2} = \frac{5}{3}(\cos 1.9 + i \sin 1.9) = \frac{5}{3} \text{cis} (1.9)$$

48.
$$\frac{z_1}{z_2} = \frac{4(\cos 1.2 + i \sin 1.2)}{8(\cos 5.2 + i \sin 5.2)} \quad [5.2]$$

$$\frac{z_1}{z_2} = \frac{1}{2} \left[\cos(1.2 - 5.2) + i \sin(1.2 - 5.2) \right]$$

$$\frac{z_1}{z_2} = \frac{1}{2}(\cos -4 + i \sin -4) = \frac{1}{2} \text{cis} (-4)$$

49.
$$[3(\cos 45° + i \sin 45°)]^5 = 3^5[\cos(5 \cdot 45°) + i \sin(5 \cdot 45°)] \quad [5.3]$$
$$= 243[\cos 225° + i \sin 225°]$$
$$= 243 \left(-\frac{\sqrt{2}}{2} - \frac{\sqrt{2}}{2} i \right)$$
$$= -\frac{243\sqrt{2}}{2} - \frac{243\sqrt{2}}{2} i$$
$$\approx -171.827 - 171.827i$$

50.
$$\left[\cos \frac{11\pi}{8} + i \sin \frac{11\pi}{8} \right]^8 = \cos \left(8 \cdot \frac{11\pi}{8} \right) + i \sin \left(8 \cdot \frac{11\pi}{8} \right) \quad [5.3]$$
$$= \cos(11\pi) + i \sin(11\pi)$$
$$= \cos \pi + i \sin \pi$$
$$= -1 + 0i$$
$$= -1$$

51.
$$z = 1 - i\sqrt{3} \quad [5.3]$$

$$r = \sqrt{1^2 + (-\sqrt{3})^2} \qquad \alpha = \tan^{-1} \left| \frac{-\sqrt{3}}{1} \right| = 60°$$
$$r = 2 \qquad\qquad \theta = 300°$$

$$z = 2(\cos 300° + i \sin 300°)$$
$$(1 - i\sqrt{3})^7 = [2(\cos 300° + i \sin 300°)]^7$$
$$= 128[\cos(7 \cdot 300°) + i \sin(7 \cdot 300°)]$$
$$= 128(\cos 2100° + i \sin 2100°)$$
$$= 128(\cos 300° + i \sin 300°)$$
$$= 64 - 64i\sqrt{3}$$
$$\approx 64 - 110.851i$$

52.
$$z = -2 - 2i \quad [5.3]$$

$$r = \sqrt{(-2)^2 + (-2)^2} \qquad \alpha = \tan^{-1} \left| \frac{-2}{-2} \right| = 45°$$
$$r = 2\sqrt{2} \qquad\qquad \theta = 225°$$

$$z = 2\sqrt{2}(\cos 225° + i \sin 225°)$$
$$(-2 - 2i)^{10} = [2\sqrt{2}(\cos 225° + i \sin 225°)]^{10}$$
$$= 32,768[\cos(10 \cdot 225°) + i \sin(10 \cdot 225°)]$$
$$= 32,768(\cos 2250° + i \sin 2250°)$$
$$= 32,768(\cos 90° + i \sin 90°)$$
$$= 0 + 32,768i$$

53.
$$z = \sqrt{2} - i\sqrt{2} \quad [5.3]$$

$$r = \sqrt{(\sqrt{2})^2 + (-\sqrt{2})^2} \qquad \alpha = \tan^{-1} \left| \frac{-\sqrt{2}}{\sqrt{2}} \right| = 45°$$
$$r = 2 \qquad\qquad \theta = 315°$$

$$z = 2(\cos 315° + i \sin 315°)$$
$$(\sqrt{2} - i\sqrt{2})^5 = [2(\cos 315° + i \sin 315°)]^5$$
$$= 32[\cos(5 \cdot 315°) + i \sin(5 \cdot 315°)]$$
$$= 32(\cos 1575° + i \sin 1575°)$$
$$= 32(\cos 135° + i \sin 135°)$$
$$= -16\sqrt{2} + 16i\sqrt{2}$$
$$\approx -22.627 + 22.627i$$

54.
$$z = 3 - 4i \quad [5.3]$$

$$r = \sqrt{3^2 + (-4)^2} \qquad \alpha = \tan^{-1} \left| \frac{-4}{3} \right| = 53.13°$$
$$r = 5 \qquad\qquad \theta = 306.87°$$

$$z = 5(\cos 306.87° + i \sin 306.87°)$$
$$(3 - 4i)^5 = [5(\cos 306.87° + i \sin 306.87°)]^5$$
$$= 3125[\cos(5 \cdot 306.87°) + i \sin(5 \cdot 306.87°)]$$
$$= 3125(\cos 1534.35° + i \sin 1534.35°)$$
$$= 3125(\cos 94.35° + i \sin 94.35°)$$
$$= -237 + 3116i$$

55. $27i = 27\left(\cos 90° + i\sin 90°\right)$ [5.3]

$$w_k = 27^{1/3}\left(\cos\frac{90° + 360°k}{3} + i\sin\frac{90° + 360°k}{3}\right) \quad k = 0, 1, 2$$

$$w_0 = 3\left(\cos\frac{90°}{3} + i\sin\frac{90°}{3}\right) \qquad w_1 = 3\left(\cos\frac{90° + 360°}{3} + i\sin\frac{90° + 360°}{3}\right) \qquad w_2 = 3\left(\cos\frac{90° + 360°\cdot 2}{3} + i\sin\frac{90° + 360°\cdot 2}{3}\right)$$

$w_0 = 3(\cos 30° + i\sin 30°)$ $\qquad\quad$ $w_1 = 3(\cos 150° + i\sin 150°)$ $\qquad\qquad$ $w_2 = 3(\cos 270° + i\sin 270°)$

$w_0 = 3 \text{ cis } 30°$ $\qquad\qquad\qquad$ $w_1 = 3 \text{ cis } 150°$ $\qquad\qquad\qquad\qquad$ $w_2 = 3 \text{ cis } 270°$

56. $8i = 8\left(\cos 90° + i\sin 90°\right)$ [5.3]

$$w_k = 8^{1/4}\left(\cos\frac{90° + 360°k}{4} + i\sin\frac{9° + 360°k}{4}\right) \quad k = 0, 1, 2, 3$$

$$w_0 = \sqrt[4]{8}\left(\cos\frac{90°}{4} + i\sin\frac{90°}{4}\right) \qquad\qquad w_1 = \sqrt[4]{8}\left(\cos\frac{90° + 360°}{4} + i\sin\frac{90° + 360°}{4}\right)$$

$w_0 = \sqrt[4]{8}(\cos 22.5° + i\sin 22.5°)$ $\qquad\qquad$ $w_1 = \sqrt[4]{8}(\cos 112.5° + i\sin 112.5°)$

$w_0 = \sqrt[4]{8} \text{ cis } 22.5°$ $\qquad\qquad\qquad\qquad$ $w_1 = \sqrt[4]{8} \text{ cis } 112.5°$

$$w_2 = \sqrt[4]{8}\left(\cos\frac{90° + 360°\cdot 2}{4} + i\sin\frac{90° + 360°\cdot 2}{4}\right) \qquad w_3 = \sqrt[4]{8}\left(\cos\frac{90° + 360°\cdot 3}{4} + i\sin\frac{90° + 360°\cdot 3}{4}\right)$$

$w_2 = \sqrt[4]{8}(\cos 202.5° + i\sin 202.5°)$ $\qquad\qquad$ $w_3 = \sqrt[4]{8}(\cos 292.5° + i\sin 292.5°)$

$w_2 = \sqrt[4]{8} \text{ cis } 202.5°$ $\qquad\qquad\qquad\qquad$ $w_3 = \sqrt[4]{8} \text{ cis } 292.5°$

57. $256 = 256\left(\cos 0° + i\sin 0°\right)$ [5.3]

$$w_k = 81^{1/4}\left(\cos\frac{0° + 360°k}{4} + i\sin\frac{0° + 360°k}{4}\right) \quad k = 0, 1, 2, 3$$

$w_0 = 3(\cos 0° + i\sin 0°)$ $\qquad\qquad\qquad\qquad$ $w_1 = 3\left(\cos\dfrac{0° + 360°}{4} + i\sin\dfrac{0° + 360°}{4}\right)$

$w_0 = 3 \text{ cis } 0°$ $\qquad\qquad\qquad\qquad\qquad\qquad$ $w_1 = 3(\cos 90° + i\sin 90°)$

$\qquad\qquad\qquad\qquad\qquad\qquad\qquad\qquad$ $w_1 = 3 \text{ cis } 90°$

$w_2 = 3\left(\cos\dfrac{0° + 360°\cdot 2}{4} + i\sin\dfrac{0° + 360°\cdot 2}{4}\right)$ $\qquad$ $w_3 = 3\left(\cos\dfrac{0° + 360°\cdot 3}{4} + i\sin\dfrac{0° + 360°\cdot 3}{4}\right)$

$w_2 = 3(\cos 180° + i\sin 180°)$ $\qquad\qquad\qquad$ $w_3 = 3(\cos 270° + i\sin 270°)$

$w_2 = 3 \text{ cis } 180°$ $\qquad\qquad\qquad\qquad\qquad$ $w_3 = 3 \text{ cis } 270°$

58. $-16\sqrt{2}-16i\sqrt{2}=32(\cos 225°+i\sin 225°)$ [5.3]

$$w_k = 32^{1/5}\left(\cos\frac{225°+360°k}{5}+i\sin\frac{225°+360°k}{5}\right) \quad k=0,1,2,3,4$$

$w_0 = 2\left(\cos\frac{225°}{5}+i\sin\frac{225°}{5}\right)$

$w_0 = 2(\cos 45°+i\sin 45°)$

$w_0 = 2$ cis $45°$

$w_1 = 2\left(\cos\frac{225°+360°}{5}+i\sin\frac{225°+360°}{5}\right)$

$w_1 = 2(\cos 117°+i\sin 117°)$

$w_1 = 2$ cis $117°$

$w_2 = 2\left(\cos\frac{225°+360°\cdot 2}{5}+i\sin\frac{225°+360°\cdot 2}{5}\right)$

$w_2 = 2(\cos 189°+i\sin 189°)$

$w_2 = 2$ cis $189°$

$w_3 = 2\left(\cos\frac{225°+360°\cdot 3}{5}+i\sin\frac{225°+360°\cdot 3}{5}\right)$

$w_3 = 2(\cos 261°+i\sin 261°)$

$w_3 = 2$ cis $261°$

$w_4 = 2\left(\cos\frac{225°+360°\cdot 4}{5}+i\sin\frac{225°+360°\cdot 4}{5}\right)$

$w_4 = 2(\cos 333°+i\sin 333°)$

$w_4 = 2$ cis $333°$

59. $81 = 81\left(\cos 0°+i\sin 0°\right)$ [5.3]

$$w_k = 256^{1/4}\left(\cos\frac{0°+360°k}{4}+i\sin\frac{0°+360°k}{4}\right) \quad k=0,1,2,3$$

$w_0 = 4(\cos 0°+i\sin 0°)$

$w_0 = 4$ cis $0°$

$w_1 = 4\left(\cos\frac{0°+360°}{4}+i\sin\frac{0°+360°}{4}\right)$

$w_1 = 4(\cos 90°+i\sin 90°)$

$w_1 = 4$ cis $90°$

$w_2 = 4\left(\cos\frac{0°+360°\cdot 2}{4}+i\sin\frac{0°+360°\cdot 2}{4}\right)$

$w_2 = 4(\cos 180°+i\sin 180°)$

$w_2 = 4$ cis $180°$

$w_3 = 4\left(\cos\frac{0°+360°\cdot 3}{4}+i\sin\frac{0°+360°\cdot 3}{4}\right)$

$w_3 = 4(\cos 270°+i\sin 270°)$

$w_3 = 4$ cis $270°$

60. $-125 = 125\left(\cos 180°+i\sin 180°\right)$ [5.3]

$$w_k = 125^{1/3}\left(\cos\frac{180°+360°k}{3}+i\sin\frac{180°+360°k}{3}\right) \quad k=0,1,2$$

$w_0 = 5\left(\cos\frac{180°}{3}+i\sin\frac{180°}{3}\right)$

$w_0 = 5(\cos 60°+i\sin 60°)$

$w_0 = 5$ cis $60°$

$w_1 = 5\left(\cos\frac{180°+360°}{3}+i\sin\frac{180°+360°}{3}\right)$

$w_1 = 5(\cos 180°+i\sin 180°)$

$w_1 = 5$ cis $180° = -5$

$w_2 = 5\left(\cos\frac{180°+360°\cdot 2}{3}+i\sin\frac{180°+360°\cdot 2}{3}\right)$

$w_2 = 5(\cos 300°+i\sin 300°)$

$w_2 = 5$ cis $300°$

••

Quantitative Reasoning

QR1.

QR3. -2 is an element of the Mendelbrot set because all of its iterates equal 2.

QR5. $(z^2+s) \to z$

Chapter Test

1. $6 + \sqrt{-9} = 6 + 3i$ [5.1]

2. $\sqrt{-18} = 3i\sqrt{2}$ [5.1]

3. $(3 + \sqrt{-4}) + (7 - \sqrt{-9}) = 3 + 2i + 7 - 3i$
$$= 10 - i$$
[5.1]

4. $(-1 + \sqrt{-25}) - (8 - \sqrt{-16}) = -1 + 5i - 8 + 4i$ [5.1]
$$= -9 + 9i$$

5. $(\sqrt{-12})(\sqrt{-3}) = (2i\sqrt{3})(i\sqrt{3})$ [5.1]
$$= 6i^2 = -6$$

6. $i^{263} = i^3 = -i$ [5.1]

7. $(3 + 7i) - (-2 - 9i) = 5 + 16i$ [5.1]

8. $(-6 - 9i)(4 + 3i) = -24 - 54i - 27i^2$
$$= 3 - 54i$$
[5.1]

9. $(3 - 5i)(-3 + 5i) = -9 + 30i - 25i^2$ [5.1]
$$= 16 + 30i$$

10. $\dfrac{4 - 5i}{i} \cdot \dfrac{-i}{-i} = \dfrac{-4i + 5i^2}{-i^2} = -5 - 4i$ [5.1]

11. $\dfrac{2 - 7i}{4 + 3i} \cdot \dfrac{4 - 3i}{4 - 3i} = \dfrac{8 - 34i + 21i^2}{16 + 9} = -\dfrac{13}{25} - \dfrac{34}{25}i$ [5.1]

12. $\dfrac{6 + 2i}{1 - i} \cdot \dfrac{1 + i}{1 + i} = \dfrac{6 + 8i + 2i^2}{1 + 1} = \dfrac{4}{2} + \dfrac{8}{2}i = 2 + 4i$ [5.1]

13. $|z| = \sqrt{(3)^2 + (-5)^2}$ [5.2]
$$= \sqrt{9 + 25} = \sqrt{34}$$

14. $r = \sqrt{(3)^2 + (-3)^2}$ [5.2]
$r = 3\sqrt{2}$

$\alpha = \tan^{-1}\left|\dfrac{-3}{3}\right|$

$= \tan^{-1} 1 = 45°$

$\alpha = 360° - 45° = 315°$

$z = 3\sqrt{2} \text{ cis } 315°$

15. $r = \sqrt{(0)^2 + (-6)^2}$ [5.2]
$r = 6$

$\alpha = \tan^{-1}\left|\dfrac{-6}{0}\right| = 90°$

$\alpha = 360° - 90° = 270°$

$z = 6 \text{ cis } 270°$

16. $z = 4(\cos 120° + i \sin 120°)$ [5.2]
$z = 4\left(-\dfrac{1}{2} + \dfrac{\sqrt{3}}{2}i\right)$
$z = -2 + 2i\sqrt{3}$

17. $z = 5(\cos 225° + i \sin 225°)$ [5.2]
$z = 5\left(-\dfrac{\sqrt{2}}{2} - \dfrac{\sqrt{2}}{2}i\right)$
$z = -\dfrac{5\sqrt{2}}{2} - \dfrac{5\sqrt{2}}{2}i$

18. $z_1 z_2 = 3(\cos 28° + i \sin 28°) \cdot 4(\cos 17° + i \sin 17°)$ [5.2]
$z_1 z_2 = 12[\cos(28° + 17°) + i \sin(28° + 17°)]$
$z_1 z_2 = 12(\cos 45° + i \sin 45°)$
$z_1 z_2 = 12\left(\dfrac{\sqrt{2}}{2} + \dfrac{\sqrt{2}}{2}i\right)$
$z_1 z_2 = 6\sqrt{2} + 6i\sqrt{2}$

19. $z_1 z_2 = 5(\cos 115° + i \sin 115°) \cdot 4(\cos 10° + i \sin 10°)$ [5.2]
$z_1 z_2 = 20[\cos(115° + 10°) + i \sin(115° + 10°)]$
$z_1 z_2 = 20(\cos 125° + i \sin 125°)$
$z_1 z_2 \approx 20(-0.5736 + 0.81915i)$
$z_1 z_2 \approx -11.472 + 16.383i$

20. $\dfrac{z_1}{z_2} = \dfrac{24(\cos 258° + i \sin 258°)}{6(\cos 78° + i \sin 78°)}$ [5.2]

$\dfrac{z_1}{z_2} = 4[\cos(258° - 78°) + i \sin(258° - 78°)]$

$\dfrac{z_1}{z_2} = 4(\cos 180° + i \sin 180°) = -4 + 0i = -4$

21. $\dfrac{z_1}{z_2} = \dfrac{18(\cos 50° + i \sin 50°)}{3(\cos 140° + i \sin 140°)}$ [5.2]

$\dfrac{z_1}{z_2} = 6[\cos(50° - 140°) + i \sin(50° - 140°)]$

$\dfrac{z_1}{z_2} = 6(\cos -90° + i \sin -90°) = 0 - 6i = -6i$

22. $z = 2 - 2i\sqrt{3}$ [5.3]

$r = \sqrt{2^2 + (-2\sqrt{3})^2}$ $\qquad$ $\alpha = \tan^{-1}\left|\dfrac{-2\sqrt{3}}{2}\right| = 60°$

$r = 4$

$\qquad\qquad\qquad\qquad\qquad$ $\theta = 300°$

$z = 4(\cos 300° + i\sin 300°)$

$(2 - 3i\sqrt{3})^{12} = [4(\cos 300° + i\sin 300°)]^{12}$

$\qquad\qquad = 16,777,216[\cos(12 \cdot 300°) + i\sin(10 \cdot 300°)]$

$\qquad\qquad = 16,777,216(\cos 3600° + i\sin 3600°)$

$\qquad\qquad = 16,777,216(\cos 0° + i\sin 0°)$

$\qquad\qquad = 16,777,216 + 0i$

23. $64 = 64(\cos 0° + i\sin 0°)$ [5.3]

$w_k = 64^{1/6}\left(\cos\dfrac{0° + 360°k}{6} + i\sin\dfrac{0° + 360°k}{6}\right)$ $\quad$ $k = 0, 1, 2, 3, 4, 5$

$w_0 = 2(\cos 0° + i\sin 0°)$ $\qquad\qquad\qquad\qquad$ $w_1 = 2\left(\cos\dfrac{0 + 360°}{6} + i\sin\dfrac{0 + 360°}{6}\right)$

$w_0 = 2 \text{ cis } 0°$ $\qquad\qquad\qquad\qquad\qquad\qquad$ $w_1 = 2(\cos 60° + i\sin 60°)$

$\qquad\qquad\qquad\qquad\qquad\qquad\qquad\qquad$ $w_1 = 2 \text{ cis } 60°$

$w_2 = 2\left(\cos\dfrac{0° + 360° \cdot 2}{6} + i\sin\dfrac{0° + 360° \cdot 2}{6}\right)$ $\qquad$ $w_3 = 2\left(\cos\dfrac{0° + 360° \cdot 3}{6} + i\sin\dfrac{0° + 360° \cdot 3}{6}\right)$

$w_2 = 2(\cos 120° + i\sin 120°)$ $\qquad\qquad\qquad\qquad$ $w_3 = 2(\cos 180° + i\sin 180°)$

$w_2 = 2 \text{ cis } 120°$ $\qquad\qquad\qquad\qquad\qquad\qquad$ $w_3 = 2 \text{ cis } 180°$

$w_4 = 2\left(\cos\dfrac{0° + 360° \cdot 4}{6} + i\sin\dfrac{0° + 360° \cdot 4}{6}\right)$ $\qquad$ $w_5 = 2\left(\cos\dfrac{0° + 360° \cdot 5}{6} + i\sin\dfrac{0° + 360° \cdot 5}{6}\right)$

$w_4 = 2(\cos 240° + i\sin 240°)$ $\qquad\qquad\qquad\qquad$ $w_5 = 2(\cos 300° + i\sin 300°)$

$w_4 = 2 \text{ cis } 240°$ $\qquad\qquad\qquad\qquad\qquad\qquad$ $w_5 = 2 \text{ cis } 300°$

24. $-1 + \sqrt{3}i = 2(\cos 120° + i\sin 120°)$ [5.3]

$\qquad w_k = 2^{1/3}\left(\cos\dfrac{120° + 360°k}{3} + i\sin\dfrac{120° + 360°k}{3}\right)$ $\quad$ $k = 0, 1, 2$

$w_0 = \sqrt[3]{2}\left(\cos\dfrac{120°}{3} + i\sin\dfrac{120°}{3}\right)$ $\quad$ $w_1 = \sqrt[3]{2}\left(\cos\dfrac{120° + 360°}{3} + i\sin\dfrac{120° + 360°}{3}\right)$ $\quad$ $w_2 = \sqrt[3]{2}\left(\cos\dfrac{120° + 360° \cdot 2}{3} + i\sin\dfrac{120° + 360° \cdot 2}{3}\right)$

$w_0 = \sqrt[3]{2}(\cos 40° + i\sin 40°)$ $\qquad\quad$ $w_1 = \sqrt[3]{2}(\cos 160° + i\sin 160°)$ $\qquad\quad$ $w_2 = \sqrt[3]{2}(\cos 280° + i\sin 280°)$

$w_0 = \sqrt[3]{2} \text{ cis } 40°$ $\qquad\qquad\qquad$ $w_1 = \sqrt[3]{2} \text{ cis } 160°$ $\qquad\qquad\qquad$ $w_2 = \sqrt[3]{2} \text{ cis } 280°$

25. $x^5 + 32 = 0$ [5.3]

$\qquad x^5 = -32$

Find the five fifth roots of -32.

$-32 = 32(\cos 180° + i\sin 180°)$

$w_k = 32^{1/5}\left(\cos\dfrac{180° + 360°k}{5} + i\sin\dfrac{180° + 360°k}{5}\right)$ $\quad$ $k = 0, 1, 2, 3, 4$

$w_0 = 2 \text{ cis } \dfrac{180°}{5}$ $\qquad\qquad\quad$ $w_1 = 2 \text{ cis } \dfrac{180° + 360°}{5}$ $\qquad\qquad\quad$ $w_2 = 2 \text{ cis } \dfrac{180° + 360° \cdot 2}{5}$

$w_0 = 2 \text{ cis } 36°$ $\qquad\qquad\qquad$ $w_1 = 2 \text{ cis } 108°$ $\qquad\qquad\qquad\qquad$ $w_2 = 2 \text{ cis } 180°$

$w_3 = 2 \text{ cis } \dfrac{180° + 360° \cdot 3}{5}$ $\qquad\qquad$ $w_4 = 2 \text{ cis } \dfrac{180° + 360° \cdot 4}{5}$

$w_3 = 2 \text{ cis } 252°$ $\qquad\qquad\qquad\qquad$ $w_4 = 2 \text{ cis } 324°$

1.
$$x^2 - x - 6 \le 0 \quad [1.1$$
$$(x-3)(x+2) \le 0$$

The product is negative or zero.
The critical values are –2 and 3.

$(x-3)(x+2)$

$$[-2, 3]$$

2.
$$x^2 - 4 = 0$$
$$(x-2)(x+2) = 0$$
$$x = -2, \ 2$$

Domain: all real numbers except –2 and 2.

3.
$$f(c) = 2 \quad [1.3]$$
$$2 = \frac{c}{c+1}$$
$$2(c+1) = c$$
$$2c+2 = c$$
$$c = -2$$

4.
$$(f \circ g)(x) = f[g(x)] \quad [1.5]$$
$$= f\left[\frac{x^2-1}{3}\right]$$
$$= \sin\left(3 \cdot \frac{x^2-1}{3}\right)$$
$$= \sin(x^2-1)$$

5.
$$f(x) = \frac{x}{x-1} \quad [1.6]$$
$$y = \frac{x}{x-1}$$
$$x = \frac{y}{y-1}$$
$$x(y-1) = y$$
$$xy - x = y$$
$$xy - y = x$$
$$y(x-1) = x$$
$$y = \frac{x}{x-1}$$
$$f^{-1}(x) = \frac{x}{x-1}$$
$$f^{-1}(3) = \frac{3}{3-1} = \frac{3}{2}$$

6.
$$\frac{3\pi}{2}\left(\frac{180°}{\pi}\right) = 270° \quad [2.1]$$

7.
$$\cos 38° = \frac{20}{c} \quad [2.2]$$
$$c = \frac{20}{\cos 38°} \approx 25.4 \text{ cm}$$

8.
$$-1 \le \sin t \le 1 \quad [2.4]$$
$$a = -1, \ b = 1$$

9. $y = 3\sin\pi x \quad [2.5]$

10. $y = \frac{1}{2}\tan\frac{\pi x}{4} \quad [2.6]$

11.
$$\frac{\cos x}{1+\sin x} = \frac{\cos x}{1+\sin x} \cdot \frac{1-\sin x}{1-\sin x} \quad [3.1]$$
$$= \frac{\cos x - \sin x \cos x}{1 - \sin^2 x}$$
$$= \frac{\cos x - \sin x \cos x}{\cos^2 x}$$
$$= \frac{1}{\cos x} - \frac{\sin x}{\cos x}$$
$$= \sec x - \tan x$$

12.
$$\sin 2x \cos 3x - \cos 2x \sin 3x = \sin(2x - 3x) \quad [3.2]$$
$$= \sin(-x) \text{ or } -\sin x$$

13. $\sin\alpha = \dfrac{4}{5}, \ \cos\alpha = \dfrac{3}{5}$ [3.2]

$\cos\beta = \dfrac{12}{13}, \ \sin\beta = -\dfrac{5}{13}$

$\cos(\alpha + \beta) = \cos\alpha\cos\beta - \sin\alpha\sin\beta$

$$= \left(\dfrac{3}{5}\right)\left(\dfrac{12}{13}\right) - \left(\dfrac{4}{5}\right)\left(-\dfrac{5}{13}\right)$$

$$= -\dfrac{48}{65} + \dfrac{15}{65} = \dfrac{56}{65}$$

14. $y = \sin\left[\sin^{-1}\left(\dfrac{3}{5}\right) + \cos^{-1}\left(-\dfrac{5}{13}\right)\right]$ [3.5]

Let $\alpha = \sin^{-1}\dfrac{3}{5}, \ \sin\alpha = \dfrac{3}{5}, \ \cos\alpha = \sqrt{1 - \left(\dfrac{3}{5}\right)^2} = \dfrac{4}{5}$.

$\beta = \cos^{-1}\left(-\dfrac{5}{13}\right), \ \cos\beta = -\dfrac{5}{13}, \ \sin\beta = \sqrt{1 - \left(-\dfrac{5}{13}\right)^2} = \dfrac{12}{13}$.

$y = \sin(\alpha + \beta)$

  $= \sin\alpha\cos\beta + \cos\alpha\sin\beta$

$$= \dfrac{3}{5}\left(-\dfrac{5}{13}\right) + \dfrac{4}{5}\left(\dfrac{12}{13}\right)$$

$$= -\dfrac{15}{65} + \dfrac{48}{65} = \dfrac{33}{65}$$

15. $\sin 2x = \sqrt{3}\sin x$ [3.6]

$2\sin x\cos x - \sqrt{3}\sin x = 0$

$\sin x(2\cos x - \sqrt{3}) = 0$

$\sin x = 0 \qquad\qquad 2\cos x - \sqrt{3} = 0$

$x = 0, \ \pi \qquad\qquad\qquad \cos x = \dfrac{\sqrt{3}}{2}$

$\qquad\qquad\qquad\qquad\qquad x = \dfrac{\pi}{6}, \ \dfrac{11\pi}{6}$

The solutions are $0, \ \dfrac{\pi}{6}, \ \pi, \ \dfrac{11\pi}{6}$.

16. $c^2 = a^2 + b^2 - 2ab\cos C$ [4.2]

$\quad = (140)^2 + (130)^2 - 2(140)(130)\cos 78°$

$\quad = 36{,}500 - 36{,}400\cos 78°$

$\quad \approx 28{,}932$

$c \approx 170$ cm

17. $\cos\theta = \dfrac{\mathbf{v}\cdot\mathbf{w}}{\|\mathbf{v}\|\ \|\mathbf{w}\|}$ [4.3]

$\cos\theta = \dfrac{(3\mathbf{i} + 2\mathbf{j})\cdot(5\mathbf{i} - 3\mathbf{j})}{\sqrt{3^2 + 2^2}\sqrt{5^2 + (-3)^2}}$

$\cos\theta = \dfrac{3(5) + (2)(-3)}{\sqrt{13}\sqrt{34}}$

$\cos\theta = \dfrac{9}{\sqrt{442}} = 0$

$\theta = 64.7°$

18. $W = \mathbf{F}\cdot\mathbf{s}$ [4.3]

$W = \|\mathbf{F}\|\ \|\mathbf{s}\|\cos\alpha$

$W = (100)(15)(\cos 15°)$

$W \approx 1449$ foot-pounds

19. $r = \sqrt{(2)^2 + (2)^2}$ [5.2]

$r = 2\sqrt{2}$

$\alpha = \tan^{-1}\left|\dfrac{2}{2}\right|$

$\quad = \tan^{-1} 1 = 45°$

$z = 2\sqrt{2} \ \text{cis} \ 45°$

20. $-27 = 27(\cos 180° + i\sin 180°)$ [5.3]

$w_k = 27^{1/3}\left(\cos\dfrac{180° + 360°k}{3} + i\sin\dfrac{180° + 360°k}{3}\right) \qquad k = 0, 1, 2$

$w_0 = 3\left(\cos\dfrac{180°}{3} + i\sin\dfrac{180°}{3}\right)$

$w_0 = 3(\cos 60° + i\sin 60°)$

$w_0 = 3\left(\dfrac{1}{2} + \dfrac{\sqrt{3}}{2}i\right) = \dfrac{3}{2} + \dfrac{3\sqrt{3}}{2}$

$w_1 = 3\left(\cos\dfrac{180° + 360°}{3} + i\sin\dfrac{180° + 360°}{3}\right)$

$w_1 = 3(\cos 180° + i\sin 180°)$

$w_1 = 3(-1 + 0i) = -3$

$w_2 = 3\left(\cos\dfrac{180° + 360°\cdot 2}{3} + i\sin\dfrac{180° + 360°\cdot 2}{3}\right)$

$w_2 = 3(\cos 300° + i\sin 300°)$

$w_2 = 3\left(\dfrac{1}{2} - \dfrac{\sqrt{3}}{2}i\right) = \dfrac{3}{2} - \dfrac{3\sqrt{3}}{2}i$

Chapter 6
Topics in Analytic Geometry

Section 6.1

1.   **a.** iii   **b.** i   **c.** iv   **d.** ii

3. $x^2 = -4y$

     $4p = -4$

     $p = -1$

     vertex $= (0,\ 0)$

     focus $= (0,\ -1)$

     directrix: $y = 1$

5. $y^2 = \frac{1}{3}x$

     $4p = \frac{1}{3}$

     $p = \frac{1}{12}$

     vertex $= (0,\ 0)$

     focus $= \left(\frac{1}{12},\ 0\right)$

     directrix: $x = -\frac{1}{12}$

7. $(x-2)^2 = 8(y+3)$

     vertex $= (2, -3)$

     $4p = 8 \quad p = 2$

     $(h, k+p) = (2, -3+2) = (2, -1)$

     focus $= (2, -1)$

     $k - p = -3 - 2 = -5$

     directrix: $y = -5$

9. $(y+4)^2 = -4(x-2)$

     vertex $= (2, -4)$

     $4p = -4 \quad p = -1$

     $(h+p, k) = (2-1, -4) = (1, -4)$

     focus $= (1, -4)$

     $h - p = 2 + 1 = 3$

     directrix: $x = 3$

11. $(y-1)^2 = 2(x+4)$

     vertex $= (-4, 1)$

     $4p = 2 \quad p = \frac{1}{2}$

     $(h+p, k) = \left(-4+\frac{1}{2}, 1\right) = \left(-\frac{7}{2}, 1\right)$

     focus $= \left(-\frac{7}{2}, 1\right)$

     $h - p = -4 - \frac{1}{2} = -\frac{9}{2}$

     directrix: $x = -\frac{9}{2}$

13. $(2x-4)^2 = 8y - 16$

     $4(x-2)^2 = 8(y-2)$

     $(x-2)^2 = 2(y-2)$

     vertex $= (2,\ 2)$

     $4p = 2 \quad p = \frac{1}{2}$

     $(h, k+p) = \left(2,\ 2+\frac{1}{2}\right) = \left(2,\ \frac{5}{2}\right)$

     focus $= \left(2,\ \frac{5}{2}\right)$

     $k - p = 2 - \frac{1}{2} = \frac{3}{2}$

     directrix: $y = \frac{3}{2}$

15. $x^2 + 8x - y + 6 = 0$

$x^2 + 8x = y - 6$

$x^2 + 8x + 16 = y - 6 + 16$

$(x+4)^2 = y + 10$

vertex $= (-4, -10)$

$4p = 1, \ p = \frac{1}{4}$

focus $= \left(-4, -\frac{39}{4}\right)$

directrix: $y = -\frac{41}{4}$

17. $x + y^2 - 3y + 4 = 0$

$y^2 - 3y = -x - 4$

$y^2 - 3y + 9/4 = -x - 4 + 9/4$

$\left(y - \frac{3}{2}\right)^2 = -\left(x + \frac{7}{4}\right)$

vertex $= \left(-\frac{7}{4}, \frac{3}{2}\right)$

$4p = -1, \ p = -\frac{1}{4}$

focus $= \left(-2, \frac{3}{2}\right)$

directrix: $x = -\frac{3}{2}$

19. $2x - y^2 - 6y + 1 = 0$

$-y^2 - 6y = -2x - 1$

$y^2 + 6y = 2x + 1$

$y^2 + 6y + 9 = 2x + 1 + 9$

$(y+3)^2 = 2(x+5)$

vertex $= (-5, -3)$

$4p = 2, \ p = \frac{1}{2}$

focus $= \left(-\frac{9}{2}, -3\right)$

directrix: $x = -\frac{11}{2}$

21. $x^2 + 3x + 3y - 1 = 0$

$x^2 + 3x = -3y + 1$

$x^2 + 3x + 9/4 = -3y + 1 + 9/4$

$\left(x + \frac{3}{2}\right)^2 = -3\left(y - \frac{13}{12}\right)$

vertex $\left(-\frac{3}{2}, \frac{13}{12}\right)$

$4p = -3, \ p = -\frac{3}{4}$

focus $= \left(-\frac{3}{2}, \frac{1}{3}\right)$

directrix: $y = \frac{11}{6}$

23. $2x^2 - 8x - 4y + 3 = 0$

$2(x^2 - 4x) = 4y - 3$

$2(x^2 - 4x + 4) = 4y - 3 + 8$

$2(x-2)^2 = 4y + 5$

$(x-2)^2 = 2y + \frac{5}{2}$

$(x-2)^2 = 2\left(y + \frac{5}{4}\right)$

vertex $= \left(2, -\frac{5}{4}\right), \ 4p = 2, \ p = \frac{1}{2}$

focus $= \left(2, -\frac{3}{4}\right)$

directrix $y = -\frac{7}{4}$

25. $2x + 4y^2 + 8y - 5 = 0$

$4y^2 + 8y = -2x + 5$

$4(y^2 + 2y) = -2x + 5$

$4(y^2 + 2y + 1) = -2x + 5 + 4$

$4(y+1)^2 = -2x + 9$

$(y+1)^2 = -\frac{1}{2}x + \frac{9}{4}$

$(y+1)^2 = -\frac{1}{2}\left(x - \frac{9}{2}\right)$

vertex $= \left(\frac{9}{2}, -1\right)$

$4p = -\frac{1}{2}, \ p = -\frac{1}{8}$

focus $= \left(\frac{35}{8}, -1\right)$

directrix $x = \frac{37}{8}$

27. $3x^2 - 6x - 9y + 4 = 0$

$$3(x^2 - 2x) = 9y - 4$$
$$3(x^2 - 2x + 1) = 9y - 1$$
$$(x-1)^2 = 3\left(y - \frac{1}{9}\right)$$

vertex $= \left(1, \dfrac{1}{9}\right)$, $4p = 3$, $p = \dfrac{3}{4}$

focus $= \left(1, \dfrac{31}{36}\right)$

directrix $y = -\dfrac{23}{36}$

29. vertex $(0,0)$, focus $(0, -4)$

$$x^2 = 4py$$
$$p = -4 \text{ since focus is } (0, p)$$
$$x^2 = 4(-4)y$$
$$x^2 = -16y$$

31. vertex $(-1, 2)$, focus $(-1, 3)$

$$(x-h)^2 = 4p(y-k)$$
$$h = -1, \quad k = 2.$$

The distance p from the vertex to the focus is 1.

$$(x+1)^2 = 4(1)(y-2)$$
$$(x+1)^2 = 4(y-2)$$

33. focus $(3, -3)$, directrix $y = -5$

The vertex is the midpoint of the line segment joining $(3, -3)$ and the point $(3, -5)$ on the directrix.

$$(h, k) = \left(\frac{3+2}{2}, \frac{-3+(-5)}{2}\right) = (3, -4)$$

The distance p from the vertex to the focus is 1.

$$4p = 4(1) = 4$$
$$(x-h)^2 = 4p(y-k)$$
$$(x-3)^2 = 4(y+4)$$

35. vertex $= (-4, 1)$, point: $(-2, 2)$ on the parabola.

Axis of symmetry $x = -4$.

If $P_1 = (-2, 2)$ $(x+4)^2 = 4p(y-1)$. Since $(-2, 2)$ is on the curve, we get

$$(-2+4)^2 = 4p(2-1)$$
$$4 = 4p \Rightarrow p = 1$$

Thus, the equation in standard form is

$$(x+4)^2 = 4(y-1)$$

37. Find the vertex.

$$x = -0.325y^2 + 13y + 120$$
$$x - 120 = -0.325y^2 + 13y$$
$$x - 120 = -0.325(y^2 - 40y)$$
$$x - 120 - 130 = -0.325(y^2 - 40y + 400)$$
$$x - 250 = -0.325(y - 20)^2$$
$$-\frac{40}{13}(x - 250) = (y - 20)^2$$

The vertex is $(250, 20)$.

Find the focus.

$$4p = -\frac{40}{13}, \quad p = -\frac{10}{13}$$

The focus is $\left(\dfrac{3240}{13}, 20\right)$.

39. Place the satellite dish on an xy-coordinate system with its vertex at $(0, -1)$ as shown.

The equation of the parabola is $x^2 = 4p(y+1)$ $\quad -1 \le y \le 0$

Because $(4, 0)$ is a point on this graph, $(4, 0)$ must be a solution of the equation of the parabola. Thus,

$$16 = 4p(0+1)$$
$$16 = 4p$$
$$4 = p$$

Because p is the distance from the vertex to the focus, the focus is on the x-axis 4 feet above the vertex.

41. The focus of the parabola is $(p, 0)$ where $y^2 = 4px$.
Half of 18.75 inches is 9.375 inches.
Therefore, the point (3.66, 9.375) is on the parabola.

$$(9.375)^2 = 4p(3.66)$$
$$87.890625 = 14.64p$$
$$\frac{87.890625}{14.64} = p$$
$$p \approx 6.0 \text{ inches}$$

43.
$$S = \frac{\pi r}{6d^2}\left[\left(r^2 + 4d^2\right)^{3/2} - r^3\right]$$

a. $r = 40.5$ feet
$d = 16$ feet

$$S = \frac{\pi(40.5)}{6(16)^2}\left[\left([40.5]^2 + 4[16]^2\right)^{3/2} - (40.5)^3\right]$$
$$= \frac{40.5\pi}{1536}\left[(266.25)^{3/2} - 66430.125\right]$$
$$= \frac{40.5\pi}{1536}\left[137518.9228 - 66430.125\right]$$
$$= \frac{40.5\pi}{1536}\left[71088.79775\right]$$
$$\approx 5900 \text{ square feet}$$

b. $r = 125$ feet
$d = 52$ feet

$$S = \frac{\pi(125)}{6(52)^2}\left[\left([125]^2 + 4[52]^2\right)^{3/2} - (125)^3\right]$$
$$= \frac{125\pi}{16224}\left[(26441)^{3/2} - 1953125\right]$$
$$= \frac{125\pi}{16224}\left[4299488.724 - 1953125\right]$$
$$= \frac{125\pi}{16224}\left[2346363.724\right]$$
$$\approx 56,800 \text{ square feet}$$

45.   The equation of the mirror is given by
$$x^2 = 4py \quad -60 \le x \le 60$$
Because p is the distance from the vertex to the focus and the coordinates of the focus are (0, 600), $p = 600$.
Therefore,
$$x^2 = 4(600)y$$
$$x^2 = 2400y$$
To determine a, substitute $(60, a)$ into the equation $x^2 = 2400y$ and solve for a.
$$x^2 = 2400y$$
$$60^2 = 2400a$$
$$3600 = 2400a$$
$$1.5 = a.$$
The concave depth of the mirror is 1.5 inches.

47. **a.** The equation of the parabola is

$$x^2 = 4p(y-32)$$

Because $(-800, 53)$ is a point on this graph, $(-800, 53)$ must be a solution of the equation of the parabola. Thus,

$$(-800)^2 = 4p(53-32)$$

$$800^2 = 84p$$

$$\frac{800^2}{84} = p$$

The equation of the parabola is $x^2 = 4\left(\dfrac{800^2}{84}\right)(y-32)$.

b. To find the width use $x = 900$, and solve for y and then multiply by 2

$$x^2 = 4\left(\frac{800^2}{84}\right)(y-32)$$

$$900^2 = 4\left(\frac{800^2}{84}\right)(y-32)$$

$$\left(\frac{900^2}{800^2}\right)\frac{84}{4} + 32 = y$$

$$y \approx 58.58$$

$$2y \approx 117$$

The width of the ski is 117 mm.

Connecting Concepts

49.
$$x^2 = 4y$$
$$4p = 4$$
$$p = 1$$

focus $= (0, 1)$
Substituting the vertical coordinate of the focus for y to obtain x-coordinates of endpoints (x_1, y_1), (x_2, y_2), we have

$$x^2 = 4(1), \text{ or } x^2 = 4$$

$$x = \pm\sqrt{4}$$

$$x_1 = -2 \quad x_2 = 2$$

Length of latus rectum $= |x_2 - x_1|$
$$= 2 - (-2) = 4.$$

51. $(x-h)^2 = 4p(y-k)$

focus $= (h, k+p)$

Substituting the vertical coordinate of the focus for y to obtain x-coordinates of endpoints (x_1, y_1), (x_2, y_2), we have

$$(x-h)^2 = 4p(k+p-k)$$
$$(x-h)^2 = 4p^2$$
$$x - h = \pm 2p$$
$$x_1 = h - 2p \quad x_2 = h + 2p$$

Solving for $|x_2 - x_1|$, we obtain

$$\Delta x = |x_2 - x_1| = |h + 2p - h + 2p| = 4|p|$$

or $(y-k)^2 = 4p(x-h)$

focus $= (h+p, k)$

Substituting the horizontal coordinate of the focus for x to obtain the y-coordinates of endpoints (x_1, y_1), (x_2, y_2), we have

$$(y-k)^2 = 4p(h+p-h)$$
$$(y-k)^2 = 4p^2$$
$$y - k = \pm 2p$$
$$y_1 = k - 2p \quad y_2 = k + 2p$$

Solving for $|y_2 - y_1|$, we obtain $\Delta y = |y_2 - y_1|$
$$= |k + 2p - k + 2p|$$
$$= 4|p|$$

Thus, the length of the latus rectum is $4|p|$.

53.

$4p = -1$

$p = -\frac{1}{4}$

focus $\left(\frac{3}{4}, -4\right)$

one point: $\left(\frac{3}{4}, k+2p\right) = \left(\frac{3}{4}, -\frac{9}{2}\right)$

one point: $\left(\frac{3}{4}, k-2p\right) = \left(\frac{3}{4}, -\frac{7}{2}\right)$

55. Graph $y = \frac{7}{4} + \frac{1}{4}x|x|$.

57. By definition, any point on the curve (x, y) will be equidistant from both the focus $(1, 1)$ and the directrix, $\left(y_2 = -x_2 - 2\right)$.

If we let d_1 equal the distance from the focus to the point (x, y), we get $d_1 = \sqrt{(x-1)^2 + (y-1)^2}$

To determine the distance d_2 from the point (x, y) to the line $y = -x - 2$, draw a line segment from (x, y) to the directrix so as to meet the directrix at a 90° angle.

Now drop a line segment parallel to the y-axis from (x, y) to the directrix. This segment will meet the directrix at a 45° angle, thus forming a right isosceles triangle with the directrix and the line segment perpendicular to the directrix from (x, y). The length of this segment, which is the hypotenuse of the triangle, is the difference between y and the y-value of the directrix at x, or $-x - 2$. Thus, the hypotenuse has a length of $y + x + 2$, and since the right triangle is also isosceles, each leg has a length of $\dfrac{y+x+2}{\sqrt{2}}$.

But since d_2 is the length of the leg drawn from (x, y) to the directrix, $d_2 = \dfrac{y+x+2}{\sqrt{2}}$.

Thus, $d_1 = \sqrt{(x-1)^2 + (y-1)^2}$ and $d_2 = \dfrac{x+y+2}{\sqrt{2}}$.

By definition, $d_1 = d_2$. So, by substitution,

$$\sqrt{(x-1)^2 + (y-1)^2} = \frac{x+y+2}{\sqrt{2}}$$

$$\sqrt{2}\sqrt{(x-1)^2 + (y-1)^2} = x + y + 2$$

$$2\left[(x-1)^2 + (y-1)^2\right] = x^2 + y^2 + 4x + 4y + 2xy + 4$$

$$2\left(x^2 - 2x + 1 + y^2 - 2y + 1\right) = x^2 + y^2 + 4x + 4y + 2xy + 4$$

$$2x^2 - 4x + 2y^2 - 4y + 4 = x^2 + y^2 + 4x + 4y + 2xy + 4$$

$$x^2 + y^2 - 8x - 8y - 2xy = 0$$

Prepare for Section 6.2

PS1. midpoint: $\dfrac{x_1 + x_2}{2} \qquad \dfrac{y_1 + y_2}{2}$

$\dfrac{5 + -1}{2} = 2 \qquad \dfrac{1 + 5}{2} = 3$

The midpoint is $(2, 3)$.

length: $\sqrt{(x_2 - x_1)^2 + (y_2 - y_1)^2}$

$\sqrt{(-1-5)^2 + (5-1)^2} = \sqrt{36 + 16} = \sqrt{52} = 2\sqrt{13}$

The length is $2\sqrt{13}$.

PS2. $x^2 + 6x - 16 = 0$

$(x+8)(x-2) = 0$

$x + 8 = 0 \qquad x - 2 = 0$

$x = -8 \qquad x = 2$

The solutions are $-8, 2$.

PS3. $x^2 - 2x = 2$
$$x^2 - 2x + 1 = 2 + 1$$
$$(x-1)^2 = 3$$
$$x - 1 = \pm\sqrt{3}$$
$$x = 1 \pm \sqrt{3}$$

PS4. $x^2 - 8x + 16 = (x-4)^2$

PS5. $(x-2)^2 + y^2 = 4$
$$y^2 = 4 - (x-2)^2$$
$$y = \pm\sqrt{4 - (x-2)^2}$$

PS6.

$(x-2)^2 + (y+3)^2 = 16$
Center: $(2, -3)$, radius 4

Section 6.2

1.   **a.** iv   **b.** i   **c.** ii   **d.** iii

3. $\dfrac{x^2}{16} + \dfrac{y^2}{25} = 1$

$a^2 = 25 \rightarrow a = 5$
$b^2 = 16 \rightarrow b = 4$
$c = \sqrt{a^2 - b^2}$
$\quad = \sqrt{25 - 16}$
$\quad = \sqrt{9}$
$\quad = 3$

Center $(0, 0)$
Vertices $(0, \pm 5)$
Foci $(0, \pm 3)$

5. $\dfrac{x^2}{9} + \dfrac{y^2}{4} = 1$

$a^2 = 9 \rightarrow a = 3$
$b^2 = 4 \rightarrow b = 2$
$c = \sqrt{a^2 - b^2}$
$\quad = \sqrt{9 - 4}$
$\quad = \sqrt{5}$

Center $(0, 0)$
Vertices $(\pm 3, 0)$
Foci $\left(\pm\sqrt{5},\, 0\right)$

7. $\dfrac{x^2}{9} + \dfrac{y^2}{7} = 1$

$a^2 = 9 \rightarrow a = 3$
$b^2 = 7 \rightarrow b = \sqrt{7}$
$c = \sqrt{a^2 - b^2}$
$\quad = \sqrt{9 - 7}$
$\quad = \sqrt{2}$

Center $(0, 0)$
Vertices $(0, \pm 3)$
Foci $\left(0, \pm\sqrt{2}\right)$

9. $\dfrac{4x^2}{9} + \dfrac{y^2}{16} = 1$

Rewrite as

$$\dfrac{x^2}{9/4} + \dfrac{y^2}{16} = 1$$

$a^2 = 16 \rightarrow a = 4$
$b^2 = 9/4 \rightarrow b = 3/2$
$c = \sqrt{a^2 - b^2}$
$\quad = \sqrt{16 - 9/4}$
$\quad = \sqrt{55}/2$

Center $(0, 0)$
Vertices $(0, \pm 4)$
Foci $\left(0, \pm\dfrac{\sqrt{55}}{2}\right)$

11. $\dfrac{(x-3)^2}{25} + \dfrac{(y+2)^2}{16} = 1$

Center $(3, -2)$

Vertices $(3 \pm 5, -2) = (8, -2), (-2, -2)$

Foci $(3 \pm 3, -2) = (6, -2), (0, -2)$

13. $\dfrac{(x+2)^2}{9} + \dfrac{y^2}{25} = 1$

Center $(-2, 0)$

Vertices $(-2, 5), (-2, -5)$

Foci $(-2, 4), (-2, -4)$

15. $\dfrac{(x-1)^2}{21} + \dfrac{(y-3)^2}{4} = 1$

Center $(1, 3)$

Vertices $\left(1 \pm \sqrt{21}, 3\right)$

Foci $\left(1 \pm \sqrt{17}, 3\right)$

17. $\dfrac{9(x-1)^2}{16} + \dfrac{(y+1)^2}{9} = 1$

Center $(1, -1)$

Vertices $(1, -1 \pm 3) = (1, 2), (1, -4)$

Foci $\left(1, -1 \pm \dfrac{\sqrt{65}}{3}\right)$

19. $3x^2 + 4y^2 = 12$

$\dfrac{x^2}{4} + \dfrac{y^2}{3} = 1$

Center $(0, 0)$

Vertices $(\pm 2, 0)$

Foci $(\pm 1, 0)$

21. $25x^2 + 16y^2 = 400$

$\dfrac{x^2}{16} + \dfrac{y^2}{25} = 1$

Center $(0, 0)$

Vertices $(0, \pm 5)$

Foci $(0, \pm 3)$

23. $64x^2 + 25y^2 = 400$

$\dfrac{x^2}{\frac{25}{4}} + \dfrac{y^2}{16} = 1$

Center $(0, 0)$

Vertices $(0, \pm 4)$

Foci $\left(0, \pm \dfrac{\sqrt{39}}{2}\right)$

25.

$$4x^2 + y^2 - 24x - 8y + 48 = 0$$

$$4(x^2 - 6x) + (y^2 - 8y) = -48$$

$$4(x^2 - 6x + 9) + (y^2 - 8y + 16) = -48 + 36 + 16$$

$$4(x-3)^3 + (y-4)^2 = 4$$

$$\frac{(x-3)^2}{1} + \frac{(y-4)^2}{4} = 1$$

Center $(3, 4)$

Vertices $(3, 4\pm2) = (3, 6),\ (3, 2)$

Foci $(3, 4\pm\sqrt{3})$

27.

$$5x^2 + 9y^2 - 20x + 54y + 56 = 0$$

$$5(x^2 - 4x) + 9(y^2 + 6y) = -56$$

$$5(x^2 - 4x + 4) + 9(y^2 + 6y + 9) = -56 + 20 + 81$$

$$5(x-2)^2 + 9(y+3)^2 = 45$$

$$\frac{(x-2)^2}{9} + \frac{(y+3)^2}{5} = 1$$

Center $(2, -3)$

Vertices $(2\pm3, -3) = (-1, -3), (5, -3)$

Foci $(2\pm2, 3) = (0, -3), (4, -3)$

29.

$$16x^2 + 9y^2 - 64x - 80 = 0$$

$$16(x^2 - 4x) + 9y^2 = 80$$

$$16(x^2 - 4x + 4) + 9y^2 = 80 + 64$$

$$16(x-2)^2 + 9y^2 = 144$$

$$\frac{(x-2)^2}{9} + \frac{y^2}{16} = 1$$

Center $(2, 0)$

Vertices $(2, \pm4) = (2, 4),\ (2, -4)$

Foci $(2\pm\sqrt{7})$

31.

$$25x^2 + 16y^2 + 50x - 32y - 359 = 0$$

$$25(x^2 + 2x) + 16(y^2 - 2y) = 359$$

$$25(x^2 + 2x + 1) + 16(y^2 - 2y + 1) = 359 + 25 + 16$$

$$25(x+1)^2 + 16(y-1)^2 = 400$$

$$\frac{(x+1)^2}{16} + \frac{(y-1)^2}{25} = 1$$

Center $(-1, 1)$

Vertices $(-1, 1\pm5) = (-1, 6), (-1, -4)$

Foci $(-1, 1\pm3) = (-1, 4), (-1, -2)$

33.

$$8x^2 + 25y^2 - 48x + 50y + 47 = 0$$

$$8(x^2 - 6x) + 25(y^2 + 2y) = -47$$

$$8(x^2 - 6x + 9) + 25(y^2 + 2y + 1) = -47 + 72 + 25$$

$$8(x-3)^2 + 25(y+1)^2 = 50$$

$$\frac{(x-3)^2}{25/4} + \frac{(y+1)^2}{2} = 1$$

Center: $(3, -1)$

Vertices: $\left(3\pm\frac{5}{2}, -1\right) = \left(\frac{11}{2}, -1\right), \left(\frac{1}{2}, -1\right)$

Foci: $\left(3\pm\frac{\sqrt{17}}{2}, -1\right)$

35.

$$2a = 10$$

$$a = 5$$

$$a^2 = 25$$

$$c = 4$$

$$c^2 = a^2 - b^2$$

$$16 = 25 - b^2$$

$$b^2 = 9$$

$$\frac{x^2}{25} + \frac{y^2}{9} = 1$$

37.
$$a = 6$$
$$a^2 = 36$$
$$b = 4$$
$$b^2 = 16$$
$$\frac{x^2}{36} + \frac{y^2}{16} = 1$$

39.
$$2a = 12$$
$$a = 6$$
$$a^2 = 36$$
$$\frac{x^2}{36} + \frac{y^2}{b^2} = 1$$
$$\frac{(2)^2}{36} + \frac{(-3)^2}{b^2} = 1$$
$$\frac{4}{36} + \frac{9}{b^2} = 1$$
$$\frac{9}{b^2} = \frac{8}{9}$$
$$8b^2 = 81$$
$$b^2 = \frac{81}{8}$$
$$\frac{x^2}{36} + \frac{y^2}{81/8} = 1$$

41.
$$c = 3$$
$$2a = 8$$
$$a = 4$$
$$a^2 = 16$$
$$c^2 = a^2 - b^2$$
$$9 = 16 - b^2$$
$$b^2 = 7$$
$$\frac{(x+2)^2}{16} + \frac{(y-4)^2}{7} = 1$$

43.
$$2a = 10$$
$$a = 5$$
$$a^2 = 25$$
Since the center of the ellipse is (2, 4) and the point (3, 3) is on the ellipse, we have
$$\frac{(x-2)^2}{b^2} + \frac{(y-4)^2}{a^2} = 1$$
$$\frac{(3-2)^2}{b^2} + \frac{(3-4)^2}{25} = 1$$
$$\frac{1}{b^2} = 1 - \frac{1}{25}$$
$$b^2 = \frac{25}{24}$$
$$\frac{(x-2)^2}{\frac{25}{24}} + \frac{(y-4)^2}{25} = 1$$

45.
center (5, 1)
$$c = 3$$
$$2a = 10$$
$$a = 5$$
$$a^2 = 25$$
$$c^2 = a^2 - b^2$$
$$9 = 25 - b^2$$
$$b^2 = 16$$
$$\frac{(x-5)^2}{16} + \frac{(y-1)^2}{25} = 1$$

47.
$$2a = 10$$
$$a = 5$$
$$a^2 = 25$$
$$\frac{c}{a} = \frac{2}{5}$$
$$\frac{c}{5} = \frac{2}{5}$$
$$c = 2$$
$$c^2 = a^2 - b^2$$
$$4 = 25 - b^2$$
$$b^2 = 21$$
$$\frac{x^2}{25} + \frac{y^2}{21} = 1$$

49.
center (0, 0)
$$c = 4$$
$$\frac{c}{a} = \frac{2}{3}$$
$$\frac{4}{a} = \frac{2}{3}$$
$$a = 6$$
$$c^2 = a^2 - b^2$$
$$16 = 36 - b^2$$
$$b^2 = 20$$
$$\frac{x^2}{20} + \frac{y^2}{36} = 1$$

51.
center (1, 3)
$$c = 2$$
$$\frac{c}{a} = \frac{2}{5}$$
$$\frac{2}{a} = \frac{2}{5}$$
$$a = 5$$
$$c^2 = a^2 - b^2$$
$$4 = 25 - b^2$$
$$b^2 = 21$$
$$\frac{(x-1)^2}{25} + \frac{(y-3)^2}{21} = 1$$

53.
$$2a = 24$$
$$a = 12$$
$$\frac{c}{a} = \frac{2}{3}$$
$$\frac{c}{12} = \frac{2}{3}$$
$$c = 8$$
$$c^2 = a^2 - b^2$$
$$64 = 144 - b^2$$
$$b^2 = 80$$
$$\frac{x^2}{80} + \frac{y^2}{144} = 1$$

55.
$$484 = 64 + c^2$$
$$c^2 = 420$$
$$c = 20.494$$
$$2c = 40.9878 \approx 41$$
The emitter should be placed
41 cm away.

57.
$$\text{Aphelion} = 2a - \text{perihelion}$$
$$934.34 = 2a - 835.14$$
$$a = 884.74 \text{ million miles}$$
$$\text{Aphelion} = a + c = 934.34$$
$$884.74 + c = 934.34$$
$$c = 49.6 \text{ million miles}$$

$$b = \sqrt{a^2 - c^2}$$
$$= \sqrt{884.74^2 - 49.6^2}$$
$$\approx 883.35 \text{ million miles}$$

An equation of the orbit of Saturn is
$$\frac{x^2}{884.74^2} + \frac{y^2}{883.35^2} = 1$$

59.
$a = $ semimajor axis $= 50$ feet
$b = $ height $= 30$ feet
$$c^2 = a^2 - b^2$$
$$c^2 = 50^2 - 30^2$$
$$c = \sqrt{1600} = 40$$
The foci are located 40 feet to
the right and to the left of center.

61.
$$2a = 36 \qquad 2b = 9$$
$$a = 18 \qquad b = \frac{9}{2}$$
$$c^2 = a^2 - b^2$$
$$c^2 = 18^2 - \left(\frac{9}{2}\right)^2$$
$$c^2 = 324 - \frac{81}{4}$$
$$c^2 = \frac{1215}{4}$$
$$c = \frac{9\sqrt{15}}{2}$$

Since one focus is at (0, 0), the center of the ellipse is at
$(9\sqrt{15}/2, 0)$
(17.43, 0). The equation of the path of Halley's Comet in
astronomical units is
$$\frac{\left(x - 9\sqrt{15}/2\right)^2}{324} + \frac{y^2}{81/4} = 1$$

63.
$$\frac{x^2}{75^2} + \frac{y^2}{34^2} = 1$$
Solve for y, where $x = 55$.
$$\frac{55^2}{75^2} + \frac{y^2}{34^2} = 1$$
$$\frac{y^2}{34^2} = 1 - \frac{55^2}{75^2}$$
$$y^2 = 34^2\left(1 - \frac{55^2}{75^2}\right)$$
$$y = \sqrt{34^2\left(1 - \frac{55^2}{75^2}\right)}$$
$$y \approx 23 \text{ ft}$$
$$h = y + 1 = 23 + 1 = 24 \text{ ft}$$

65. a.
$$c^2 = a^2 - b^2$$
$$c^2 = 4^2 - 3^2$$
$$c^2 = 7$$
$$c = \sqrt{7}$$
$\sqrt{7}$ ft to the right and left of O.

b.
$$2a = 2(4) = 8 \text{ ft}$$

67. $9y^2 + 36y + 16x^2 - 108 = 0$

$$y = \frac{-36 \pm \sqrt{36^2 - 4(9)(16x^2 - 108)}}{2(9)}$$

$$= \frac{-36 \pm \sqrt{1296 - 36(16x^2 - 108)}}{18}$$

$$= \frac{-36 \pm \sqrt{1296 - 576x^2 + 3888}}{18}$$

$$= \frac{-36 \pm \sqrt{-576x^2 + 5184}}{18}$$

$$= \frac{-36 \pm \sqrt{576(-x^2 + 9)}}{18}$$

$$= \frac{-36 \pm 24\sqrt{(-x^2 + 9)}}{18}$$

$$= \frac{-6 \pm 4\sqrt{(-x^2 + 9)}}{3}$$

69. $9y^2 + 18y + 4x^2 + 24x + 44 = 0$

$$y = \frac{-18 \pm \sqrt{18^2 - 4(9)(4x^2 + 24x + 44)}}{2(9)}$$

$$= \frac{-18 \pm \sqrt{324 - 36(4x^2 + 24x + 44)}}{18}$$

$$= \frac{-18 \pm \sqrt{324 - 144x^2 - 864x - 1584}}{18}$$

$$= \frac{-18 \pm \sqrt{-144x^2 - 864x - 1260}}{18}$$

$$= \frac{-18 \pm \sqrt{36(-4x^2 - 24x - 35)}}{18}$$

$$= \frac{-18 \pm 6\sqrt{-4x^2 - 24x - 35}}{18}$$

$$= \frac{-3 \pm \sqrt{-4x^2 - 24x - 35}}{3}$$

Connecting Concepts

71. The sum of the distances between the two foci and a point on the ellipse is $2a$.

$$2a = \sqrt{\left(\frac{9}{2} - 0\right)^2 + (3-3)^2} + \sqrt{\left(\frac{9}{2} - 0\right)^2 + (3+3)^2}$$

$$= \sqrt{\left(\frac{9}{2}\right)^2} + \sqrt{\frac{225}{4}}$$

$$= \frac{9}{2} + \frac{15}{2}$$

$$= 12$$

$$a = 6$$

$$c = 3$$

$$c^2 = a^2 - b^2$$

$$9 = 36 - b^2$$

$$b^2 = 27$$

$$\frac{x^2}{36} + \frac{y^2}{27} = 1$$

73. The sum of the distances between the two foci and a point on the ellipse is $2a$.

$$2a = \sqrt{(5-2)^2 + (3+1)^2} + \sqrt{(5-2)^2 + (3-3)^2}$$

$$= \sqrt{25} + \sqrt{3^2}$$

$$= 5 + 3$$

$$= 8$$

$$a = 4$$

$$c = 2$$

$$c^2 = a^2 - b^2$$

$$4 = 16 - b^2$$

$$b^2 = 12$$

$$\frac{(x-1)^2}{16} + \frac{(y-2)^2}{12} = 1$$

75. Center $(1,-1)$

$c^2 = a^2 - b^2$

$c^2 = 16 - 9$

$c^2 = 7$

$c = \sqrt{7}$

The latus rectum is on the graph of $y = -1 + \sqrt{7}$, or $y = -1 - \sqrt{7}$

$$\frac{(x-1)^2}{9} + \frac{(y+1)^2}{16} = 1$$

$$\frac{(x-1)^2}{9} + \frac{\left(-1+\sqrt{7}+1\right)^2}{16} = 1 \quad \text{or} \quad \frac{(x-1)^2}{9} + \frac{\left(-1-\sqrt{7}+1\right)^2}{16} = 1$$

$$\frac{(x-1)^2}{9} + \frac{7}{16} = 1$$

$$\frac{(x-1)^2}{9} = \frac{9}{16}$$

$$16(x-1)^2 = 81$$

$$(x-1)^2 = \frac{81}{16}$$

$$x-1 = \pm\sqrt{\frac{81}{16}}$$

$$x-1 = \pm\frac{9}{4}$$

$$x = \frac{13}{4} \quad \text{and} \quad -\frac{5}{4}$$

The x-coordinates of the endpoints of the latus rectum are $\frac{13}{4}$ and $-\frac{5}{4}$.

$$\left|\frac{13}{4} - \left(-\frac{5}{4}\right)\right| = \frac{9}{2}$$

The length of the latus rectum is $\frac{9}{2}$.

77. Let us transform the general equation of an ellipse into an $x'y'$-coordinate system where the center is at the origin by replacing $(x-h)$ by x' and $(y-k)$ by y'.

We have $\dfrac{x'^2}{a^2} + \dfrac{y'^2}{b^2} = 1$.

Letting $x' = c$ and solving for y' yields

$$\frac{(c)^2}{a^2} + \frac{y'^2}{b^2} = 1$$

$$b^2 c^2 + a^2 y'^2 = a^2 b^2$$

$$a^2 y'^2 = a^2 b^2 - b^2 c^2$$

$$a^2 y'^2 = b^2(a^2 - c^2)$$

But since $c^2 = a^2 - b^2$, $b^2 = a^2 - c^2$, we can substitute to obtain

$$a^2 y'^2 = b^2(b^2)$$

$$y'^2 = \frac{b^4}{a^2}$$

$$y' = \pm\sqrt{\frac{b^4}{a^2}} = \pm\frac{b^2}{a}$$

The endpoints of the latus rectum, then, are $\left(c, \dfrac{b^2}{a}\right)$ and $\left(c, -\dfrac{b^2}{a}\right)$.

The distance between these points is $\dfrac{2b^2}{a}$.

• •

Prepare for Section 6.3

PS1. $\dfrac{4 + -2}{2} = 1$

$\dfrac{-3 + 1}{2} = -1$

Midpoint: $(1, -1)$

$\sqrt{(-2-4)^2 + (1--3)^2} = \sqrt{52} = 2\sqrt{13}$

Length: $2\sqrt{13}$

PS2. $(x-1)(x+3) = 5$

$x^2 + 2x - 3 = 5$

$x^2 + 2x - 8 = 0$

$(x+4)(x-2) = 0$

$x + 4 = 0 \qquad x - 2 = 0$

$x = -4 \qquad x = 2$

PS3. $\dfrac{4}{\sqrt{8}} = \dfrac{4\sqrt{8}}{8} = \dfrac{8\sqrt{2}}{8} = \sqrt{2}$

PS4. $4x^2 + 24x = 4(x^2 + 6x)$

$= 4(x^2 + 6x + 9)$

$= 4(x+3)^2$

PS5. $\dfrac{x^2}{4} - \dfrac{y^2}{9} = 1$

$-\dfrac{y^2}{9} = 1 - \dfrac{x^2}{4}$

$y^2 = \dfrac{9x^2}{4} - 9$

$y = \pm\sqrt{\dfrac{9x^2}{4} - 9}$

$y = \pm\dfrac{3}{2}\sqrt{x^2 - 4}$

PS6.

Section 6.3

1. a. iii   b. ii   c. i   d. iv

3. $\dfrac{x^2}{16} - \dfrac{y^2}{25} = 1$

    Center $(0, 0)$

    Vertices $(\pm 4, 0)$

    Foci $\left(\pm\sqrt{41},\ 0\right)$

    Asymptotes $y = \pm\dfrac{5}{4}x$

5. $\dfrac{y^2}{4} - \dfrac{x^2}{25} = 1$

    Center $(0, 0)$

    Vertices $(0,\ \pm 2)$

    Foci $\left(0,\ \pm\sqrt{29}\right)$

    Asymptotes $y = \pm\dfrac{2}{5}x$

7. $\dfrac{x^2}{7} - \dfrac{y^2}{9} = 1$

    Center $(0, 0)$

    Vertices $\left(\pm\sqrt{7}, 0\right)$

    Foci $(\pm 4, 0)$

    Asymptotes $y = \pm\dfrac{3\sqrt{7}}{7}x$

9. $\dfrac{4x^2}{9} - \dfrac{y^2}{16} = 1$

    Center $(0, 0)$

    Vertices $\left(\pm\dfrac{3}{2}, 0\right)$

    Foci $\left(\pm\dfrac{\sqrt{73}}{2},\ 0\right)$

    Asymptotes $y = \pm\dfrac{8}{3}x$

11. $\dfrac{(x-3)^2}{16} - \dfrac{(y+4)^2}{9} = 1$

    Center $(3, -4)$

    Vertices $(3 \pm 4, -4) = (7, -4), (-1, -4)$

    Foci $(3 \pm 5, -4) = (8, -4), (-2, -4)$

    Asymptotes $y + 4 = \pm\dfrac{3}{4}(x-3)$

13. $\dfrac{(y+2)^2}{4} - \dfrac{(x-1)^2}{16} = 1$

    Center $(1, -2)$

    Vertices $(1, -2 \pm 2) = (1, 0), (1, -4)$

    Foci $\left(1, -2 \pm 2\sqrt{5}\right) = \left(1, -2 + 2\sqrt{5}\right), \left(1, -2 - 2\sqrt{5}\right)$

    Asymptotes $y + 2 = \pm\dfrac{1}{2}(x-1)$

15. $\dfrac{(x+2)^2}{9} - \dfrac{y^2}{25} = 1$

Center $(-2, 0)$

Vertices $(-2 \pm 3, 0) = (1, 0), (-5, 0)$

Foci $\left(-2 \pm \sqrt{34}, 0\right)$

Asymptotes $y = \pm \dfrac{5}{3}(x+2)$

17. $\dfrac{9(x-1)^2}{16} - \dfrac{(y+1)^2}{9} = 1$

$\dfrac{(x-1)^2}{16/9} - \dfrac{(y+1)^2}{9}$

Center $(1, -1)$

Vertices $\left(1 \pm \dfrac{4}{3}, -1\right) = \left(\dfrac{7}{3}, -1\right), \left(-\dfrac{1}{3}, -1\right)$

Foci $\left(1 \pm \dfrac{\sqrt{97}}{3}, -1\right)$

Asymptotes $(y+1) = \pm \dfrac{9}{4}(x-1)$

19. $x^2 - y^2 = 9$

$\dfrac{x^2}{9} - \dfrac{y^2}{9} = 1$

Center $(0, 0)$

Vertices $(\pm 3, 0)$

Foci $\left(\pm 3\sqrt{2}, 0\right)$

Asymptotes $y = \pm x$

21. $16y^2 - 9x^2 = 144$

$\dfrac{y^2}{9} - \dfrac{x^2}{16} = 1$

Center $(0, 0)$

Vertices $(0, \pm 3)$

Foci (0 ± 5)

Asymptotes $y = \pm \dfrac{3}{4}x$

23. $9y^2 - 36x^2 = 4$

$\dfrac{y^2}{4/9} - \dfrac{x^2}{1/9} = 1$

Center $(0, 0)$

Vertices $\left(0, \pm \dfrac{2}{3}\right)$

Foci $\left(0, \pm \dfrac{\sqrt{5}}{3}\right)$

Asymptotes $y = \pm 2x$

25.
$$x^2 - y^2 - 6x + 8y = 3$$
$$(x^2 - 6x) - (y^2 - 8y) = 3$$
$$(x^2 - 6x + 9) - (y^2 - 8y + 16) = 3 + 9 - 16$$
$$(x-3)^2 - (y-4)^2 = -4$$
$$\frac{(y-4)^2}{4} - \frac{(x-3)^2}{4} = 1$$

Center $(3, 4)$

Vertices $(3, 4 \pm 2) = (3, 6), (3, 2)$

Foci $(3, 4 \pm 2\sqrt{2}) = (3, 4 + 2\sqrt{2}), (3, 4 - 2\sqrt{2})$

Asymptotes $y - 4 = \pm(x - 3)$

27.
$$9x^2 - 4y^2 + 36x - 8y + 68 = 0$$
$$9x^2 + 36x - 4y^2 - 8y = -68$$
$$9(x^2 + 4x) - 4(y^2 + 2y) = -68$$
$$9(x^2 + 4x + 4) - 4(y^2 + 2y + 1) = -68 + 36 - 4$$
$$9(x+2)^2 - 4(y+1)^2 = -36$$
$$\frac{(y+1)^2}{9} - \frac{(x+2)^2}{4} = 1$$

Center $(-2, -1)$

Vertices $(-2, -1 \pm 3) = (-2, 2), (-2, -4)$

Foci $(-2, -1 \pm \sqrt{13}) = (-2, -1 + \sqrt{13}), (-2, -1 - \sqrt{13})$

Asymptotes $y + 1 = \pm\frac{3}{2}(x + 2)$

29.
$$y = \frac{-6 \pm \sqrt{6^2 - 4(-1)(4x^2 + 32x + 39)}}{2(-1)}$$
$$= \frac{-6 \pm \sqrt{36 + 4(4x^2 + 32x + 39)}}{-2}$$
$$= \frac{-6 \pm \sqrt{16x^2 + 128x + 192}}{-2}$$
$$= \frac{-6 \pm \sqrt{16(x^2 + 8x + 12)}}{-2}$$
$$= \frac{-6 \pm 4\sqrt{x^2 + 8x + 12}}{-2}$$
$$= 3 \pm 2\sqrt{x^2 + 8x + 12}$$

31.
$$y = \frac{64 \pm \sqrt{(-64)^2 - 4(-16)(9x^2 - 36x + 116)}}{2(-16)}$$
$$= \frac{64 \pm \sqrt{4096 + 64(9x^2 - 36x + 116)}}{-32}$$
$$= \frac{64 \pm \sqrt{64(9x^2 - 36x + 116 + 64)}}{-32}$$
$$= \frac{64 \pm 8\sqrt{(9x^2 - 36x + 180)}}{-32}$$
$$= \frac{64 \pm 8\sqrt{9(x^2 - 4x + 20)}}{-32}$$
$$= \frac{64 \pm 24\sqrt{x^2 - 4x + 20}}{-32}$$
$$= \frac{-8 \pm 3\sqrt{x^2 - 4x + 20}}{4}$$

33.
$$y = \frac{18 \pm \sqrt{(-18)^2 - 4(-9)(4x^2 + 8x - 6)}}{2(-9)}$$

$$= \frac{18 \pm \sqrt{324 + 36(4x^2 + 8x - 6)}}{-18}$$

$$= \frac{18 \pm \sqrt{36(4x^2 + 8x - 6 + 9)}}{-18}$$

$$= \frac{18 \pm 6\sqrt{(4x^2 + 8x + 3)}}{-18}$$

$$= \frac{-3 \pm \sqrt{(4x^2 + 8x + 3)}}{3}$$

35. vertices $(3, 0)$ and $(-3, 0)$, foci $(4, 0)$ and $(-4, 0)$

Traverse axis is on x-axis. For a standard hyperbola, the vertices are at $(h + a, k)$ and $(h - a, k)$, $h + a = 3$, $h - a = -3$, and $k = 0$..

If $h + a = 3$ and $h - a = -3$, then $h = 0$ and $a = 3$.

The foci are located at $(4, 0)$ and $(-4, 0)$. Thus, $h = 0$ and $c = 4$.

Since $c^2 = a^2 + b^2$, $b^2 = c^2 - a^2$

$$b^2 = (4)^2 - (3)^2 = 16 - 9 = 7$$

$$\frac{(x - h)^2}{a^2} - \frac{(y - k)^2}{b^2} = 1$$

$$\frac{(x - 0)^2}{(3)^2} - \frac{(y - 0)^2}{7} = 1$$

$$\frac{x^2}{9} - \frac{y^2}{7} = 1$$

37. foci $(0, 5)$ and $(0, -5)$, asymptotes $y = 2x$ and $y = -2x$

Transverse axis is on y-axis. Since foci are at $(h, k + c)$ and $(h, k - c)$, $k + c = 5$, $k - c = -5$, and $h = 0$.

Therefore, $k = 0$ and $c = 5$.

Since one of the asymptotes is $y = \frac{a}{b}x$, $\frac{a}{b} = 2$ and $a = 2b$.

$a^2 + b^2 = c^2$; then substituting $a = 2b$ and $c = 5$ yields $(2b)^2 + b^2 = (5)^2$, or $5b^2 = 25$.

Therefore, $b^2 = 5$ and $b = \sqrt{5}$.

Since $a = 2b$, $a = 2(\sqrt{5}) = 2\sqrt{5}$.

$$\frac{(y - k)^2}{a^2} - \frac{(x - h)^2}{b^2} = 1$$

$$\frac{y^2}{(2\sqrt{5})^2} - \frac{x^2}{5} = 1$$

$$\frac{y^2}{20} - \frac{x^2}{5} = 1$$

39. vertices $(0, 3)$ and $(0, -3)$, point $(2, 4)$

The distance between the two vertices is the length of the transverse axis, which is $2a$.

$$2a = |3 - (-3)| = 6 \text{ or } a = 3.$$

Since the midpoint of the transverse axis is the center of the hyperbola, the center is given by $\left(\frac{0 + 0}{2}, \frac{3 + (-3)}{2}\right)$, or $(0, 0)$.

Since both vertices lie on the y-axis, the transverse axis must be on the y-axis.

Taking the standard form of the hyperbola, we have

$$\frac{y^2}{a^2} - \frac{x^2}{b^2} = 1$$

Substituting the point $(2, 4)$ for x and y, and 3 for a, we have

$$\frac{16}{9} - \frac{4}{b^2} = 1$$

Solving for b^2 yields $b^2 = \frac{36}{7}$.

Therefore, the equation is

$$\frac{y^2}{9} - \frac{x^2}{36/7} = 1$$

41. vertices $(0, 4)$ and $(0, -4)$, asymptotes $y = \frac{1}{2}x$ and $y = -\frac{1}{2}x$.

The length of the transverse axis, or the distance between the vertices, is equal to $2a$.

$$2a = 4 - (-4) = 8 \text{, or } a = 4$$

The center of the hyperbola, or the midpoint of the line segment joining the vertices, is

$\left(\dfrac{0+0}{2}, \dfrac{4+(-4)}{2}\right)$, or $(0, 0)$

Since both vertices lie on the y-axis, the transverse axis must lie on the y-axis. Therefore, the asymptotes are given by

$y = \frac{a}{b}x$ and $y = -\frac{a}{b}x$. One asymptote is $y = \frac{1}{2}x$. Thus $\frac{a}{b} = \frac{1}{2}$ or $b = 2a$.

Since $b = 2a$ and $a = 4$, $b = 2(4) = 8$.

Thus, the equation is

$$\frac{y^2}{4^2} - \frac{x^2}{8^2} = 1 \text{ or } \frac{y^2}{16} - \frac{x^2}{64} = 1$$

43. vertices $(6, 3)$ and $(2, 3)$, foci $(7, 3)$ and $(1, 3)$

Length of transverse axis = distance between vertices

$$2a = |6 - 2|$$
$$a = 2$$

The center of the hyperbola (h, k) is the midpoint of the line segment joining the vertices, or the point $\left(\dfrac{6+2}{2}, \dfrac{3+3}{2}\right)$.

Thus, $h = \dfrac{6+2}{2}$, or 4, and $k = \dfrac{3+3}{3}$, or 3.

Since both vertices lie on the horizontal line $y = 3$, the transverse axis is parallel to the x-axis. The location of the foci is given by $(h + c, k)$ and $(h - c, k)$, or specifically $(7, 3)$ and $(1, 3)$. Thus $h + c = 7$, $h - c = 1$, and $k = 3$. Solving for h and c simultaneously yields $h = 4$ and $c = 3$.

Since $c^2 = a^2 + b^2$, $b^2 = c^2 - a^2$.

Substituting, we have $b^2 = 3^2 - 2^2 = 9 - 4 = 5$.

Substituting $a = 2, b^2 = 5, h = 4$, and $k = 3$ in the standard equation $\dfrac{(x-h)^2}{a^2} - \dfrac{(y-k)^2}{b^2} = 1$ yields $\dfrac{(x-4)^2}{4} - \dfrac{(y-3)^2}{5} = 1$.

45. foci $(1, -2)$ and $(7, -2)$, slope of an asymptote $= \frac{5}{4}$

Both foci lie on the horizontal line $y = -2$; therefore, the transverse axis is parallel to the x-axis.

The foci are given by $(h + c, k)$ and $(h - c, k)$.

Thus, $h - c = 1, h + c = 7$, and $k = -2$. Solving simultaneously for h and c yields $h = 4$ and $c = 3$.

Since $y - k = \frac{b}{a}(x - h)$ is the equation for an asymptote, and the slope of an asymptote is given as $\frac{5}{4}$, $\frac{b}{a} = \frac{5}{4}$, $b = \frac{5a}{4}$,

and $b^2 = \frac{25a^2}{16}$.

Because $a^2 + b^2 = c^2$, substituting $c = 3$ and $b^2 = \dfrac{25a^2}{16}$ yields $a^2 = \dfrac{144}{41}$.

Therefore, $b^2 = \dfrac{3600}{656} = \dfrac{225}{41}$.

Substituting in the standard equation for a hyperbola yields $\dfrac{(x-4)^2}{144/41} - \dfrac{(y+2)^2}{225/41} = 1$

47. Because the transverse axis is parallel to the y-axis and the center is $(7, 2)$, the equation of the hyperbola is

$$\frac{(y-2)^2}{a^2} - \frac{(x-7)^2}{b^2} = 1$$

Because $(9, 4)$ is a point on the hyperbola,

$$\frac{(4-2)^2}{a^2} - \frac{(9-7)^2}{b^2} = 1$$

The slope of the asymptote is $\frac{1}{2}$. Therefore $\frac{1}{2} = \frac{a}{b}$ or $b = 2a$.

Substituting, we have

$$\frac{4}{a^2} - \frac{4}{4a^2} = 1$$

$$\frac{4}{a^2} - \frac{1}{a^2} = 1, \text{ or } a^2 = 3$$

Since $b = 2a$, $b^2 = 4a^2$, or $b^2 = 12$. The equation is $\dfrac{(y-2)^2}{3} - \dfrac{(x-7)^2}{12} = 1$.

49. vertices $(1, 6)$ and $(1, 8)$, eccentricity $= 2$

Length of transverse axis $=$ distance between vertices

$$2a = |6 - 8| = 2$$

$$a = 1 \text{ and } a^2 = 1$$

Center (midpoint of transverse axis) is $\left(\dfrac{1+1}{2}, \dfrac{6+8}{2}\right)$, or $(1, 7)$.

Therefore, $h = 1$ and $k = 7$.

Since both vertices lie on the vertical line $x = 1$, the transverse axis is parallel to the y-axis.

Since $e = \dfrac{c}{a}$, $c = ae = (1)(2) = 2$.

Because $b^2 = c^2 - a^2$, $b^2 = (2)^2 - (1)^2 = 4 - 1 = 3$.

Substituting h, k, a^2, and b^2 into the standard equation yields $\dfrac{(y-7)^2}{1} - \dfrac{(x-1)^2}{3} = 1$

51. foci $(4, 0)$ and $(-4, 0)$, eccentricity $= 2$

Center (midpoint of line segment joining foci) is $\left(\dfrac{4+(-4)}{2}, \dfrac{0+0}{2}\right)$, or $(0, 0)$

Thus, $h = 0$ and $k = 0$.

Since both foci lie on the horizontal line $y = 0$, the transverse axis is parallel to the x-axis. The locations of the foci are given by $(h + c, k)$ and $(h - c, k)$, or specifically $(4, 0)$ and $(-4, 0)$

Since $h = 0$, $c = 4$.

Because $e = \dfrac{c}{a}$, $a = \dfrac{c}{e} = \dfrac{4}{2} = 2$ and $a^2 = 4$.

Because $b^2 = c^2 - a^2$, $b^2 = 4^2 - 2^2 = 16 - 4 = 12$.

Substituting h, k, a^2 and b^2 into the standard formula for a hyperbola yields $\dfrac{x^2}{4} - \dfrac{y^2}{12} = 1$

53. conjugate axis length = 4, center (4, 1), eccentricity = $\frac{4}{3}$

$2b$ = conjugate axis length = 4

$b = 2$ and $b^2 = 4$

Since

$e = \dfrac{c}{a} = \dfrac{4}{3}$, $c = \dfrac{4a}{3}$ and $c^2 = \dfrac{16a^2}{9}$. Since $a^2 + b^2 = c^2$, substituting $b^2 = 4$ and $c^2 = \dfrac{16a^2}{9}$ and solving for a^2 yields $a^2 = \dfrac{36}{7}$.

Substituting into the two standard equations of a hyperbola yields $\dfrac{(x-4)^2}{36/7} - \dfrac{(y-1)^2}{4} = 1$ and $\dfrac{(y-1)^2}{36/7} - \dfrac{(x-4)^2}{4} = 1$

55. **a.** Because the transmitters are 250 miles apart,

$2c = 250$ and $c = 125$.

$2a$ = rate × time

$2a = 0.186 \times 500 = 93$

Thus, $a = 46.5$ miles.

$b = \sqrt{c^2 - a^2} = \sqrt{125^2 - 46.5^2} = \sqrt{13,462.75}$ miles

The ship is located on the hyperbola given by

$\dfrac{x^2}{2,162.25} - \dfrac{y^2}{13,462.75} = 1$

b. $x = 100$

$\dfrac{10,000}{2,162.25} - \dfrac{y^2}{13,462.75} = 1$

$\dfrac{-y^2}{13,462.75} \approx -3.6248121$

$y^2 \approx 48,799.939$

$y \approx 221$

The ship is 221 miles from the coastline.

57. When the wave hits Earth, $z = 0$.

$y^2 = x^2 + (z - 10,000)^2$

$y^2 = x^2 + (0 - 10,000)^2$

$y^2 - x^2 = 10,000^2$

It is a hyperbola.

59. **a.** Using the eccentricity, and $a = 2$,

$\dfrac{c}{2} = \dfrac{\sqrt{17}}{4} \Rightarrow c = \dfrac{\sqrt{17}}{2}$

Solve for b.

$a^2 + b^2 = c^2$

$b^2 = c^2 - a^2$

$b^2 = \left(\dfrac{\sqrt{17}}{2}\right)^2 - 2^2$

$b^2 = \dfrac{17}{4} - \dfrac{16}{4}$

$b^2 = \dfrac{1}{4}$

$b = \dfrac{1}{2} = 0.5$

$\dfrac{x^2}{2^2} - \dfrac{y^2}{0.5^2} = 1$

b. For *FG*, $y = 0.6$.

$\dfrac{x^2}{2^2} - \dfrac{0.6^2}{0.5^2} = 1$

$\dfrac{x^2}{2^2} = 1 + \dfrac{0.6^2}{0.5^2}$

$x^2 = 2^2\left(1 + \dfrac{0.6^2}{0.5^2}\right)$

$x = \sqrt{2^2\left(1 + \dfrac{0.6^2}{0.5^2}\right)}$

$x \approx 3.1241$

$FG = 2x \approx 6.25$ in.

61.
$$4x^2 + 9y^2 - 16x - 36y + 16 = 0$$
$$4(x^2 - 4x) + 9(y^2 - 4y) = -16$$
$$4(x^2 - 4x + 4) + 9(y^2 - 4y + 4) = -16 + 16 + 36$$
$$4(x-2)^2 + 9(y-2)^2 = 36$$
$$\frac{(x-2)^2}{9} + \frac{(y-2)^2}{4} = 1$$

ellipse

center (2, 2)

vertices $(2 \pm 3, 2) = (5, 2), (-1, 2)$

foci $(2 \pm \sqrt{5}, 2) = (2 + \sqrt{5}, 2), (2 - \sqrt{5}, 2)$

63.
$$5x - 4y^2 + 24y - 11 = 0$$
$$-4(y^2 - 6y) = -5x + 11$$
$$-4(y^2 - 6y + 9) = -5x + 11 - 36$$
$$-4(y-3)^2 = -5(x - 25)$$
$$-4(y-3)^2 = -5(x + 5)$$
$$(y-3)^2 = \frac{5}{4}(x + 5)$$

parabola

vertex $(-5, 3)$

focus $\left(-5 + \dfrac{5}{16}, 3\right) = \left(-\dfrac{75}{16}, 3\right)$

directrix $x = -5 - \dfrac{5}{16}$, or $x = \dfrac{-85}{16}$

65.
$$x^2 + 2y - 8x = 0$$
$$x^2 - 8x = -2y$$
$$x^2 - 8x + 16 = -2y + 16$$
$$(x-4)^2 = -2(y - 8)$$

parabola

vertex (4, 8)

foci $\left(4, 8 - \dfrac{1}{2}\right) = \left(4, \dfrac{15}{2}\right)$

directrix $y = 8 + \dfrac{1}{2}$, or $y = \dfrac{17}{2}$

67.
$$25x^2 + 9y^2 - 50x - 72y - 56 = 0$$
$$25(x^2 - 2x) + 9(y^2 - 8y) = 56$$
$$25(x^2 - 2x + 1) + 9(y^2 - 8y + 16) = 56 + 25 + 144$$
$$25(x-1)^2 + 9(y-4)^2 = 225$$
$$\frac{(x-1)^2}{9} + \frac{(y-4)^2}{25} = 1$$

ellipse

center (1, 4)

vertices $(1, 4 \pm 5) = (1, 9), (1, -1)$

foci $(1, 4 \pm 4) = (1, 8), (1, 0)$

•••

69. foci $F_1(2,0)$, $F_2(-2,0)$ passing through $P_1(2,3)$

$$d(P_1,F_2)-d(P_1,F_1)=\sqrt{(2+2)^2+3^2}-\sqrt{(2-2)^2+3^2}=5-3=2$$

Let $P(x,y)$ be any point on the hyperbola. Since the difference between F_1P and F_2P is the same as the difference between F_1P_1 and

F_2P_1, we have

$$\sqrt{(x-2)^2+y^2}-\sqrt{(x+2)^2+y^2}=2$$
$$\sqrt{(x-2)^2+y^2}=2+\sqrt{(x+2)^2+y^2}$$
$$x^2-4x+4+y^2=4+4\sqrt{(x+2)^2+y^2}+x^2+4x+4+y^2$$
$$-8x-4=4\sqrt{(x+2)^2+y^2}$$
$$-2x-1=\sqrt{(x+2)^2+y^2}$$
$$4x^2+4x+1=x^2+4x+4+y^2$$
$$3x^2-y^2=3$$
$$\frac{x^2}{1}-\frac{y^2}{3}=1$$

71. foci $(0,4)$ and $(0,-4)$, point $\left(\frac{7}{3},4\right)$

Difference in distances from (x,y) to foci = difference of distances from $\left(\frac{7}{3},4\right)$ to foci

$$\sqrt{(x-0)^2+(y-4)^2}-\sqrt{(x-0)^2+(y+4)^2}=\sqrt{\left(\frac{7}{3}-0\right)^2+(4-4)^2}-\sqrt{\left(\frac{7}{3}-0\right)^2+(4+4)^2}$$
$$\sqrt{x^2+y^2-8y+16}-\sqrt{x^2+y^2+8y+16}=\frac{7}{3}-\frac{25}{3}=-6$$
$$\sqrt{x^2+y^2-8y+16}=\sqrt{x^2+y^2+8y+16}-6$$
$$x^2+y^2-8y+16=x^2+y^2+8y+16-12\sqrt{x^2+y^2+8y+16}+36$$
$$-16y-36=-12\sqrt{x^2+y^2+8y+16}$$
$$4y+9=3\sqrt{x^2+y^2+8y+16}$$
$$16y^2+72y+81=9x^2+9y^2+72y+144$$
$$7y^2-9x^2=63$$
$$\frac{y^2}{9}-\frac{x^2}{7}=1$$

73.

PS1. $\cos(\alpha + \beta) = \cos\alpha\cos\beta - \sin\alpha\sin\beta$

PS2. $\sin(\alpha + \beta) = \sin\alpha\cos\beta + \cos\alpha\sin\beta$

PS3. $\cot 2\alpha = \dfrac{\sqrt{3}}{3}$

$\tan 2\alpha = \dfrac{3}{\sqrt{3}}$

$2\alpha = \tan^{-1}\left(\dfrac{3}{\sqrt{3}}\right) = \dfrac{\pi}{3}$

$\alpha = \dfrac{\pi}{6}$

PS4. $\sin\alpha = \dfrac{1}{2}, \ \alpha = 30° \text{ or } 150°$

$\cos\alpha = -\dfrac{\sqrt{3}}{2} \ \alpha = 150° \text{ or } 210°$

$\alpha = 150°$

PS5. $4x^2 - 6y^2 + 9x + 16y - 8 = 0$
$A = 4, B = 0, C = -6$
Since $B^2 - 4AC = 0^2 - 4(4)(-6) = 96 > 0$,
the graph is a hyperbola.

PS6.

Section 6.4

1. $xy = 3$

$A = 0, B = 1, C = 0$

$\cot 2\alpha = \dfrac{A - C}{B}$

$\cot 2\alpha = \dfrac{0 - 0}{1}$

$\cot 2\alpha = 0$

$2\alpha = 90°$

$\alpha = 45°$

3. $9x^2 - 24xy + 16y^2 - 320x - 240y = 0$

$A = 9, B = -24, C = 16$

$\cot 2\alpha = \dfrac{A - C}{B}$

$\cot 2\alpha = \dfrac{9 - 16}{-24}$

$\cot 2\alpha = \dfrac{-7}{-24}$

$\cot 2\alpha = \dfrac{7}{24}$

$2\alpha \approx 73.74°$

$\alpha \approx 36.9°$

5. $5x^2 - 6\sqrt{3}xy - 11y^2 + 4x - 3y + 2 = 0$

$A = 5, B = -6\sqrt{3}, C = -11$

$\cot 2a = \dfrac{A - C}{B}$

$\cot 2a = \dfrac{5 - (-11)}{-6\sqrt{3}}$

$\cot 2a = \dfrac{5 + 11}{-6\sqrt{3}}$

$\cot 2a = \dfrac{16}{-6\sqrt{3}}$

$\cot 2a = -\dfrac{8}{3\sqrt{3}}$

$\cot 2a = -\dfrac{8\sqrt{3}}{9}$

$2\alpha \approx 147°$

$\alpha \approx 73.5°$

7. $2x^2 + xy + y^2 - 4 = 0$

$A = 2, B = 1, C = 1$

$\cot 2a = \dfrac{A - C}{B}$

$\cot 2a = \dfrac{2 - 1}{1}$

$\cot 2a = 1$

$2\alpha \approx 45°$

$\alpha \approx 22.5°$

9.
$$xy = 4$$
$$xy - 4 = 0$$

$$A = 0, B = 1, C = 0, F = -4$$

$$\cot 2\alpha = \frac{A-C}{B} = \frac{0-0}{1} = 0$$

$$\csc^2 2\alpha = \cot^2 2\alpha + 1$$
$$\csc^2 2\alpha = 0^2 + 1 = 1$$
$$\csc 2\alpha = +1 \quad (2\alpha \text{ is in the first quadrant.})$$

$$\sin 2\alpha = \frac{1}{\csc 2\alpha} = \frac{1}{1} = 1$$

$$\sin^2 2\alpha + \cos^2 2\alpha = 1$$
$$\cos^2 2\alpha = 1 - \sin^2 2\alpha$$
$$\cos^2 2\alpha = 1 - (1)^2$$
$$\cos^2 2\alpha = 0$$
$$\cos 2\alpha = 0$$

$$\sin \alpha = \sqrt{\frac{1-(0)}{2}} = \frac{\sqrt{2}}{2} \qquad \cos \alpha = \sqrt{\frac{1+(0)}{2}} = \frac{\sqrt{2}}{2}$$

$$\alpha = 45°$$

$$A' = A\cos^2 \alpha + B\cos \alpha \sin \alpha + C\sin^2 \alpha = 0\left(\frac{\sqrt{2}}{2}\right)^2 + 1\left(\frac{\sqrt{2}}{2}\right)\left(\frac{\sqrt{2}}{2}\right) + 0\left(\frac{\sqrt{2}}{2}\right)^2 = \frac{1}{2}$$

$$C' = A\sin^2 \alpha - B\cos \alpha \sin \alpha + C\cos^2 \alpha = 0\left(\frac{\sqrt{2}}{2}\right)^2 - 1\left(\frac{\sqrt{2}}{2}\right)\left(\frac{\sqrt{2}}{2}\right) + 0\left(\frac{\sqrt{2}}{2}\right)^2 = -\frac{1}{2}$$

$$F' = F = -4$$

$$\frac{1}{2}x'^2 - \frac{1}{2}y'^2 - 4 = 0 \text{ or } \frac{x'^2}{8} - \frac{y'^2}{8} = 1$$

11. $6x^2 - 6xy + 14y^2 - 45 = 0$

$A = 6, B = -6, C = 14, F = -45$

$$\cot 2\alpha = \frac{A - C}{B} = \frac{6 - 14}{-6} = \frac{4}{3}$$

$$\csc^2 2\alpha = \cot^2 2\alpha + 1$$

$$\csc^2 2\alpha = \left(\frac{4}{3}\right)^2 + 1 = \frac{25}{9}$$

$$\csc 2\alpha = +\sqrt{\frac{25}{9}} = \frac{5}{3} \qquad (2\alpha \text{ is in the first quadrant.})$$

$$\sin 2\alpha = \frac{1}{\csc 2\alpha} = \frac{3}{5}$$

$$\sin^2 \alpha + \cos^2 2\alpha = 1$$

$$\cos^2 2\alpha = 1 - \sin^2 2\alpha$$

$$\cos^2 2\alpha = 1 - \left(\frac{3}{5}\right)^2 = \frac{16}{25}$$

$$\cos 2\alpha = +\sqrt{\frac{16}{25}} = \frac{4}{5} \qquad (2\alpha \text{ is in the first quadrant.})$$

$$\sin \alpha = \sqrt{\frac{1 - \left(\frac{4}{5}\right)}{2}} = \frac{\sqrt{10}}{10} \qquad \cos \alpha = \sqrt{\frac{1 + \left(\frac{4}{5}\right)}{2}} = \frac{3\sqrt{10}}{10}$$

$$\alpha = 18.4°$$

$$A' = A\cos^2 \alpha + B\cos \alpha \sin \alpha + C\sin^2 \alpha = 6\left(\frac{3\sqrt{10}}{10}\right)^2 - 6\left(\frac{3\sqrt{10}}{10}\right)\left(\frac{\sqrt{10}}{10}\right) + 14\left(\frac{\sqrt{10}}{10}\right)^2 = 5$$

$$C' = A\sin^2 \alpha - B\cos \alpha \sin \alpha + C\cos^2 \alpha = 6\left(\frac{\sqrt{10}}{10}\right)^2 + 6\left(\frac{3\sqrt{10}}{10}\right)\left(\frac{\sqrt{10}}{10}\right) + 14\left(\frac{3\sqrt{10}}{10}\right)^2 = 15$$

$$F' = F = -45$$

$$5x'^2 + 15y'^2 - 45 = 0 \text{ or } \frac{(x')^2}{9} + \frac{(y')^2}{3} = 1$$

13. $x^2 - 4xy + 2y^2 - 1 = 0$

$A = 1, B = 4, C = -2, F = -1$

$$\cot 2\alpha = \frac{A - C}{B} = \frac{1 - (-2)}{4} = \frac{3}{4}$$

$$\csc^2 2\alpha = \cot^2 2\alpha + 1$$

$$\csc^2 2\alpha = \left(\frac{3}{4}\right)^2 + 1 = \frac{25}{16}$$

$$\csc 2\alpha = +\sqrt{\frac{25}{16}} = \frac{5}{4} \qquad (2\alpha \text{ is in the first quadrant.})$$

$$\sin 2\alpha = \frac{1}{\csc 2\alpha} = \frac{4}{5}$$

$$\sin^2 \alpha + \cos^2 2\alpha = 1$$

$$\cos^2 2\alpha = 1 - \sin^2 \alpha$$

$$\cos^2 2\alpha = 1 - \left(\frac{4}{5}\right)^2 = \frac{9}{25}$$

$$\cos 2\alpha = +\sqrt{\frac{9}{25}} = \frac{3}{5} \qquad (2\alpha \text{ is in first quadrant.})$$

$$\sin\alpha = \sqrt{\frac{1 - \left(\frac{3}{5}\right)}{2}} = \frac{\sqrt{5}}{5} \qquad\qquad \cos\alpha = \sqrt{\frac{1 + \left(\frac{3}{5}\right)}{2}} = \frac{2\sqrt{5}}{5}$$

$\alpha \approx 26.6°$

$$A' = A\cos^2\alpha + B\cos\alpha\sin\alpha + C\sin^2\alpha = 1\left(\frac{2\sqrt{5}}{5}\right)^2 + 4\left(\frac{2\sqrt{5}}{5}\right)\left(\frac{\sqrt{5}}{5}\right) - 2\left(\frac{\sqrt{5}}{5}\right)^2 = 2$$

$$C' = A\sin^2\alpha - B\cos\alpha\sin\alpha + C\cos^2\alpha = 1\left(\frac{\sqrt{5}}{5}\right)^2 - 4\left(\frac{2\sqrt{5}}{5}\right)\left(\frac{\sqrt{5}}{5}\right) - 2\left(\frac{2\sqrt{5}}{5}\right)^2 = -3$$

$F' = F = -1$

$$2(x')^2 + 3(y')^2 = 1$$

15. $3x^2 + 2\sqrt{3}xy + y^2 + 2x - 2\sqrt{3}y + 16 = 0$

$A = 3, B = 2\sqrt{3}, C = 1, D = 2, E = -2\sqrt{3}, F = 16$

$\cot 2\alpha = \dfrac{A - C}{B} = \dfrac{3 - 1}{2\sqrt{3}} = \dfrac{\sqrt{3}}{3}$

$\csc^2 2\alpha = \cot^2 2\alpha + 1$

$\csc^2 2\alpha = \left(\dfrac{\sqrt{3}}{3}\right)^2 + 1 = \dfrac{4}{3}$

$\csc 2\alpha = +\sqrt{\dfrac{4}{3}} = \dfrac{2\sqrt{3}}{3}$ $\qquad$ (2α is in the first quadrant.)

$\sin 2\alpha = \dfrac{1}{\csc 2\alpha} = \dfrac{\sqrt{3}}{2}$

$\sin^2 \alpha + \cos^2 2\alpha = 1$

$\cos^2 2\alpha = 1 - \sin^2 2\alpha$

$\cos^2 2\alpha = 1 - \left(\dfrac{\sqrt{3}}{2}\right)^2 = \dfrac{1}{4}$

$\cos 2\alpha = +\sqrt{\dfrac{1}{4}} = \dfrac{1}{2}$ $\qquad$ (2α is in the first quadrant.)

$\sin \alpha = \sqrt{\dfrac{1 - \left(\frac{1}{2}\right)}{2}} = \dfrac{1}{2}$ $\qquad\qquad$ $\cos \alpha = \sqrt{\dfrac{1 + \left(\frac{1}{2}\right)}{2}} = \dfrac{\sqrt{3}}{2}$

$\alpha = 30°$

$A' = A\cos^2 \alpha + B\cos\alpha\sin\alpha + C\sin^2\alpha = 3\left(\dfrac{\sqrt{3}}{2}\right)^2 + 2\sqrt{3}\left(\dfrac{\sqrt{3}}{2}\right)\left(\dfrac{1}{2}\right) + 1\left(\dfrac{1}{2}\right)^2 = 4$

$C' = A\sin^2\alpha - B\cos\alpha\sin\alpha + C\cos^2\alpha = 3\left(\dfrac{1}{2}\right)^2 - 2\sqrt{3}\left(\dfrac{\sqrt{3}}{2}\right)\left(\dfrac{1}{2}\right) + 1\left(\dfrac{\sqrt{3}}{2}\right)^2 = 0$

$D' = D\cos\alpha + E\sin\alpha = 2\left(\dfrac{\sqrt{3}}{2}\right) - 2\sqrt{3}\left(\dfrac{1}{2}\right) = 0$

$E' = -D\sin\alpha + E\cos\alpha = -2\left(\dfrac{1}{2}\right) - 2\sqrt{3}\left(\dfrac{\sqrt{3}}{2}\right) = -4$

$F' = F = 16$

$4(x')^2 - 4y' + 16 = 0$ or $y' = (x')^2 + 4$

17. $\quad 9x^2 + 24xy + 16y^2 - 40x - 30y + 100 = 0$

$A = 9, B = -24, C = 16, D = -40, E = -30 \ F = 100$

$$\cot 2\alpha = \frac{A-C}{B} = \frac{9-16}{-24} = \frac{7}{24}$$

$$\csc^2 2\alpha = \cot^2 2\alpha + 1$$

$$\csc^2 2\alpha = \left(\frac{7}{24}\right)^2 + 1 = \frac{625}{576}$$

$$\csc 2\alpha = +\sqrt{\frac{625}{576}} = \frac{25}{24} \qquad (2\alpha \text{ is in the first quadrant.})$$

$$\sin 2\alpha = \frac{1}{\csc 2\alpha} = \frac{24}{25}$$

$$\sin^2 \alpha + \cos^2 2\alpha = 1$$

$$\cos^2 2\alpha = 1 - \sin^2 2\alpha$$

$$\cos^2 2\alpha = 1 - \left(\frac{24}{25}\right)^2 = \frac{49}{625}$$

$$\cos 2\alpha = +\sqrt{\frac{49}{625}} = \frac{7}{25} \qquad (2\alpha \text{ is in first quadrant.})$$

$$\sin \alpha = \sqrt{\frac{1-\left(\frac{7}{25}\right)}{2}} = \frac{3}{5} \qquad \cos \alpha = \sqrt{\frac{1+\left(\frac{7}{25}\right)}{2}} = \frac{4}{5}$$

$\alpha \approx 36.9°$

$$A' = A\cos^2 \alpha + B\cos\alpha\sin\alpha + C\sin^2 \alpha = 9\left(\frac{4}{5}\right)^2 - 24\left(\frac{4}{5}\right)\left(\frac{3}{5}\right) + 16\left(\frac{3}{5}\right)^2 = 0$$

$$C' = A\sin^2 \alpha - B\cos\alpha\sin\alpha + C\cos^2 \alpha = 9\left(\frac{3}{5}\right)^2 + 24\left(\frac{4}{5}\right)\left(\frac{3}{5}\right) + 16\left(\frac{4}{5}\right)^2 = 25$$

$$D' = D\cos\alpha + E\sin\alpha = -40\left(\frac{4}{5}\right) - 30\left(\frac{3}{5}\right) = -50$$

$$E' = -D\sin\alpha + E\cos\alpha = 40\left(\frac{3}{5}\right) - 30\left(\frac{4}{5}\right) = 0$$

$F' = F = 100$

$$25(x')^2 - 50x' + 100 = 0 \text{ or } (y')^2 = 2(x-2)$$

19. $6x^2 + 24xy - y^2 - 12x + 26y + 11 = 0$

$A = 6, B = 24, C = -1, D = -12, E = 26 \; F = 11$

$$\cot 2\alpha = \frac{A-C}{B} = \frac{6-(-1)}{24} = \frac{7}{24}$$

$$\csc^2 2\alpha = \cot^2 2\alpha + 1$$

$$\csc^2 2\alpha = \left(\frac{7}{24}\right)^2 + 1 = \frac{625}{576}$$

$$\csc 2\alpha = +\sqrt{\frac{625}{576}} = \frac{25}{24} \qquad (2\alpha \text{ is in the first quadrant.})$$

$$\sin 2\alpha = \frac{1}{\csc 2\alpha} = \frac{24}{25}$$

$$\sin^2 2\alpha + \cos^2 2\alpha = 1$$

$$\cos^2 2\alpha = 1 - \sin^2 2\alpha$$

$$\cos^2 2\alpha = 1 - \left(\frac{24}{25}\right)^2 = \frac{49}{625}$$

$$\cos 2\alpha = +\sqrt{\frac{49}{625}} = \frac{7}{25} \qquad (2\alpha \text{ is in the first quadrant.})$$

$$\sin \alpha = \sqrt{\frac{1-\left(\frac{7}{25}\right)}{2}} = \frac{3}{5} \qquad\qquad \cos \alpha = \sqrt{\frac{1+\left(\frac{7}{25}\right)}{2}} = \frac{4}{5}$$

$\alpha \approx 36.9°$

$$A' = A\cos^2 \alpha + B\cos \alpha \sin \alpha + C\sin^2 \alpha = 6\left(\frac{4}{5}\right)^2 + 24\left(\frac{4}{5}\right)\left(\frac{3}{5}\right) - 1\left(\frac{3}{5}\right)^2 = 15$$

$$C' = A\sin^2 \alpha - B\cos \alpha \sin \alpha + C\cos^2 \alpha = 6\left(\frac{3}{5}\right)^2 - 24\left(\frac{4}{5}\right)\left(\frac{3}{5}\right) - 1\left(\frac{4}{5}\right)^2 = -10$$

$$D' = D\cos\alpha + E\sin\alpha = -12\left(\frac{4}{5}\right) + 26\left(\frac{3}{5}\right) = 6$$

$$E' = -D\sin\alpha + E\cos\alpha = 12\left(\frac{3}{5}\right) + 26\left(\frac{4}{5}\right) = 28$$

$F' = F = 11$

$$15(x')^2 - 10(y')^2 + 6x' + 28y' + 11 = 0$$

21. $A = 6, B = -1, C = 2, D = 4, E = -12, F = 7$

Graph $y = \dfrac{-(-x-12) \pm \sqrt{(-x-12)^2 - 8(6x^2 + 4x + 7)}}{4}$

The graph will appear disconnected at the endpoints of
the minor axes on a graphing utility.

23. $A = 1, B = -6, C = 1, D = -2, E = -5, F = 4$

Graph $y = \dfrac{-(-6x-5) \pm \sqrt{(-6x-5)^2 - 4(x^2 - 2x + 4)}}{2}$

25. $A = 3, B = -6, C = 3, D = 10, E = -8, F = -2$

Graph $y = \dfrac{-(-6x-8) \pm \sqrt{(-6x-8)^2 - 12(3x^2 + 10x - 2)}}{6}$

27. $\dfrac{2x'^2}{1} - \dfrac{3y'^2}{1} = 1 \qquad \sin\alpha = \dfrac{\sqrt{5}}{5} \qquad \cos\alpha = \dfrac{2\sqrt{5}}{5}$

$a^2 = \dfrac{1}{2}; \quad a = \dfrac{\sqrt{2}}{2}$

$b^2 = \dfrac{1}{3}; \quad b = \dfrac{\sqrt{3}}{3}$

Asymptotes: $y' = \pm\dfrac{b}{a}x'$ or $y' = \pm\dfrac{\sqrt{6}}{3}x'$

Using the transformation formulas for x' and y' yields

$y\cos\alpha - x\sin\alpha = \pm\dfrac{\sqrt{6}}{3}(x\cos\alpha + y\sin\alpha)$

$\dfrac{2\sqrt{5}}{5}y - \dfrac{\sqrt{5}}{5}x = \pm\dfrac{\sqrt{6}}{3}\left(\dfrac{2\sqrt{5}}{5}x + \dfrac{\sqrt{5}}{5}y\right)$

$\dfrac{2\sqrt{5}}{5}y - \dfrac{\sqrt{5}}{5}x = \pm\left(\dfrac{2\sqrt{30}}{15} + \dfrac{\sqrt{30}}{15}y\right)$

Multiplying both sides of the equation by $15/\sqrt{5}$ yields

$6y - 3x = \pm\left(2\sqrt{6}x + \sqrt{6}y\right)$

$6y - 3x = 2\sqrt{6}x + \sqrt{6}y$ and $\qquad 6y - 3x = -(2\sqrt{6}x + \sqrt{6}y)$

$6y - \sqrt{6}y = 3x + 2\sqrt{6}x \qquad\qquad 6y + \sqrt{6}y = 3x - 2\sqrt{6}x$

$y = \dfrac{3 + 2\sqrt{6}}{6 - \sqrt{6}}x \qquad\qquad\qquad y = \dfrac{3 - 2\sqrt{6}}{6 + \sqrt{6}}x$

Rationalizing the denominators, we obtain

$y = \dfrac{2 + \sqrt{6}}{2}x \qquad$ and $\qquad y = \dfrac{2 - \sqrt{6}}{2}x$

29. From Exercise 11, $\dfrac{x'^2}{9} + \dfrac{y'^2}{3} = 1$, $\sin\alpha = \dfrac{\sqrt{10}}{10}$, $\cos\alpha = \dfrac{3\sqrt{10}}{10}$, $a^2 = 9$, $b^2 = 3$, $c^2 = 9 - 3 = 6$.

Thus $c = \sqrt{6}$.

Foci in $x'y'$-coordinates are $(\pm\sqrt{6}, 0)$. Thus $x' = \pm\sqrt{6}$, $y' = 0$.

$$x = x'\cos\alpha - y'\sin\alpha \qquad\qquad y = y'\cos\alpha + x'\sin\alpha$$

$$= \pm\sqrt{6}\left(\dfrac{3\sqrt{10}}{10}\right) - 0\cdot\dfrac{\sqrt{10}}{10} \qquad\qquad = 0\cdot\cos\alpha \pm \sqrt{6}\left(\dfrac{\sqrt{10}}{10}\right)$$

$$= \pm\dfrac{3\sqrt{15}}{5} \qquad\qquad\qquad\qquad = \pm\dfrac{\sqrt{15}}{5}$$

Foci in the xy-coordinate system are $\left(\dfrac{3\sqrt{15}}{5}, \dfrac{\sqrt{15}}{5}\right)$ and $\left(-\dfrac{3\sqrt{15}}{5}, -\dfrac{\sqrt{15}}{5}\right)$.

31. $x^2 + xy - y^2 - 40 = 0$
$A = 1$, $B = 1$, $C = -1$
Since $B^2 - 4AC = 1^2 - 4(1)(-1) = 5 > 0$,
the graph is a hyperbola.

33. $3x^2 + 2\sqrt{3}xy + y^2 - 3x + 2y + 20 = 0$
$A = 3$, $B = 2\sqrt{3}$, $C = 1$
Since $B^2 - 4AC = (2\sqrt{3})^2 - 4(3)(1) = 0$,
the graph is a parabola.

35. $4x^2 - 4xy + y^2 - 12y - 20 = 0$
$A = 4$, $B = -4$, $C = 1$
Since $B^2 - 4AC = (-4)^2 - 4(4)(1) = 0$, the graph is a parabola.

37. $5x^2 - 6\sqrt{3}xy - 11y^2 + 4x - 3y + 2 = 0$
$A = 5$, $B = -6\sqrt{3}$, $C = -11$
Since $B^2 - 4AC = (-6\sqrt{3})^2 - 4(5)(-11) = 328 > 0$,
the graph is a hyperbola.

39. $6x^2 + 2\sqrt{3}xy + 5y^2 - 3x + 2y - 20 = 0$
$A = 6$, $B = 2\sqrt{3}$, $C = 5$
Since $B^2 - 4AC = (2\sqrt{3})^2 - 4(6)(5) = -108 < 0$,
the graph is an ellipse.

Connecting Concepts

41. $x^2 + y^2 = r^2$

Substitute $x = x'\cos\alpha - y'\sin\alpha$ and $y = y'\cos\alpha + x'\sin\alpha$.

$$(x'\cos\alpha - y'\sin\alpha)^2 + (y'\cos\alpha + x'\sin\alpha)^2 = r^2$$

$$x'^2\cos^2\alpha - 2x'y'\cos\alpha\sin\alpha + y'^2\sin^2\alpha + y'^2\cos^2\alpha + 2x'y'\cos\alpha\sin\alpha + x'^2\sin^2\alpha = r^2$$

$$x'^2(\cos^2\alpha + \sin^2\alpha) + x'y'(2\cos\alpha\sin\alpha - 2\cos\alpha\sin\alpha) + y'^2(\sin^2\alpha + \cos^2\alpha) = r^2$$

$$x'^2(1) + x'y'(0) + y'^2(1) = r^2$$

$$x'^2 + y'^2 = r^2$$

43. Vertices $(2, 4)$ and $(-2, -4)$ imply that the major axis is on the line $y = 2x$. Consider an $x'y'$-coordinate system rotated an angle α,

where $\tan \alpha = 2$. From this equation, using identities, $\cos \alpha = \dfrac{\sqrt{5}}{5}$ and $\sin \alpha = \dfrac{2\sqrt{5}}{5}$.

The equation of the ellipse in the $x'y'$-coordinate system is

$$\frac{x'^2}{20} + \frac{y'^2}{10} = 1 \qquad (1)$$

This follows from the fact that $(2, 4)$ in xy-coordinates is $\left(\sqrt{20}, \, 0\right)$ in $x'y'$-coordinates.

Therefore, $a = \sqrt{20}$. Also, $\left(\sqrt{2}, \, 2\sqrt{2}\right)$ in xy-coordinates is $\left(\sqrt{10}, \, 0\right)$ in $x'y'$-coordinates.

Therefore, $c = \sqrt{10}$. Thus $b = \sqrt{10}$.

Now let $x' = \dfrac{\sqrt{5}}{5} x + \dfrac{2\sqrt{5}}{5} y$ and $y' = \dfrac{\sqrt{5}}{5} y - \dfrac{2\sqrt{5}}{5} x$ and substitute into Equation (1):

$$\frac{\left(\frac{\sqrt{5}}{5} x + \frac{2\sqrt{5}}{5} y\right)^2}{20} + \frac{\left(\frac{\sqrt{5}}{5} y - \frac{2\sqrt{5}}{5} x\right)^2}{10} = 1$$

Simplifying, we have

$$\frac{\left(\frac{1}{5} x^2 + \frac{4}{5} xy + \frac{4}{5} y^2\right)}{20} + \frac{\left(\frac{4}{5} x^2 - \frac{4}{5} xy + \frac{1}{5} y^2\right)}{10} = 1$$

$$\frac{\frac{9}{5} x^2 - \frac{4}{5} xy + \frac{6}{5} y^2}{20} = 1$$

$$\frac{9x^2 - 4xy + 6y^2}{100} = 1$$

$$\text{or } 9x^2 - 4xy + 6y^2 = 100$$

45. $A' + C' = A\cos^2\alpha + B\cos\alpha\sin\alpha + C\sin^2\alpha + A\sin^2\alpha - B\cos\alpha\sin\alpha + C\cos^2\alpha$

$\qquad\quad = A(\cos^2\alpha + \sin^2\alpha) + B(\cos\alpha\sin\alpha - \cos\alpha\sin\alpha) + C(\sin^2\alpha + \cos^2\alpha)$

$\qquad\quad = A + C$

47. Ellipse with major axis parallel to x-axis:

$$\frac{(x-h)^2}{a^2} + \frac{(y-k)^2}{b^2} = 1$$

$$b^2(x-h)^2 + a^2(y-k)^2 = a^2b^2$$
$$b^2(x^2 - 2hx + h^2) + a^2(y^2 - 2ky + k^2) = a^2b^2$$
$$b^2x^2 - 2b^2hx + b^2h^2 + a^2y^2 - 2a^2ky + a^2k^2 = a^2b^2$$
$$b^2x^2 + a^2y^2 - 2b^2hx - 2a^2ky + b^2h^2 + a^2k^2 - a^2b^2 = 0$$

$A = b^2, B = 0, C = a^2$

$B^2 - 4AC = 0^2 - 4b^2a^2 = 4a^2b^2 < 0$

$B^2 - 4AC < 0$ for an ellipse whose major axis is parallel to the x-axis.

Ellipse with major axis parallel to y-axis:

$$\frac{(y-k)^2}{a^2} + \frac{(x-h)^2}{b^2} = 1$$

$$b^2(y-k)^2 + a^2(x-h)^2 = a^2b^2$$
$$b^2(y^2 - 2ky + k^2) + a^2(x^2 - 2hx + h^2) = a^2b^2$$
$$b^2y^2 - 2b^2ky + b^2k^2 + a^2x^2 - 2a^2hx + a^2h^2 = a^2b^2$$
$$b^2y^2 + a^2x^2 - 2b^2ky - 2a^2hx + b^2k^2 + a^2h^2 - a^2b^2 = 0$$

$A = b^2, B = 0, C = a^2$

$B^2 - 4AC = 0^2 - 4b^2a^2 = -4a^2b^2 < 0$

$B^2 - 4AC < 0$ for an ellipse whose major axis is parallel to the y-axis.

Parabola with axis of symmetry parallel to y-axis:

$$(x-h)^2 = 4p(y-k)$$
$$x^2 - 2hx + h^2 = 4py - 4pk$$
$$x^2 - 2hx - 4py + h^2 + 4pk = 0$$

$A = 1, B = 0, C = 0$

$B^2 - 4AC = 0^2 - 4(1)(0) = 0$

$B^2 - 4AC = 0$ for a parabola with axis of symmetry parallel to the y-axis.

Parabola with axis of symmetry parallel to x-axis:

$$(y-k)^2 = 4p(x-h)$$
$$y^2 - 2ky + k^2 = 4px - 4ph$$
$$y^2 - 2ky - 4px + k^2 + 4ph = 0$$

$A = 1, B = 0, C = 0$

$B^2 - 4AC = 0^2 - 4(1)(0) = 0$

$B^2 - 4AC = 0$ for a parabola with axis of symmetry parallel to the x-axis.

Hyperbola with the transverse axis parallel to x-axis:

$$\frac{(x-h)^2}{a^2} - \frac{(y-k)^2}{b^2} = 1$$

$$b^2(x-h)^2 - a^2(y-k)^2 = a^2b^2$$
$$b^2(x^2 - 2hx + h^2) - a^2(y^2 - 2ky + k^2) = a^2b^2$$
$$b^2x^2 - 2b^2hx + b^2h^2 - a^2y^2 + 2a^2ky - a^2k^2 = a^2b^2$$
$$b^2x^2 - a^2y^2 - 2b^2hx + 2a^2ky + b^2h^2 - a^2k^2 - a^2b^2 = 0$$

$A = b^2, B = 0, C = -a^2$

$B^2 - 4AC = 0^2 - 4b^2(-a^2) = 4a^2b^2 > 0$

$B^2 - 4AC > 0$ for a hyperbola whose transverse axis is parallel to the x-axis.

Hyperbola with transverse axis parallel to y-axis:

$$\frac{(y-k)^2}{a^2} - \frac{(x-h)^2}{b^2} = 1$$

$$b^2(y-k)^2 - a^2(x-h)^2 = a^2b^2$$
$$b^2(y^2 - 2ky + k^2) - a^2(x^2 - 2hx + h^2) = a^2b^2$$
$$b^2y^2 - 2b^2ky + b^2k^2 - a^2x^2 + 2a^2hx - a^2h^2 = a^2b^2$$
$$b^2y^2 - a^2x^2 - 2b^2ky + 2a^2hx + b^2k^2 - a^2h^2 - a^2b^2 = 0$$

$A = -a^2, B = 0, C = b^2$

$B^2 - 4AC = 0^2 - 4b^2(-a^2) = 4a^2b^2 > 0$

$B^2 - 4AC > 0$ for a hyperbola whose transverse axis is parallel to the y-axis.

●●   **P**repare for **S**ection **6.5**

PS1. $\sin(-x) = -\sin x$ odd function

PS2. $\cos(-x) = \cos x$ even function

PS3. $\tan \alpha = -\sqrt{3}$
$$\alpha = \frac{2\pi}{3}, \frac{5\pi}{3}$$

PS4. $\sin \alpha = -\frac{\sqrt{3}}{2}$, $\alpha = 240°$ or $300°$
$$\cos \alpha = -\frac{1}{2}\quad \alpha = 120° \text{ or } 240°$$
$$\alpha = 240°$$

PS5. $(r\cos\theta)^2 + (r\sin\theta)^2 = r^2\cos^2\theta + r^2\sin^2\theta$
$$= r^2(\cos^2\theta + \sin^2\theta)$$
$$= r^2$$

PS6. $\sin 32° = \dfrac{y}{5}$ $y = 5\sin 32° \approx 2.6$
$$\cos 32° = \frac{x}{5}\quad x = 5\cos 32° \approx 4.2$$
$$(4.2, 2.6)$$

Section 6.5

25. Graph for $0 \le \theta \le 2\pi$.

27. Graph for $0 \le \theta \le \pi$.

29. Graph for $0 \le \theta \le \pi$.

31. Graph for $0 \le \theta \le \pi$.

33. Graph for $0 \le \theta \le 4\pi$.

35. Graph for $0 \le \theta \le 6\pi$.

37. Graph for $0 \le \theta \le 2\pi$.

39. Graph for $0 \le \theta \le 2\pi$ with $\theta_{step} = \pi/200$.

(Some graphing utilities may draw a false asymptote in "connected" mode.)

41.
$$r = \sqrt{x^2 + y^2}$$
$$= \sqrt{1^2 + (-\sqrt{3})^2}$$
$$= \sqrt{1 + 3}$$
$$= \sqrt{4}$$
$$= 2$$

$$\theta = \tan^{-1}\frac{y}{x}$$
$$= \tan^{-1}\left(\frac{-\sqrt{3}}{1}\right)$$
$$= \tan^{-1}\left(-\frac{\sqrt{3}}{1}\right)$$
$$= -60°$$

The polar coordinates of the point are $(2, -60°)$.

43.
$$x = r\cos\theta$$
$$= (-3)\left(\cos\frac{2\pi}{3}\right)$$
$$= (-3)\left(-\frac{1}{2}\right)$$
$$= \frac{3}{2}$$

$$y = r\sin\theta$$
$$= (-3)\left(\sin\frac{2\pi}{3}\right)$$
$$= (-3)\left(\frac{\sqrt{3}}{2}\right)$$
$$= -\frac{3\sqrt{3}}{2}$$

The rectangular coordinates of the point are $\left(\frac{3}{2}, -\frac{3\sqrt{3}}{2}\right)$.

45.
$$x = r\cos\theta$$
$$= 0\cos\left(-\frac{\pi}{2}\right)$$
$$= 0$$

$$y = r\sin\theta$$
$$= 0\sin\left(-\frac{\pi}{2}\right)$$
$$= 0$$

The rectangular coordinates of the point are $(0, 0)$

47.
$$r = \sqrt{x^2 + y^2}$$
$$= \sqrt{(3)^2 + (4)^2}$$
$$= \sqrt{9 + 16}$$
$$= \sqrt{25}$$
$$= 5$$

$$\theta = \tan^{-1}\frac{y}{x}$$
$$= \tan^{-1}\frac{4}{3}$$
$$\approx 53.1°$$

The approximate polar coordinates of the point are $(5, 53.1°)$.

49.
$$r = 3\cos\theta$$
$$r - 3\cos\theta = 0$$
$$r^2 - 3r\cos\theta = 0$$
$$x^2 + y^2 - 3x = 0$$

51.
$$r = 3\sec\theta$$
$$r = \frac{3}{\cos\theta}$$
$$r\cos\theta = 3$$
$$x = 3$$

53.
$$r = 4$$
$$\sqrt{x^2 + y^2} = 4$$
$$x^2 + y^2 = 16$$

55.
$$\theta = \frac{\pi}{6}$$
$$\tan\theta = \frac{y}{x}$$
$$\tan\frac{\pi}{6} = \frac{y}{x}$$
$$\frac{\sqrt{3}}{3} = \frac{y}{x}$$
$$y = \frac{\sqrt{3}}{3}x$$

57.
$$r = \tan\theta$$
$$r = \frac{\sin\theta}{\cos\theta}$$
$$r\cos\theta = \sin\theta$$
$$r\cos\theta - \sin\theta = 0$$
$$r^2\cos\theta - r\sin\theta = 0$$
$$\sqrt{x^2+y^2}(x) - y = 0$$
$$\sqrt{x^2+y^2} = \frac{y}{x}$$
$$x^2 + y^2 = \frac{y^2}{x^2}$$
$$x^4 - y^2 + x^2y^2 = 0$$

59.
$$r = \frac{2}{1+\cos\theta}$$
$$r + r\cos\theta = 2$$
$$\sqrt{x^2+y^2} + x = 2$$
$$\sqrt{x^2+y^2} = 2 - x$$
$$x^2 + y^2 = 4 - 4x + x^2$$
$$y^2 + 4x - 4 = 0$$

61.
$$r(\sin\theta - 2\cos\theta) = 6$$
$$r\sin\theta - 2r\cos\theta = 6$$
$$y - 2x = 6$$
$$y = 2x + 6$$

63.
$$y = 2$$
$$r\sin\theta = 2$$
$$r = 2\csc\theta$$

65.
$$y = \sqrt{3}x$$
$$r\sin\theta = \sqrt{3}(r\cos\theta)$$
$$\tan\theta = \sqrt{3}$$
$$\theta = \frac{\pi}{3}$$

67.
$$x = 3$$
$$r\cos\theta = 3$$
$$r = 3\sec\theta$$

69.
$$x^2 + y^2 = 4$$
$$r^2 = 4$$
$$r = 2$$

71.
$$x^2 = 8y$$
$$r^2\cos^2\theta = 8r\sin\theta$$
$$r\cos^2\theta = 8\sin\theta$$

73.
$$x^2 - y^2 = 25$$
$$r^2\cos^2\theta - r^2\sin^2\theta = 25$$
$$r^2(\cos^2\theta - \sin^2\theta) = 25$$
$$r^2(\cos 2\theta) = 25$$

75.

77.

79.

81.

83. $\cos\theta = \pm\sqrt{\cos^2\theta}$ is *not* an identity.

85. Enter as $r = \sqrt{4\cos 2\theta}$ and
$r = -\sqrt{4\cos 2\theta}$ for $0 \leq \theta \leq 4\pi$.

87. Graph for $0 \leq \theta \leq 2\pi$ with
$\theta_{\text{step}} = \pi/200$.

(Some graphing utilities may produce a
false asymptote in "connected" mode.)

89. Graph $r = 2/\theta$ for $-4\pi < \theta < 4\pi$.   **91.** Graph for $-30 \le \theta \le 30$.

93. a. $0 \le \theta \le 5\pi$ **b.** $0 \le \theta \le 20\pi$

Prepare for Section 6.6

PS1. $\dfrac{x^2}{25} + \dfrac{y^2}{16} = 1$

$a^2 = 25, \ a = 5$

$b^2 = 16$

$c^2 = a^2 - b^2 = 25 - 16 = 9$

$c = 3$

$e = \dfrac{c}{a} = \dfrac{3}{5}$

The eccentricity is $\dfrac{3}{5}$.

PS2. $y^2 = 4x$

$4p = 4$

$p = 1$

directrix: $x = -1$

PS3.
$$y = 2(1 + yx)$$
$$y = 2 + 2yx$$
$$y - 2yx = 2$$
$$y(1 - 2x) = 2$$
$$y = \dfrac{2}{1 - 2x}$$

PS4. $1 + \sin x = 0$

$\sin x = -1$

$x = \dfrac{3\pi}{2}$

PS5. For a hyperbola, $e > 1$.

PS6. $\dfrac{4 \sec x}{2 \sec x - 1} = \dfrac{4 \dfrac{1}{\cos x}}{2 \dfrac{1}{\cos x} - 1}$

$= \dfrac{\cos x}{\cos x} \cdot \dfrac{4 \dfrac{1}{\cos x}}{2 \dfrac{1}{\cos x} - 1}$

$= \dfrac{4}{2 - \cos x}$

Section 6.6

1. $r = \dfrac{12}{3-6\cos\theta} = \dfrac{4}{1-2\cos\theta}$

$e = 2$ The graph is a hyperbola.

The transverse axis is on the polar axis.

Let $\theta = 0$.

$\qquad r = \dfrac{12}{3-6\cos 0} = \dfrac{12}{3-6} = -4$

Let $\theta = \pi$.

$\qquad r = \dfrac{12}{3-6\cos\pi} = \dfrac{12}{3+6} = \dfrac{4}{3}$

The vertices are at $(-4,\ 0)$ and $\left(\dfrac{4}{3},\ \pi\right)$.

3. $r = \dfrac{8}{4+3\sin\theta} = \dfrac{2}{1+\frac{3}{4}\sin\theta}$

$e = \dfrac{3}{4}$ The graph is an ellipse.

The major axis is on the line $\theta = \dfrac{\pi}{2}$.

Let $\theta = \dfrac{\pi}{2}$.

$\qquad r = \dfrac{8}{4+3\sin\frac{\pi}{2}} = \dfrac{8}{4+3} = \dfrac{8}{7}$

Let $\theta = \dfrac{3\pi}{2}$.

$\qquad r = \dfrac{8}{4+3\sin\frac{3\pi}{2}} = \dfrac{8}{4-3} = 8$

Vertices on major axis are at $\left(\dfrac{8}{7},\ \dfrac{\pi}{2}\right)$ and $\left(8,\ \dfrac{3\pi}{2}\right)$.

Let $\theta = 0$.

$\qquad r = \dfrac{8}{4+3\sin 0} = \dfrac{8}{4+0} = 2$

Let $\theta = \pi$.

$\qquad r = \dfrac{8}{4+3\sin\pi} = \dfrac{8}{4+0} = 2$

The curve also goes through $(2, 0)$ and $(2, \pi)$.

5. $r = \dfrac{9}{3-3\sin\theta} = \dfrac{3}{1-\sin\theta}$

$e = 1$ The graph is a parabola.

The axis of symmetry is $\theta = \dfrac{\pi}{2}$.

When $\theta = \dfrac{\pi}{2}$, r is undefined.

Let $\theta = \dfrac{3\pi}{2}$.

$r = \dfrac{9}{3-3\sin\dfrac{3\pi}{2}} = \dfrac{9}{3+3} = \dfrac{3}{2}$

Vertex is at $\left(\dfrac{3}{2}, \dfrac{3\pi}{2}\right)$.

7. $r = \dfrac{10}{5+6\cos\theta} = \dfrac{2}{1+\dfrac{6}{5}\cos\theta}$

$e = \dfrac{6}{5}$ The graph is a hyperbola.

The transverse axis is on the polar axis.

Let $\theta = 0$.

$r = \dfrac{10}{5+6\cos 0} = \dfrac{10}{5+6} = \dfrac{10}{11}$

Let $\theta = \pi$.

$r = \dfrac{10}{5+6\cos\pi} = \dfrac{10}{5-6} = -10$

The vertices are at $\left(\dfrac{10}{11}, 0\right)$ and $(-10, \pi)$.

226

Chapter 6: Topics in Analytic Geometry

9. $r = \dfrac{4\sec\theta}{2\sec\theta - 1}$

$= \dfrac{\dfrac{4}{\cos\theta}}{\dfrac{2}{\cos\theta} - 1} = \dfrac{4}{2 - \cos\theta}$

$= \dfrac{2}{1 - \frac{1}{2}\cos\theta}$

$e = \frac{1}{2}$ The graph is an ellipse.

The major axis is on the polar axis.

However, the original equation is undefined at $\frac{\pi}{2}$ and at $\frac{3\pi}{2}$.

Thus, the ellipse will have holes at those angles.

Let $\theta = 0$.

$r = \dfrac{4}{2 - \cos 0} = \dfrac{4}{2 - 1} = 4$

Let $\theta = \pi$.

$r = \dfrac{4}{2 - \cos\pi} = \dfrac{4}{2 + 1} = \dfrac{4}{3}$

Vertices on major axis are at $(4, 0)$ and $\left(\frac{4}{3}, \pi\right)$.

$\theta = \dfrac{\pi}{2}$.

$r = \dfrac{4}{2 - \cos\frac{\pi}{2}} = \dfrac{4}{2 - 0} = 2$

Let $\theta = \dfrac{3\pi}{2}$.

$r = \dfrac{4}{2 - \cos\frac{3\pi}{2}} = \dfrac{4}{2 - 0} = 2$

Vertices on minor axis of $\dfrac{2}{1 - \frac{1}{2}\cos\theta}$ are at

$\left(2, \frac{\pi}{2}\right)$ and $\left(2, \frac{3\pi}{2}\right)$.

Thus, the equation $r = \dfrac{4\sec\theta}{2\sec\theta - 1}$ will have holes at

$\left(2, \frac{\pi}{2}\right)$ and $\left(2, \frac{3\pi}{2}\right)$

11. $r = \dfrac{12\csc\theta}{6\csc\theta - 2}$

$= \dfrac{\dfrac{12}{\sin\theta}}{\dfrac{6}{\sin\theta} - 2} = \dfrac{12}{6 - 2\sin\theta}$

$= \dfrac{2}{1 - \frac{1}{3}\sin\theta}$

$e = \frac{1}{3}$ The graph is an ellipse.

The major axis is on $\theta = \frac{\pi}{2}$.

Let $\theta = \dfrac{\pi}{2}$.

$r = \dfrac{12}{6 - 2\sin\frac{\pi}{2}} = \dfrac{12}{6 - 2} = 3$

Let $\theta = \dfrac{3\pi}{2}$.

$r = \dfrac{12}{6 - 2\sin\frac{3\pi}{2}} = \dfrac{12}{6 + 2} = \dfrac{3}{2}$

Vertices on major axis are at $\left(3, \frac{\pi}{2}\right)$ and $\left(\frac{3}{2}, \frac{3\pi}{2}\right)$.

The equation $r = \dfrac{12\csc\theta}{6\csc\theta - 2}$ has holes at $(2, 0)$ and $(2, \pi)$.

13. $r = \dfrac{3}{\cos\theta - 1}$

$ = \dfrac{-3}{1 - \cos\theta}$

$e = 1$ The graph is a parabola.
The axis of symmetry is the polar axis.

Let $\theta = \pi$.

$r = \dfrac{-3}{1 - \cos\pi} = \dfrac{-3}{1 - (-1)} = \dfrac{-3}{1 + 1} = -\dfrac{3}{2}$

Vertex is at $\left(-\dfrac{3}{2}, \pi\right)$.

15. $r = \dfrac{12}{3 - 6\cos\theta}$

$r(3 - 6\cos\theta) = 12$

$3r - 6r\cos\theta = 12$

$3\sqrt{x^2 + y^2} - 6x = 12$

$3\sqrt{x^2 + y^2} = 6x + 12$

$\sqrt{x^2 + y^2} = 2x + 4$

$x^2 + y^2 = 4x^2 + 16x + 16$

$3x^2 - y^2 + 16x + 16 = 0$

17. $r = \dfrac{8}{4 + 3\sin\theta}$

$r(4 + 3\sin\theta) = 8$

$4r + 3r\sin\theta = 8$

$4\sqrt{x^2 + y^2} + 3y = 8$

$4\sqrt{x^2 + y^2} = -3y + 8$

$16x^2 + 16y^2 = 9y^2 - 48y + 64$

$16x^2 + 7y^2 + 48y - 64 = 0$

19. $r = \dfrac{9}{3 - 3\sin\theta}$

$r(3 - 3\sin\theta) = 9$

$3r - 3r\sin\theta = 9$

$3\sqrt{x^2 + y^2} - 3y = 9$

$3\sqrt{x^2 + y^2} = 3y + 9$

$3\sqrt{x^2 + y^2} = 3(y + 3)$

$\sqrt{x^2 + y^2} = y + 3$

$x^2 + y^2 = y^2 + 6y + 9$

$x^2 - 6y - 9 = 0$

21. $e = 2,\ r\cos\theta = -1,$

$d = \left|-1\right| = 1$

$r = \dfrac{ed}{1 - e\cos\theta}$

$ = \dfrac{(2)(1)}{1 - (2)\cos\theta}$

$ = \dfrac{2}{1 - 2\cos\theta}$

23. $e = 1,\ r\sin\theta = 2,\ d = \left|2\right| = 2$

$r = \dfrac{ed}{1 + e\sin\theta}$

$ = \dfrac{(1)(2)}{1 + (1)\sin\theta}$

$ = \dfrac{2}{1 + \sin\theta}$

25. $e = \dfrac{2}{3},\ r\sin\theta = -4,$

$d = |-4| = 4$

$r = \dfrac{ed}{1 - e\sin\theta}$

$ = \dfrac{\left(\frac{2}{3}\right)(4)}{1 - \left(\frac{2}{3}\right)\sin\theta}$

$ = \dfrac{\frac{8}{3}}{1 - \frac{2}{3}\sin\theta}$

$ = \dfrac{8}{3 - 2\sin\theta}$

27. $e = \dfrac{3}{2},\ r = 2\sec\theta$

$r = \dfrac{2}{\cos\theta}$

$r\cos\theta = 2,\quad d = |2| = 2$

$r = \dfrac{ed}{1 + e\cos\theta}$

$ = \dfrac{\left(\frac{3}{2}\right)(2)}{1 + \left(\frac{3}{2}\right)\cos\theta}$

$ = \dfrac{3}{1 + \frac{3}{2}\cos\theta}$

$ = \dfrac{6}{2 + 3\cos\theta}$

29. vertex: $(2, \pi)$, curve: parabola

$r = \dfrac{ed}{1 - e\cos\theta}\qquad e = 1$ (by definition of a parabola)

When $\theta = \pi, r = 2$. Substituting into

$r = \dfrac{ed}{1 - e\cos\theta},$

we have

$2 = \dfrac{1 \cdot d}{1 - 1 \cdot \cos(\pi)} = \dfrac{d}{2}$

Therefore, $d = 4$. Substituting $e = 1$ and $d = 4$ yields

$r = \dfrac{(1)(4)}{1 - (1)\cos\theta}$ or $r = \dfrac{4}{1 - \cos\theta}$

31. vertex: $(1, 3\pi/2),\ e = 2$

$r = \dfrac{ed}{1 - e\sin\theta}$

When $\theta = \dfrac{3\pi}{2}, r = 1$. Substituting into

$r = \dfrac{ed}{1 - e\sin\theta},$

we have

$1 = \dfrac{2d}{1 - 2\sin\left(\frac{3\pi}{2}\right)} = \dfrac{2d}{3}$

Therefore $d = \dfrac{3}{2}$. Substituting $e = 2$ and $d = \dfrac{3}{2}$ yields

$r = \dfrac{(2)\left(\frac{2}{3}\right)}{1 - (2)\sin\theta}$

$ = \dfrac{3}{1 - 2\sin\theta}$

33.

Rotate the graph of Exercise 1 $\dfrac{\pi}{6}$ radians counterclockwise about the pole.

35.

Rotate the graph of Exercise 3 π radians counterclockwise about the pole.

37.

Rotate the graph of Exercise 5 $\dfrac{\pi}{6}$ radians clockwise about the pole.

39.

Rotate the graph of Exercise 7 π radians clockwise about the pole.

41.   **43.**   **45.** $0 \le \theta \le 12\pi$

47. Convert the equations for the conic and the directrix to rectangular form.

$$r = \frac{ed}{1 - e\cos\theta} \qquad\qquad d = -r\cos\theta \text{ (by definition)}$$

$$r(1 - e\cos\theta) = ed \qquad\qquad\qquad = -x$$

$$r - er\cos\theta = ed \cdot \qquad\qquad\qquad x = -d$$

$$\sqrt{x^2 + y^2} - ex = ed$$

$$\sqrt{x^2 + y^2} = ex + ed = e(x + d)$$

$$x^2 + y^2 = e^2(x^2 + 2dx + d^2)$$

$$x^2 + y^2 - e^2x^2 - 2e^2dx - e^2d^2 = 0$$

$$(1 - e^2)x^2 + y^2 - (2e^2d)x - (e^2d^2) = 0$$

Solving for y^2 yields $y^2 = (e^2 - 1)x^2 + (2ed)x + (e^2d^2)$.

Now, with $y^2 = (e^2 - 1)x^2 + (2ed)x + (e^2d^2)$ and a directrix of $x = -d$, let $k = \dfrac{d(P, F)}{d(P, D)}$, where the focus is at the origin (by

definition).

$$k = \frac{d(P, F)}{d(P, D)} = \frac{\sqrt{x^2 + y^2}}{|x + d|}$$

We can substitute $y^2 = (e^2 - 1)x^2 + (2ed)x + (e^2d^2)$ to obtain

$$k = \frac{\sqrt{x^2 + (e^2 - 1)x^2 + (2ed)x + (e^2d^2)}}{|x + d|} = \frac{\sqrt{e^2x^2 + 2e^2dx + e^2d^2}}{|x + d|}$$

Solving for k^2 gives us $k^2 = \dfrac{e^2x^2 + 2e^2dx + e^2d^2}{x^2 + 2dx + d^2} = \dfrac{e^2(x^2 + 2dx + d^2)}{x^2 + 2dx + d^2} = e^2$.

Since $k^2 = e^2$, $k = \pm\sqrt{e^2} = \pm e$. But, since $k = \dfrac{d(P, F)}{d(P, D)}$, and the ratio of two distances must be positive, k cannot be negative.

Therefore, $k = e$, or $\dfrac{d(P, F)}{d(P, D)} = e$.

PS1. $y^2 + 3y + \left(\dfrac{3}{2}\right)^2 = \left(y + \dfrac{3}{2}\right)^2$ **PS2.** $y = x^2 = (2t + 1)^2 = 4t^2 + 4t + 1$

PS3. ellipse   **PS4.** $x^2 + y^2 = (\sin t)^2 + (\cos t)^2 = 1$

PS5. $y = \ln t$ **PS6.** Domain: $(-\infty, \infty)$

     $e^y = t$ Range: $[-3, 3]$

Section 6.7

1. A table of five arbitrarily chosen values of t and the corresponding values of x and y are shown in the table below.

t	$x = 2t$	$y = -t$	(x, y)
-2	-4	2	(-4, 2)
-1	-2	1	(-2, 1)
0	0	0	(0, 0)
1	2	-1	(2, -1)
2	4	-2	(4, -2)

Plotting points for several values of t yields the following graph.

3. A table of five arbitrarily chosen values of t and the corresponding values of x and y are shown in the table below.

t	$x = -t$	$y = t^2 - 1$	(x, y)
-2	2	3	(2, 3)
-1	1	0	(1, 0)
0	0	-1	(0, -1)
1	-1	0	(-1, 0)
2	-2	3	(-2, 3)

Plotting points for several values of t yields the following graph.

5. A table of five arbitrarily chosen values of t and the corresponding values of x and y are shown the table below.

t	$x = t^2$	$y = t^3$	(x, y)
-2	4	-8	(4, -8)
-1	1	-1	(1, -1)
0	0	0	(0, 0)
1	1	1	(1, 1)
2	4	8	(4, 8)

Plotting points for several values of t yields the following graph.

7. A table of eight values of t in the specified interval and the corresponding values of x and y are shown the table below.

t	$x = 2\cos t$	$y = 3\sin t$	(x, y)
0	2	0	(2, 0)
$\pi/4$	$\sqrt{2}$	$3\sqrt{2}/2$	$(\sqrt{2}, 3\sqrt{2}/2)$
$\pi/2$	0	3	(0, 3)
$3\pi/4$	$-\sqrt{2}$	$3\sqrt{2}/2$	$(-\sqrt{2}, 3\sqrt{2}/2)$
π	-2	0	(-2, 0)
$5\pi/4$	$-\sqrt{2}$	$-3\sqrt{2}/2$	$(-\sqrt{2}, -3\sqrt{2}/2)$
$3\pi/2$	0	-3	(0, -3)
$7\pi/4$	$\sqrt{2}$	$-3\sqrt{2}/2$	$(\sqrt{2}, -3\sqrt{2}/2)$

Plotting points for several values of t yields the following graph.

9. A table of five arbitrarily chosen values of t and the corresponding values of x and y are shown the table below.

t	$x = 2^t$	$y = 2^{t+1}$	(x, y)
-2	1/4	1/2	(1/4, 1/2)
-1	1/2	1	(1/2, 1)
0	1	2	(1, 2)
1	2	4	(2, 4)
2	4	8	(4, 8)

Plotting points for several values of t yields the following graph.

11.
$$x = \sec t \qquad -\frac{\pi}{2} < t < \frac{\pi}{2}$$
$$y = \tan t$$

$$\tan^2 t + 1 = \sec^2 t$$
$$y^2 + 1 = x^2$$
$$x^2 - y^2 - 1 = 0 \qquad x \geq 1, y \in R$$

13.
$$x = 2 - t^2 \qquad t \in R$$
$$y = 3 + 2t^2$$

$$x = 2 - t^2 \rightarrow t^2 = 2 - x$$
$$y = 3 + 2(2 - x)$$
$$y = -2x + 7$$

Because $x = 2 - t^2$ and $t^2 \geq 0$ for all real numbers t, $x \leq 2$ for all t. Similarly, $y \geq 3$ for all t.

15.
$$x = \cos^3 t \qquad 0 \leq t < 2\pi$$
$$y = \sin^3 t$$

$$\cos^2 t = x^{2/3}$$
$$\sin^2 t = y^{2/3}$$

$$\cos^2 t + \sin^2 t = 1 \qquad -1 \leq x \leq 1$$
$$x^{2/3} + y^{2/3} = 1 \qquad -1 \leq y \leq 1$$

17.
$$x = \sqrt{t+1} \qquad t \geq -1$$
$$y = t$$

$$x = \sqrt{y+1} \qquad x \geq 0$$
$$y = x^2 - 1 \qquad y \geq -1$$

19.

$$x = 2 + 3\cos t \qquad\qquad y = 3 + 2\sin t$$

$$x - 2 = 3\cos t \qquad\qquad y - 3 = 2\sin t$$

$$\frac{x-2}{3} = \cos t \qquad\qquad \frac{y-3}{2} = \sin t$$

$$\left(\frac{x-2}{3}\right)^2 + \left(\frac{y-3}{2}\right)^2 = \cos^2 t + \sin^2 t$$

$$\left(\frac{x-2}{3}\right)^2 + \left(\frac{y-3}{2}\right)^2 = 1$$

At $t = 0$,

$$x = 2 + 3\cos 0 = 5 \qquad\qquad y = 3 + 2\sin 0 = 3$$

At $t = \pi$,

$$x = 2 + 3\cos\pi = -1 \qquad\qquad y = 3 + 2\sin\pi = 3$$

The point traces the top half of the ellipse

$\left(\dfrac{x-2}{3}\right)^2 + \left(\dfrac{y-3}{2}\right)^2 = 1$, as shown in the figure. The point

starts at (5, 3) and moves counterclockwise along the
ellipse until it reaches the point (–1, 3) at time $t = \pi$.

21.

$$y = t + 1 \implies t = y - 1$$

$$x = 2t - 1$$

$$x = 2(y - 1) - 1$$

$$x = 2y - 3$$

$$x + 3 = 2y$$

$$y = \frac{1}{2}x + 3$$

At $t = 0$,

$$x = 2(0) - 1 = -1 \qquad\qquad y = 0 + 1 = 1$$

At $t = 3$,

$$x = 2(3) - 1 = 5 \qquad y = 3 + 1 = 4$$

The point traces a line segment, as shown in the figure.
The point starts at (–1, 1) and moves along the line
segment until it reaches the point (5, 4) at time $t = 3$.

23.

$$x = \tan\left(\frac{\pi}{4} - t\right) \qquad\qquad y = \sec\left(\frac{\pi}{4} - t\right)$$

$$y^2 - x^2 = \sec^2\left(\frac{\pi}{4} - t\right) - \tan^2\left(\frac{\pi}{4} - t\right)$$

$$y^2 - x^2 = 1 \quad \text{Since } 1 + \tan^2\theta = \sec^2\theta$$

At $t = 0$,

$$x = \tan\left(\frac{\pi}{4} - 0\right) = 1 \qquad\qquad y = \sec\left(\frac{\pi}{4} - 0\right) = \sqrt{2}$$

At $t = \dfrac{\pi}{2}$,

$$x = \tan\left(\frac{\pi}{4} - \frac{\pi}{2}\right) = -1 \qquad\qquad y = \sec\left(\frac{\pi}{4} - \frac{\pi}{2}\right) = \sqrt{2}$$

The point traces a portion of the top branch of the

hyperbola $y^2 - x^2 = 1$, as shown in the figure. The point

starts at $\left(1, \sqrt{2}\right)$ and moves along the hyperbola until it

reaches the point $\left(-1, \sqrt{2}\right)$ at time $t = \dfrac{\pi}{2}$.

25.

$$C_1: x = 2 + t^2$$

$$y = 1 - 2t^2$$

$$x = 2 + t^2 \to t^2 = x - 2$$

$$y = 1 - 2(x - 2)$$

$$y = -2x + 5 \quad x \geq 2, y \leq 1$$

$$C_2: x = 2 + t$$

$$y = 1 - 2t$$

$$x = 2 + t \to t = x - 2$$

$$y = 1 - 2(x - 2)$$

$$y = -2x + 5 \quad x \in R, y \in R$$

The graph of C_1 is a ray beginning at (2, 1) with slope –2.

The graph of C_2 is a line passing through (2, 1) with slope –2.

27. $x = \sin t$
$y = \csc t$

$$\csc t = \frac{1}{\sin t}$$

$$y = \frac{1}{x}$$

Range for graph 1: $0 < t \le \dfrac{\pi}{2}$

$0 < x \le 1$

$y \ge 1$

Range for graph 2: $\pi \le t \le \dfrac{3\pi}{2}$

$-1 \le x \le 0$

$y \le -1$

29.

Xscl $= 2\pi$

31.   **a.**   For the Hummer,
$x = 6$
$y = 60t$ for $t \ge 0$

b.   Using the graphing calculator in SIMUL and PAR mode, the Hummer is the first to reach the intersection.

33.

Maximum height (to the nearest foot) of 462 feet is attained when $t \approx 5.38$ seconds.

The projectile has a range (to the nearest foot) of 1295 feet and hits the ground in about 10.75 seconds.

35.

Maximum height (to the nearest foot) of 694 feet is attained when $t \approx 6.59$ seconds.

The projectile has a range (to the nearest foot) of 3084 feet and hits the ground in about 13.17 seconds.

37.

39.

41. Let $P_1(x_1, y_1)$ and $P_2(x_2, y_2)$ be two distinct points on a line.

If $P(x, y)$ is any other point on the line, then

$\dfrac{y - y_1}{x - x_1} = \dfrac{y_2 - y_1}{x_2 - x_1}$ (Slope is constant along entire line.) This

equation can be rewritten as

$\dfrac{y - y_1}{y_2 - y_1} = \dfrac{x - x_1}{x_2 - x_1}$ Let this value equal t.

Thus, $\dfrac{x - x_1}{x_2 - x_1} = t$ and $\dfrac{y - y_1}{y_2 - y_1} = t$.

Solving for x and y, respectively, we have

$x = x_1 + t(x_2 - x_1)$ and $y = y_1 + t(y_2 - y_1)$

43. radius $= a$, $\theta = \angle TOR$

The x-coordinate of $P(x, y)$ is given by $x = OR + QP$.
The y-coordinate is given by $y = TR - TQ$.

From the figure,

$OR = a\cos\theta$ and $QP = a\theta\sin\theta$. Thus,

$x = a\cos\theta + a\theta\sin\theta$

$TR = a\sin\theta$ and $TQ = a\theta\cos\theta$. Thus,

$y = a\sin\theta - a\theta\cos\theta$

The parametric equations are

$x = a\cos\theta + a\theta\sin\theta$

$y = a\sin\theta - a\theta\cos\theta$

45. Because the circle moves without slipping, $b\theta = a\alpha$.

Therefore, $\alpha = \dfrac{b\theta}{a}$. Let $P(x, y)$ be the coordinates of the moving point.

Angle $\phi = \dfrac{\pi}{2} - \left(\dfrac{b-a}{a}\right)\theta$

Thus, $x = (b-a)\cos\theta + a\sin\left[\dfrac{\pi}{2} - \left(\dfrac{b-a}{a}\right)\theta\right]$

$y = (b-a)\sin\theta - a\cos\left[\dfrac{\pi}{2} - \left(\dfrac{b-a}{a}\right)\theta\right]$

Simplifying, we have

$x = (b-a)\cos\theta + a\cos\left(\dfrac{b-a}{a}\theta\right)$

$y = (b-a)\sin\theta - a\sin\left(\dfrac{b-a}{a}\theta\right)$

Using a Graphing Calculator to Find the nth Roots of z

1. The procedure for a TI-83 calculator is illustrated below.
    $z = -27 = 27(\cos 180° + i \sin 180°)$, thus, $r = 27$, $\theta = 180°$.

    cube roots $\Rightarrow n = 3$

Use the TRACE feature and the arrow keys to display and move to each of the vertices of the polygon.

Thus, the three cube roots of -27 are
$1.5 + 2.598076i$, $1.5 - 2.598076i$, and -3.

3. Here is the procedure for a TI-83 graphing calculator. Be sure the calculator is in parametric and degree mode. In the WINDOW menu, set Tmin=0, Tmax=360, and, since $360/4 = 90$, set Tstep=90. Set Xmin, Xmax, Ymin, and Ymax to appropriate values that will allow the roots to be seen. Since $z = \sqrt{8} + \sqrt{8}i = 4(\cos 45° + i \sin 45°)$, in the Y- menu, enter X_{1T}=4^(1/4)cos(45/4+T) and Y_{1T}=4^(1/4)sin(45/4+T). Press GRAPH to display a polygon. The x- and y-coordinates of each vertex represent a root of z in the rectangular form $x + yi$. Use the TRACE feature and the arrow keys to display and move to each of the vertices of the polygon.

    The fourth roots of $\sqrt{8} + \sqrt{8}i$ are $1.38704 + 0.2758994i$, $-0.2758994 + 1.38704i$, $-1.38704 - 0.2758994i$, and $0.2758994 - 1.38704i$.

1.	d	2.	b	3.	e	4.	c
5.	a	6.	f	7.	g	8.	i
9.	h	10.	j	11.	k	12.	k

•••

1. $x^2 - y^2 = 4$ [6.3]

$$\frac{x^2}{4} - \frac{y^2}{4} = 1$$

hyperbola
center: $(0, 0)$
vertices: $(\pm 2, 0)$
foci: $(\pm 2\sqrt{2}, 0)$
asymptotes: $y = \pm x$

2. $y^2 = 16x$ [6.1]

parabola
vertex: $(0, 0)$
focus: $(4, 0)$
directrix: $x = -4$

3. $x^2 + 4y^2 - 6x + 8y - 3 = 0$ [6.2]

$$x^2 - 6x + 4(y^2 + 2y) = 3$$

$$(x^2 - 6x + 9) + 4(y^2 + 2y + 1) = 3 + 9 + 4$$

$$(x - 3)^2 + 4(y + 1)^2 = 16$$

$$\frac{(x - 3)^2}{16} + \frac{(y + 1)^2}{4} = 1$$

ellipse
center: $(3, -1)$
vertices: $(3 \pm 4, -1) = (7, -1), (-1, -1)$
foci: $(3 \pm 2\sqrt{3}, -1) = (3 + 2\sqrt{3}, -1), (3 - 2\sqrt{3}, -1)$

4. $3x^2 - 4y^2 + 12x - 24y - 36 = 0$ [6.3]

$$3(x^2 + 4x) - 4(y^2 + 6y) = 36$$

$$3(x^2 + 4x + 4) - 4(y^2 + 6y + 9) = 36 + 12 - 36$$

$$3(x + 2)^2 - 4(y + 3)^2 = 12$$

$$\frac{(x + 2)^2}{4} - \frac{(y + 3)^2}{3} = 1$$

hyperbola
center: $(-2, -3)$
vertices: $(-2 \pm 2, -3) = (0, -3), (-4, -3)$
foci: $(-2 \pm \sqrt{7}, -3) = (-2 + \sqrt{7}, -3), (-2 - \sqrt{7}, -3)$
asymptotes: $y + 3 = \pm \frac{\sqrt{3}}{2}(x + 2)$

5. $3x - 4y^2 + 8y + 2 = 0$ [6.1]

$$-4(y^2 - 2y) = -3x - 2$$
$$-4(y^2 - 2y + 1) = -3x - 2 - 4$$
$$-4(y - 1)^2 = -3(x + 2)$$
$$(y - 1)^2 = \frac{3}{4}(x + 2)$$

parabola

vertex: $(-2, 1)$

focus: $\left(-2 + \frac{3}{16}, 1\right) = \left(-\frac{29}{16}, 1\right)$

directrix: $x = -2 - \frac{3}{16}$, or $x = -\frac{35}{16}$

6. $3x + 2y^2 - 4y - 7 = 0$ [6.1]

$$2(y^2 - 2y) = -3x + 7$$
$$2(y^2 - 2y + 1) = -3x + 7 + 2$$
$$2(y - 1)^2 = -3(x - 3)$$
$$(y - 1)^2 = -\frac{3}{2}(x - 3)$$

parabola

vertex: $(3, 1)$

focus: $\left(3 - \frac{3}{8}, 1\right) = \left(\frac{21}{8}, 1\right)$

directrix: $x = 3 + \frac{3}{8}$, or $x = \frac{27}{8}$

7. $9x^2 + 4y^2 + 36x - 8y + 4 = 0$ [6.2]

$$9(x^2 + 4x) + 4(y^2 - 2y) = -4$$
$$9(x^2 + 4x + 4) + 4(y^2 - 2y + 1) = -4 + 36 + 4$$
$$9(x + 2)^2 + 4(y - 1)^2 = 36$$
$$\frac{(x + 2)^2}{4} + \frac{(y - 1)^2}{9} = 1$$

ellipse

center: $(-2, 1)$

vertices: $(-2, 1 \pm 3) = (-2, 4), (-2, -2)$

foci: $(-2, 1 \pm \sqrt{5}) = (-2, 1 + \sqrt{5}), (-2, 1 - \sqrt{5})$

8. $11x^2 - 25y^2 - 44x - 50y - 256 = 0$ [6.3]

$$11(x^2 - 4x) - 25(y^2 + 2y) = 256$$
$$11(x^2 - 4x + 4) - 25(y^2 + 2y + 1) = 256 + 44 - 25$$
$$11(x - 2)^2 - 25(y + 1)^2 = 275$$
$$\frac{(x - 2)^2}{25} - \frac{(y + 1)^2}{11} = 1$$

hyperbola

center: $(2, -1)$

vertices: $(2 \pm 5, -1) = (7, -1), (-3, -1)$

foci: $(2 \pm 6, -1) = (8, -1), (-4, -1)$

asymptotes: $y + 1 = \pm \frac{\sqrt{11}}{5}(x - 2)$

9. $4x^2 - 9y^2 - 8x + 12y - 144 = 0$ [6.3]

$$4(x^2 - 2x) - 9\left(y^2 - \frac{4}{3}y\right) = 144$$

$$4(x^2 - 2x + 1) - \left(9y^2 - \frac{4}{3}y + \frac{4}{9}\right) = 144 + 4 - 4$$

$$4(x-1)^2 - 9\left(y - \frac{2}{3}\right)^2 = 144$$

$$\frac{(x-1)^2}{36} - \frac{\left(y - \frac{2}{3}\right)^2}{16} = 1$$

hyperbola

center: $\left(1, \dfrac{2}{3}\right)$

vertices: $\left(1 \pm 6, \dfrac{2}{3}\right) = \left(7, \dfrac{2}{3}\right), \left(-5, \dfrac{2}{3}\right)$

foci: $\left(1 \pm 2\sqrt{13}, \dfrac{2}{3}\right) = \left(1 + 2\sqrt{13}, \dfrac{2}{3}\right), \left(1 - 2\sqrt{13}, \dfrac{2}{3}\right)$

asymptotes: $y - \dfrac{2}{3} = \pm \dfrac{2}{3}(x-1)$

10. $9x^2 + 16y^2 + 36x - 16y - 104 = 0$ [6.2]

$$9(x^2 + 4x) + 16(y^2 - y) = 104$$

$$9(x^2 + 4x + 4) + 16\left(y^2 - y + \frac{1}{4}\right) = 104 + 36 + 4$$

$$9(x+2)^2 + 16\left(y - \frac{1}{2}\right)^2 = 144$$

$$\frac{(x+2)^2}{16} + \frac{\left(y - \frac{1}{2}\right)^2}{9} = 1$$

ellipse

center: $\left(-2, \dfrac{1}{2}\right)$

vertices: $\left(-2 \pm 4, \dfrac{1}{2}\right) = \left(2, \dfrac{1}{2}\right), \left(-6, \dfrac{1}{2}\right)$

foci: $\left(-2 \pm \sqrt{7}, \dfrac{1}{2}\right) = \left(-2 + \sqrt{7}, \dfrac{1}{2}\right), \left(-2 - \sqrt{7}, \dfrac{1}{2}\right)$

11. $4x^2 + 28x + 32y + 81 = 0$ [6.1]

$$4(x^2 + 7x) = -32y - 81$$

$$4\left(x^2 + 7x + \frac{49}{4}\right) = -32y - 81 + 49$$

$$4\left(x + \frac{7}{2}\right)^2 = -32(y+1)$$

$$\left(x + \frac{7}{2}\right)^2 = -8(y+1)$$

parabola

$4p = -8 \Rightarrow p = -2$

vertex: $\left(-\dfrac{7}{2}, -1\right)$

focus: $\left(-\dfrac{7}{2}, -1-2\right) = \left(-\dfrac{7}{2}, -3\right)$

directrix: $y = 1$

12. $x^2 - 6x - 9y + 27 = 0$ [6.1]

$$x^2 - 6x = 9y - 27$$

$$x^2 - 6x + 9 = 9y - 27 + 9$$

$$(x-3)^2 = 9(y-2)$$

parabola

$4p = 9 \Rightarrow p = \dfrac{9}{4}$

vertex: $(3, 2)$

focus: $\left(3, 2 + \dfrac{9}{4}\right) = \left(3, \dfrac{17}{4}\right)$

directrix: $y = 3 - \dfrac{9}{4}$, or $y = -\dfrac{1}{4}$

13. $2a = |7-(-3)| = 10$ [6.2]

$a = 5$

$a^2 = 25$

$2b = 8$

$b = 4$

$b^2 = 16$

center (2, 3)

$$\frac{(x-2)^2}{25} + \frac{(y-3)^2}{16} = 1$$

14. $2a = |4-(-2)| = 6$ [6.3]

$a = 3$

$a^2 = 9$

$e = \dfrac{c}{a} = \dfrac{4}{3}$

$\dfrac{c}{3} = \dfrac{4}{3}$

$c = 4$

$c^2 = a^2 + b^2$

$16 = 9 + b^2$

$b^2 = 7$

center (1, 1)

$$\frac{(x-1)^2}{9} - \frac{(y-1)^2}{7} = 1$$

15. center $(-2, 2)$, $c = 3$ [6.3]

$2a = 4$

$a = 2$

$a^2 = 4$

$c^2 = a^2 + b^2$

$9 = 4 + b^2$

$b^2 = 5$

$$\frac{(x+2)^2}{4} - \frac{(y-2)^2}{5} = 1$$

16. $(h, k) = \left(\dfrac{6+2}{2}, \dfrac{-3-3}{2}\right) = (4, -3)$ [6.1]

$p = 2 - 4$

$p = -2$

$4p = -8$

$(y+3)^2 = -8(x-4)$

17. $(x-h)^2 = 4p(y-k)$ or $(y-k)^2 = 4p(x-h)$

$(3-0)^2 = 4p(4+2)$ $(4+2)^2 = 4p(3-0)$

$9 = 4p(6)$ $36 = 4p(3)$

$p = \dfrac{3}{8}$ $p = 3$

Thus, there are two parabolas that satisfy the given conditions:

$$x^2 = \frac{3}{2}(y+2) \ \text{ or } \ (y+2)^2 = 12x \quad [6.1]$$

18. center $(-2, -1)$

$c = 2$

$e = \dfrac{c}{a} = \dfrac{2}{3}$

$\dfrac{2}{a} = \dfrac{2}{3}$

$a = 3$

$c^2 = a^2 - b^2$

$4 = 9 - b^2$

$b^2 = 5$

$$\frac{(x+2)^2}{9} + \frac{(y+1)^2}{5} = 1 \quad [6.2]$$

19. $a = 6$ and the transverse axis is on the x-axis.

$\pm\dfrac{b}{a} = \pm\dfrac{1}{9}$

$\dfrac{b}{6} = \dfrac{1}{9}$

$b = \dfrac{2}{3}$

$$\frac{x^2}{36} - \frac{y^2}{4/9} = 1 \quad [6.3]$$

20. $(x-h)^2 = 4p(y-k)$

$(1-h)^2 = 4p(0-k)$

$(2-h)^2 = 4p(1-k)$

$(0-h)^2 = 4p(1-k)$

In the last two equations, by substitution:

$$(2-h)^2 = (0-h)^2$$

$$4 - 4h + h^2 = h^2$$

$$4 - 4h = 0$$

$$4h = 4$$

$$h = 1$$

Thus:

$(1-1)^2 = 4p(0-k)$

$0 = -4pk$

$k = 0$

$(2-1)^2 = 4p(1-k)$

$1 = 4p(1)$

$p = \dfrac{1}{4}$

The equation is $y = (x-1)^2$ [6.1]

21. $A = 11$, $B = -6$, $C = 19$, $D = 0$, $E = 0$, $F = -40$

$\cot 2\alpha = \dfrac{11-19}{-6} = \dfrac{-8}{-6} = \dfrac{4}{3}$ (2α is in quadrant I.)

$\csc^2 2\alpha = 1 + \cot^2 2\alpha = 1 + \dfrac{16}{9} = \dfrac{25}{9}$

$\csc 2\alpha = \dfrac{5}{3}$

Thus, $\sin 2\alpha = \dfrac{3}{5}$ and $\cos 2\alpha = \dfrac{4}{5}$.

$\sin \alpha = \sqrt{\dfrac{1 - \frac{4}{5}}{2}} = \dfrac{\sqrt{10}}{10}$ $\cos \alpha = \dfrac{3\sqrt{10}}{10}$

$A' = 11\left(\dfrac{3\sqrt{10}}{10}\right)^2 - 6\left(\dfrac{\sqrt{10}}{10}\right)\left(\dfrac{3\sqrt{10}}{10}\right) + 19\left(\dfrac{\sqrt{10}}{10}\right)^2$

$\quad = \dfrac{99}{10} - \dfrac{18}{10} + \dfrac{19}{10} = \dfrac{100}{10} = 10$

$B' = 0$

$C' = 11\left(\dfrac{\sqrt{10}}{10}\right)^2 + 6\left(\dfrac{\sqrt{10}}{10}\right)\left(\dfrac{3\sqrt{10}}{10}\right) + 19\left(\dfrac{3\sqrt{10}}{10}\right)^2$

$\quad = \dfrac{11}{10} + \dfrac{18}{10} + \dfrac{171}{10} = \dfrac{200}{10} = 20$

$F' = F$

$10(x')^2 + 20(y')^2 - 40 = 0$ or $(x')^2 + 2(y')^2 - 4 = 0$

The graph is an ellipse.   [6.4]

22. $A = 3$, $B = 6$, $C = 3$, $D = -4$, $E = 5$, $F = -12$

$\cot 2\alpha = \dfrac{3-3}{6} = 0$

$1 + \cot^2 2\alpha = \csc^2 2\alpha$

$\qquad 1 = \csc^2 2\alpha$

$\quad \csc 2\alpha = 1$

Thus, $\sin 2\alpha = 1$ so $2\alpha = 90°$ or $\alpha = 45°$.

Therefore, $\cos \alpha = \dfrac{\sqrt{2}}{2}$, $\sin \alpha = \dfrac{\sqrt{2}}{2}$.

$A' = 3\left(\dfrac{\sqrt{2}}{2}\right)^2 + 6\left(\dfrac{\sqrt{2}}{2}\right)\left(\dfrac{\sqrt{2}}{2}\right) + 3\left(\dfrac{\sqrt{2}}{2}\right)^2 = \dfrac{3}{2} + \dfrac{6}{2} + \dfrac{3}{2} = 6$

$B' = 0$

$C' = 3\left(\dfrac{\sqrt{2}}{2}\right)^2 - 6\left(\dfrac{\sqrt{2}}{2}\right)\left(\dfrac{\sqrt{2}}{2}\right) + 3\left(\dfrac{\sqrt{2}}{2}\right)^2 = \dfrac{3}{2} - \dfrac{6}{2} + \dfrac{3}{2} = 0$

$D' = -4\left(\dfrac{\sqrt{2}}{2}\right) + 5\left(\dfrac{\sqrt{2}}{2}\right) = \dfrac{\sqrt{2}}{2}$

$E' = 4\left(\dfrac{\sqrt{2}}{2}\right) + 5\left(\dfrac{\sqrt{2}}{2}\right) = \dfrac{9\sqrt{2}}{2}$

$F' = -12$

$6(x')^2 + \dfrac{\sqrt{2}}{2}x' + \dfrac{9\sqrt{2}}{2}y' - 12 = 0$

The graph is a parabola.   [6.4]

23. $A = 1, B = 2\sqrt{3}, C = 3, D = 8\sqrt{3}, E = -8, F = 32$

$\cot 2\alpha = \dfrac{1-3}{2\sqrt{3}} = -\dfrac{1}{\sqrt{3}}$. Thus $90° < 2\alpha < 180°$.

$\cot^2 2\alpha + 1 = \csc^2 2\alpha$

$\dfrac{1}{3} + 1 = \csc^2 2\alpha$

$\dfrac{4}{3} = \csc^2 2\alpha$, or $\csc 2\alpha = \dfrac{2}{\sqrt{3}}$

Therefore, $\sin 2\alpha = \dfrac{\sqrt{3}}{2}$ and $\cos 2\alpha = -\dfrac{1}{2}$.

Since 2α is in quadrant II,

$\cos \alpha = \sqrt{\dfrac{1+(-1/2)}{2}} = \dfrac{1}{2}$ $\sin \alpha = \sqrt{\dfrac{1-(-1/2)}{2}} = \dfrac{\sqrt{3}}{2}$

$A' = 1\left(\dfrac{1}{2}\right)^2 + 2\sqrt{3}\left(\dfrac{1}{2}\right)\left(\dfrac{\sqrt{3}}{2}\right) + 3\left(\dfrac{\sqrt{3}}{2}\right)^2 = \dfrac{1}{4} + \dfrac{6}{4} + \dfrac{9}{4} = 4$

$B' = 0$

$C' = 1\left(\dfrac{\sqrt{3}}{2}\right)^2 - 2\sqrt{3}\left(\dfrac{1}{2}\right)\left(\dfrac{\sqrt{3}}{2}\right) + 3\left(\dfrac{1}{2}\right)^2 = \dfrac{3}{4} - \dfrac{6}{4} + \dfrac{3}{4} = 0$

$D' = 8\sqrt{3}\left(\dfrac{1}{2}\right) - 8\left(\dfrac{\sqrt{3}}{2}\right) = 0$

$E' = 8\sqrt{3}\left(\dfrac{\sqrt{3}}{2}\right) - 8\left(\dfrac{1}{2}\right) = -12 - 4 = -16$

$F' = 32$

$4(x')^2 - 16y' + 32 = 0$ or $(x')^2 - 4y' + 8 = 0$

The graph is a parabola.   [6.4]

24. $A = 0, B = 1, C = 0, D = -1, E = -1, F = -1$

$\cot 2\alpha = \dfrac{0-0}{1} = 0$.   Thus $\alpha = 45°$.

Therefore, $\sin \alpha = \dfrac{\sqrt{2}}{2}$ and $\cos \alpha = \dfrac{\sqrt{2}}{2}$.

$A' = 0 + 1\left(\dfrac{\sqrt{2}}{2}\right)\left(\dfrac{\sqrt{2}}{2}\right) + 0 = \dfrac{1}{2}$

$B' = 0$

$C' = 0 - \dfrac{\sqrt{2}}{2}\dfrac{\sqrt{2}}{2} + 0 = -\dfrac{1}{2}$

$D' = -1\left(\dfrac{\sqrt{2}}{2}\right) - 1\left(\dfrac{\sqrt{2}}{2}\right) = -\sqrt{2}$

$E' = \dfrac{\sqrt{2}}{2} + (-1)\dfrac{\sqrt{2}}{2} = 0$

$F' = -1$

$\dfrac{1}{2}(x')^2 - \dfrac{1}{2}(y')^2 - \sqrt{2}x' - 1 = 0$

The equation is a hyperbola.   [6.4]

25.   [6.5]

26.   [6.5]

27.   [6.5]

28.   [6.5]

29.   [6.5]

30. vertical line through (3, 0).   [6.5]

31. horizontal line through (0, 4)   [6.5]

32.   [6.5]

33.   [6.5]

34.   [6.5]

35. $y^2 = 16x$ [6.5]

$(r\sin \theta)^2 = 16(r\cos \theta)$

$r^2 \sin^2 \theta = 16r\cos \theta$

$r \sin^2 \theta = 16\cos \theta$

36. $x^2 + y^2 + 4x + 3y = 0$ [6.5]

$(r\cos \theta)^2 + (r\sin \theta)^2 + 4(r\cos \theta) + 3(r\sin \theta) = 0$

$r^2(\cos^2 \theta + \sin^2 \theta) + 4r\cos \theta + 3r\sin \theta = 0$

$r + 4\cos \theta + 3\sin \theta = 0$

37. $3x - 2y = 6$ [6.6]

$3r\cos \theta - 2r\sin \theta = 6$

38.
$$xy = 4 \quad [6.5]$$
$$(r \cos \theta)(r \sin \theta) = 4$$
$$r^2 \cos \theta \, \sin \theta = 4$$
$$r^2 (2)\cos \theta \, \sin \theta = 4(2)$$
$$r^2 \sin 2\theta = 8$$

39.
$$r = \frac{4}{1 - \cos \theta} \quad [6.5]$$
$$r - r \cos \theta = 4$$
$$\sqrt{x^2 + y^2} - x = 4$$
$$\sqrt{x^2 + y^2} = x + 4$$
$$x^2 + y^2 = x^2 + 8x + 16$$
$$y^2 = 8x + 16$$

40.
$$r^2 = 3r \cos \theta - 4r \sin \theta \quad [6.5]$$
$$x^2 + y^2 = 3x - 4y$$
$$x^2 - 3x + y^2 + 4y = 0$$

41.
$$r^2 = \cos 2\theta \quad [6.5]$$
$$r^2 = \cos^2 \theta - \sin^2 \theta$$
$$r^4 = r^2 \cos^2 \theta - r^2 \sin^2 \theta$$
$$(r^2)^2 = r^2 \cos^2 \theta - r^2 \sin^2 \theta$$
$$(x^2 + y^2)^2 = x^2 - y^2$$
$$x^4 + 2x^2 y^2 + y^4 = x^2 - y^2$$
$$x^4 + y^4 + 2x^2 y^2 - x^2 + y^2 = 0$$

42.
$$\theta = 1 \quad [6.5]$$
$$\tan \theta = \tan 1$$
$$\frac{y}{x} \approx 1.5574$$
$$y = 1.5574x$$

43. $r = \dfrac{4}{3 - 6 \sin \theta}$

[6.6]

44. $r = \dfrac{2}{1 + \cos \theta}$

[6.6]

45. $r = \dfrac{2}{2 - \cos \theta}$

[6.6]

46. $r = \dfrac{6}{4 + 3 \sin \theta}$

[6.6]

47. $x = 4t - 2,\ y = 3t + 1,\ t \in R$
$$4t = x + 2$$
$$t = \frac{x + 2}{4}$$
$$y = 3t + 1$$
$$y = 3\left(\frac{x+2}{4}\right) + 1$$
$$y = \frac{3}{4}x + \frac{5}{2}$$

[6.7]

48. $x = 1 - t^2,\ y = 3 - 2t^2,\ t \in R$
$$t^2 = -x + 1$$
$$y = 3 - 2t^2$$
$$y = 3 - 2(-x + 1)$$
$$y = 3 + 2x - 2$$
$$y = 2x + 1,\ x \le 1$$

[6.7]

49.

$$x = 4 \sin t \qquad y = 3 \cos t \qquad 0 \le t < 2\pi$$

$$\frac{1}{4}x = \sin t \qquad \frac{1}{3}y = \cos t$$

$$\frac{1}{16}x^2 = \sin^2 t \qquad \frac{1}{9}y^2 = \cos t^2$$

Using the trignometric identity $\sin^2 t + \cos^2 t = 1$, we have

$$\frac{1}{16}x^2 + \frac{1}{9}y^2 = 1$$

$$\frac{x^2}{19} + \frac{y^2}{9} = 1$$

[6.7]

50.

$$x = \sec t \qquad y = 4 \tan t \qquad -\frac{\pi}{2} < t < \frac{\pi}{2}$$

$$x^2 = \sec^2 t \qquad \frac{1}{4}y = \tan t$$

$$\left(\frac{1}{4}y\right)^2 = \tan^2 t$$

$$\frac{1}{16}y^2 = \tan^2 t$$

Using the trignometric identity $1 + \tan^2 t = \sec^2 t$, we have

$$1 + \frac{y^2}{16} = x^2$$

$$\frac{x^2}{1} - \frac{y^2}{16} = 1$$

[6.7]

51.

$$x = \frac{1}{t} \qquad y = -\frac{2}{t} \qquad t > 0$$

$$y = -2\left(\frac{1}{t}\right)$$

$$y = -2x, \ x > 0$$

[6.7]

52.

$$x = 1 + \cos t \qquad y = 2 - \sin t \qquad 0 \le t < 2\pi$$

$$x - 1 = \cos t \qquad 2 - y = \sin t$$

$$-(y - 2) = \sin t$$

$$(x-1)^2 = \cos^2 t \qquad (-(y-2))^2 = \sin^2 t$$

$$(y-2)^2 = \sin^2 t$$

Using the trigonometric identity $\cos^2 t + \sin^2 t = 1$ we have

$$(x-1)^2 + (y-2)^2 = 1$$

[6.7]

53. $x = \sqrt{t}, \ y = 2^{-t}, t \ge 0$

$$t = x^2$$

$$y = 2^{-x^2}, \ x \ge 0$$

[6.7]

54.

[6.7]

55. Graph $y = \dfrac{-(4x+5) \pm \sqrt{(4x+5)^2 - 8(x^2 - 2x + 1)}}{4}$.

[6.4]

56.

[6.5]

57.

Graph in parametric mode. Use the TRACE feature to determine that the maximum height (to the nearest foot) of 278 feet is attained when $t \approx 4.17$ seconds. [6.7]

••

QR1. They appear to be the same.

QR3. 5.2 units

••

Chapter Test

1.
$y = \dfrac{1}{8}x^2$

$x^2 = 8y$

$4p = 8$

$p = 2$

vertex: $(0, 0)$ [6.1]

focus: $(0, 2)$

directrix: $y = -2$

2.

(graph)

[6.2]

3. $25x^2 - 150x + 9y^2 + 18y + 9 = 0$ [6.2]

$25(x^2 - 6x + 9) + 9(y^2 + 2y + 1) = -9 + 255 + 9$

$25(x-3)^2 + 9(y+1)^2 = 225$

$\dfrac{(x-3)^2}{9} + \dfrac{(y+1)^2}{25} = 1$

$a = 5 \quad b = 3 \quad c = 4$

vertices: $(3, 4), (3, -6)$
foci: $(3, 3), (3, -5)$

4. $2b = 6 \qquad c = 6$ [6.2]

  $b = 3$

$a^2 = 9 + 36 = 45$

center = $(0, -3)$

$\dfrac{x^2}{45} + \dfrac{(y+3)^2}{9} = 1$

5.

(graph) [6.3]

6. $\dfrac{x^2}{36} - \dfrac{y^2}{64} = 1$

vertices: $(6, 0), (-6, 0)$

foci: $(10, 0), (-10, 0)$

asymptotes: $y = \pm\dfrac{4}{3}x$ [6.3]

7. $\dfrac{(y+1)^2}{4} - \dfrac{(x+3)^2}{16} = 1$

(graph) [6.3]

8. $x^2 - 4xy - 5y^2 + 3x - 5y - 20 = 0$ [6.4]

$A = 1 \quad B = -4 \quad C = -5 \quad D = 3 \quad E = -5 \quad F = -20$

$\cot 2\alpha = \dfrac{A - C}{B} = \dfrac{1 - (-5)}{-4} = -\dfrac{3}{2}$ 2α is in quadrant II.

$\tan 2\alpha = -\dfrac{2}{3}$

$2\alpha = \tan^{-1}\left(-\dfrac{2}{3}\right)$

$2\alpha \approx (-33.69° + 180°)$

$\alpha \approx 73.15°$

9. $A = 8 \quad B = 5 \quad C = 2 \quad D = -10 \quad E = 5 \quad F = 4$

Since $B^2 - 4AC = (5)^2 - 4(8)(2) = -39 < 0$, the graph is an ellipse. [6.4]

10. $P(1, -\sqrt{3})$ $r = \sqrt{x^2 + y^2}$ $r\cos\theta = x$ $r\sin\theta = y$ θ is in quadrant IV.

$r = \sqrt{1^2 + (-\sqrt{3})^2}$ $2\cos\theta = 1$ $2\sin\theta = -\sqrt{3}$ $\theta = 300°$

$r = 2$ $\cos\theta = \dfrac{1}{2}$ $\sin\theta = -\dfrac{\sqrt{3}}{2}$

$P(1, -\sqrt{3}) = P(2,\ 300°)$ [6.5]

11. $r = 4\cos\theta$

[6.5]

12. $r = 3(1 - \sin\theta)$

[6.5]

13. $r = 2\sin 4\theta$

[6.5]

14. $x = r\cos\theta$ $y = r\sin\theta$

$x = 5\cos\dfrac{7\pi}{3}$ $y = 5\sin\dfrac{7\pi}{3}$

$x = \dfrac{5}{2}$ $y = \dfrac{5\sqrt{3}}{2}$

The rectangular coordinates of the point are $(5/2,\ 5\sqrt{3}/2)$.   [6.5]

15. $r - r\cos x = 4$ [6.5]

$\sqrt{x^2 + y^2} - x = 4$

$\sqrt{x^2 + y^2} = x + 4$

$x^2 + y^2 = x^2 + 8x + 16$

$y^2 - 8x - 16 = 0$

16. $r = \dfrac{4}{1 + \sin\theta}$ [6.6]

$r + r\sin\theta = 4$

$\sqrt{x^2 + y^2} + y = 4$

$x^2 + y^2 = 16 - 8y + y^2$

$x^2 + 8y - 16 = 0$

17. $x = t - 3$

$x + 3 = t$

$(x + 3)^2 = t^2$

$2(x + 3)^2 = 2t^2$

$2(x + 3)^2 = y$

$(x + 3)^2 = \dfrac{1}{2}y$

[6.7]

18. $x = 4\sin\theta$ $y = \cos\theta + 2$

$\sin\theta = x/4$ $\cos\theta = y - 2$

$\sin^2\theta + \cos^2\theta = 1$

$\left(\dfrac{x}{4}\right)^2 + (y - 2)^2 = 1$

$\dfrac{x^2}{16} + \dfrac{(y - 2)^2}{1} = 1$

[6.7]

19.

$Xscl = 2\pi$ [6.7]

20.

The projectile will travel $256\sqrt{3}$ feet ≈ 443 feet.   [6.7]

1. $x^2 + 4x - 6 = 0$ [1.1]

$x = \dfrac{-4 \pm \sqrt{4^2 - 4(1)(-6)}}{2(1)} = \dfrac{-4 \pm \sqrt{40}}{2}$

$= \dfrac{-4 \pm 2\sqrt{10}}{2} = -2 \pm \sqrt{10}$

The solutions are $2 \pm \sqrt{10}$.

2. $y = x^3 - 4x$ [1.4]

symmetric with
respect to the origin

3. $(g \circ f)(x) = g[f(x)]$ [1.5]
$= g[\sin x]$
$= 3\sin x - 2$

4. $240° = 240°\left(\dfrac{\pi}{180°}\right) = \dfrac{4\pi}{3}$ [2.1]

5. $v = \omega r$ [2.1]

$= \dfrac{3 \cdot 2\pi \cdot 60 \cdot 60 \cdot 10}{12 \cdot 5280}$

≈ 11 mph

6. $\sin t = -\dfrac{\sqrt{3}}{2} = \dfrac{\text{opp}}{\text{hyp}}$ [2.4]

$\text{adj} = \sqrt{2^2 - \left(\sqrt{-3}\right)^2} = \sqrt{1} = 1$

$\tan t = \dfrac{\text{opp}}{\text{adj}} = \dfrac{-\sqrt{3}}{1} = -\sqrt{3}$

7. $\tan 43° = \dfrac{a}{20}$ [2.4]

$a = 20 \tan 43°$

$a = 19$ cm

8. $y = \dfrac{1}{2}\cos\left(\dfrac{\pi x}{3}\right)$ [2.5]

period: 6, amplitude: $\dfrac{1}{2}$

9. $y = 2\tan\left(\dfrac{\pi x}{3}\right)$ [2.6]

period: 3

10. $\dfrac{\sin x}{1 - \cos x} = \dfrac{\sin x}{1 - \cos x} \cdot \dfrac{1 + \cos x}{1 + \cos x}$ [3.1]

$= \dfrac{\sin x(1 + \cos x)}{1 - \cos^2 x}$

$= \dfrac{\sin x(1 + \cos x)}{\sin^2 x}$

$= \dfrac{1}{\sin x} + \dfrac{\cos x}{\sin x}$

$= \csc x + \cot x$

11. $\sin \alpha = \dfrac{3}{5}, \ \cos \alpha = -\dfrac{4}{5}$ [3.2]

$\cos \beta = -\dfrac{5}{13}, \ \sin \beta = \dfrac{12}{13}$

$\sin(\alpha + \beta) = \sin \alpha \cos \beta + \cos \alpha \sin \beta$

$= \left(\dfrac{3}{5}\right)\left(-\dfrac{5}{13}\right) + \left(-\dfrac{4}{5}\right)\left(\dfrac{12}{13}\right)$

$= -\dfrac{15}{65} - \dfrac{48}{65} = -\dfrac{63}{65}$

12. $y = \sin\left[\cos^{-1}\left(\dfrac{1}{5}\right)\right]$ [3.5]

Let $\alpha = \cos^{-1}\dfrac{1}{5}, \ \cos \alpha = \dfrac{1}{5} = \dfrac{\text{adj}}{\text{hyp}}$

$\sin \alpha = \dfrac{\text{opp}}{\text{hyp}} = \dfrac{\sqrt{5^2 - (1)^2}}{5} = \dfrac{2\sqrt{6}}{5}$

$y = \sin\left[\cos^{-1}\left(\dfrac{1}{5}\right)\right] = \dfrac{2\sqrt{6}}{5}$

13. $\cos^{-1} x = \sin^{-1}\left(\dfrac{12}{13}\right)$ [3.6]

$\qquad x = \cos\left[\sin^{-1}\left(\dfrac{12}{13}\right)\right]$

Let $\alpha = \sin^{-1}\dfrac{12}{13},\ \sin\alpha = \dfrac{12}{13} = \dfrac{\text{opp}}{\text{hyp}}$

$\cos\alpha = \dfrac{\text{adj}}{\text{hyp}} = \dfrac{\sqrt{13^2 - 12^2}}{13} = \dfrac{5}{13}$

$x = \cos\left[\sin^{-1}\left(\dfrac{12}{13}\right)\right] = \dfrac{5}{13}$

14.

$C = 180° - A - B$

$\quad = 180° - 71° - 80°$

$\quad = 29°$

$\dfrac{b}{\sin B} = \dfrac{c}{\sin C}$

$\quad b = \dfrac{155 \sin 80°}{\sin 29°} \approx 314.9 \text{ ft}$

$\dfrac{a}{\sin A} = \dfrac{c}{\sin C}$

$\quad a = \dfrac{155 \sin 71°}{\sin 29°} \approx 302.3 \text{ ft}$

$\sin 71° = \dfrac{d}{314.9}$

$\quad d = 314.9 \sin 71°$

$\qquad \approx 298 \text{ ft} \qquad$ [4.1]

15. $b^2 = a^2 + c^2 - 2ac \cos B$ [4.2]

$\cos B = \dfrac{a^2 + c^2 - b^2}{2ac} = \dfrac{4^2 + 3.6^2 - 2.5^2}{2(4)(3.6)} = \dfrac{22.7}{28.8}$

$\quad B \approx 38°$

16. $a_1 = 30 \cos 145° \approx -24.6$ [4.3]

$a_2 = 30 \sin 145° \approx 17.2$

$\mathbf{v} = -24.6\mathbf{i} + 17.2\mathbf{j}$

17. $\mathbf{v} \cdot \mathbf{w} = (3\mathbf{i} + 2\mathbf{j}) \cdot (5\mathbf{i} - 7\mathbf{j})$ [4.3]

$\qquad = 3(5) + (2)(-7)$

$\qquad = 1 \neq 0$

No. the vectors are not orthogonal.

18. $z = -2 + 2i\sqrt{3}$ [5.2]

$r = \sqrt{(-2)^2 + (2\sqrt{3})^2} = \sqrt{16} = 4$

$\alpha = \tan^{-1}\left|\dfrac{2\sqrt{3}}{-2}\right| = \tan^{-1}\sqrt{3} = \dfrac{\pi}{3}$

$\theta = \pi - \dfrac{\pi}{3} = \dfrac{2\pi}{3}$

$z = 4 \operatorname{cis}\left(\dfrac{2\pi}{3}\right)$

19. $r = 3\sin(2\theta)$ [6.5]

20. $x = 2t - 1$ [6.7]

$x + 1 = 2t$

$t = \dfrac{x+1}{2}$

$y = 4t^2 + 1$

$y = 4\left(\dfrac{x+1}{2}\right)^2 + 1 = 4\left(\dfrac{x^2 + 2x + 1}{4}\right) + 1$

$y = x^2 + 2x + 2$

Chapter 7
Exponential and Logarithmic Functions

1. $f(0) = 3^0 = 1$
$f(4) = 3^4 = 81$

3. $g(-2) = 10^{-2} = \dfrac{1}{100}$
$g(3) = 10^3 = 1000$

5. $h(2) = \left(\dfrac{3}{2}\right)^2 = \dfrac{9}{4}$
$h(-3) = \left(\dfrac{3}{2}\right)^{-3} = \dfrac{8}{27}$

7. $j(-2) = \left(\dfrac{1}{2}\right)^{-2} = 4$
$j(4) = \left(\dfrac{1}{2}\right)^4 = \dfrac{1}{16}$

9. $f(3.2) = 2^{3.2} \approx 9.19$

11. $g(2.2) = e^{2.2} \approx 9.03$

13. $h(\sqrt{2}) = 5^{\sqrt{2}} \approx 9.74$

15. $f(x) = 5^x$ is a basic exponential graph.

$g(x) = 1 + 5^{-x}$ is the graph of $f(x)$ reflected across the y-axis and moved up 1 unit.

$h(x) = 5^{x+3}$ is the graph of $f(x)$ moved to the left 3 units.

$k(x) = 5^x + 3$ is the graph of $f(x)$ moved up 3 units.

  a. $k(x)$ **b.** $g(x)$ **c.** $h(x)$ **d.** $f(x)$

17. $f(x) = 3^x$

19. $f(x) = 10^x$

21. $f(x) = \left(\dfrac{3}{2}\right)^x$

23. $f(x) = \left(\dfrac{1}{3}\right)^x$

25. Shift the graph of f vertically upward 2 units.

27. Shift the graph of f horizontally to the right 2 units.

29. Reflect the graph of f across the y-axis.

31. Stretch the graph of f vertically away from the x-axis by a factor of 2.

33. Reflect the graph of f across the y-axis, and then shift this graph vertically upward 2 units.

35. Shift the graph of f horizontally 4 units to the right, and then reflect this graph across the x-axis.

37. Reflect the graph of f across the y-axis, and then shift this graph vertically upward 3 units.

39. $f(x) = \dfrac{3^x + 3^{-x}}{2}$

No horizontal asymptote

41. $f(x) = \dfrac{e^x - e^{-x}}{2}$

No horizontal asymptote.

43. $f(x) = -e^{(x-4)}$

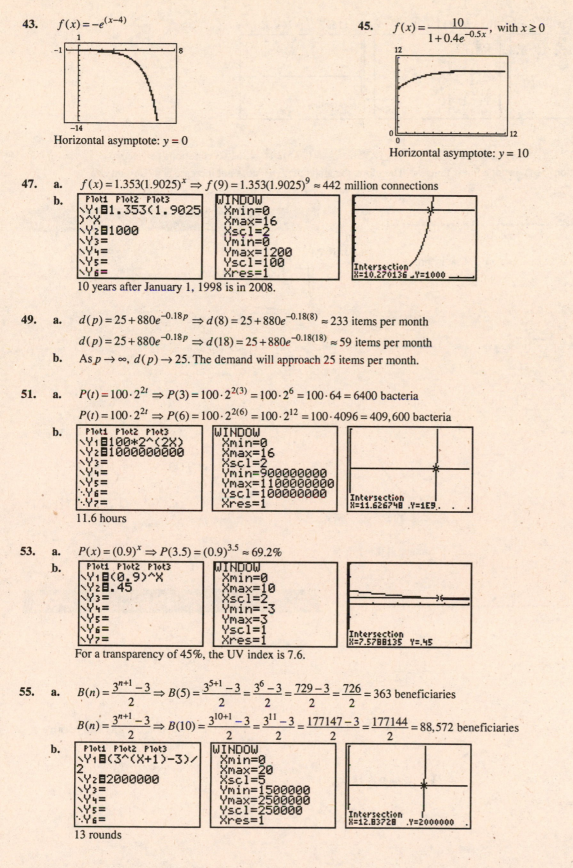

Horizontal asymptote: $y = 0$

45. $f(x) = \dfrac{10}{1 + 0.4e^{-0.5x}}$, with $x \geq 0$

Horizontal asymptote: $y = 10$

47.   **a.** $f(x) = 1.353(1.9025)^x \Rightarrow f(9) = 1.353(1.9025)^9 \approx 442$ million connections

   b.

Plot1 Plot2 Plot3	WINDOW
\Y₁◻1.353(1.9025)^X	Xmin=0
\Y₂◻1000	Xmax=16
\Y₃=	Xscl=2
\Y₄=	Ymin=0
\Y₅=	Ymax=1200
\Y₆=	Yscl=100
	Xres=1

Intersection
X=10.270136 Y=1000

10 years after January 1, 1998 is in 2008.

49.   **a.** $d(p) = 25 + 880e^{-0.18p} \Rightarrow d(8) = 25 + 880e^{-0.18(8)} \approx 233$ items per month

$d(p) = 25 + 880e^{-0.18p} \Rightarrow d(18) = 25 + 880e^{-0.18(18)} \approx 59$ items per month

   b.   As $p \to \infty$, $d(p) \to 25$. The demand will approach 25 items per month.

51.   **a.** $P(t) = 100 \cdot 2^{2t} \Rightarrow P(3) = 100 \cdot 2^{2(3)} = 100 \cdot 2^6 = 100 \cdot 64 = 6400$ bacteria

$P(t) = 100 \cdot 2^{2t} \Rightarrow P(6) = 100 \cdot 2^{2(6)} = 100 \cdot 2^{12} = 100 \cdot 4096 = 409{,}600$ bacteria

   b.

Plot1 Plot2 Plot3	WINDOW
\Y₁◻100*2^(2X)	Xmin=0
\Y₂◻1000000000	Xmax=16
\Y₃=	Xscl=2
\Y₄=	Ymin=900000000
\Y₅=	Ymax=1100000000
\Y₆=	Yscl=100000000
\Y₇=	Xres=1

Intersection
X=11.626748 Y=1E9

11.6 hours

53.   **a.** $P(x) = (0.9)^x \Rightarrow P(3.5) = (0.9)^{3.5} \approx 69.2\%$

   b.

Plot1 Plot2 Plot3	WINDOW
\Y₁◻(0.9)^X	Xmin=0
\Y₂◻.45	Xmax=10
\Y₃=	Xscl=2
\Y₄=	Ymin=-3
\Y₅=	Ymax=3
\Y₆=	Yscl=1
\Y₇=	Xres=1

Intersection
X=7.5788135 Y=.45

For a transparency of 45%, the UV index is 7.6.

55.   **a.** $B(n) = \dfrac{3^{n+1} - 3}{2} \Rightarrow B(5) = \dfrac{3^{5+1} - 3}{2} = \dfrac{3^6 - 3}{2} = \dfrac{729 - 3}{2} = \dfrac{726}{2} = 363$ beneficiaries

$B(n) = \dfrac{3^{n+1} - 3}{2} \Rightarrow B(10) = \dfrac{3^{10+1} - 3}{2} = \dfrac{3^{11} - 3}{2} = \dfrac{177147 - 3}{2} = \dfrac{177144}{2} = 88{,}572$ beneficiaries

   b.

Plot1 Plot2 Plot3	WINDOW
\Y₁◻(3^(X+1)-3)/ 2	Xmin=0
\Y₂◻2000000	Xmax=20
\Y₃=	Xscl=5
\Y₄=	Ymin=1500000
\Y₅=	Ymax=2500000
\Y₆=	Yscl=250000
	Xres=1

Intersection
X=12.83728 Y=2000000

13 rounds

57. **a.** $T(t) = 65 + 115e^{-0.042t} \Rightarrow T(10) = 65 + 115e^{-0.042(10)} = 65 + 115e^{-0.42} \approx 141° \text{ F}$

b.

Plot1 Plot2 Plot3
\Y1■65+115e^(-0.
042X)
\Y2■100
\Y3=
\Y4=
\Y5=
\Y6=

WINDOW
Xmin=0
Xmax=40
Xscl=10
Ymin=90
Ymax=110
Yscl=10
Xres=1

Intersection
X=28.32343 Y=100

28.3 minutes

59. **a.** $f(n) = (27.5)2^{(n-1)/12} \Rightarrow f(40) = (27.5)2^{(40-1)/12} = (27.5)2^{39/12} = (27.5)2^{3.25} \approx 261.63$ vibrations per second

b. No. The function $f(n)$ is not a linear function. Therefore, the graph of $f(n)$ does not increase at a constant rate.

Connecting Concepts

61. $\sinh(x) = \dfrac{e^x - e^{-x}}{2}$ is an odd function. That is, prove

$\sinh(-x) = -\sinh(x)$.

Proof: $\sinh(x) = \dfrac{e^x - e^{-x}}{2}$

$\sinh(-x) = \dfrac{e^{-x} - e^x}{2}$

$\sinh(-x) = \dfrac{-e^{-x} + e^x}{2}$

$\sinh(-x) = \dfrac{(e^x - e^{-x})}{2}$

$\sinh(-x) = -F(x)$

63.

65.

domain: $(-\infty, \infty)$

67.

domain: $[0, \infty)$

69. Let $f(x) = e^x$ and $g(x) = 2x - 5$.
Then

$h(x) = e^{(2x-5)} = f[2x - 5] = f[g(x)] =$

Prepare for Section 7.2

PS1. $2^x = 16$

$2^x = 2^4$

$x = 4$

PS2. $3^{-x} = \dfrac{1}{3^x} = \dfrac{1}{27}$

$\dfrac{1}{3^x} = \dfrac{1}{3^3}$

$x = 3$

PS3. $x^4 = 625$

$x^4 = 5^4$

$x = 5$

PS4. $f(x) = \dfrac{2x}{x+3}$

$x = \dfrac{2y}{y+3}$

$xy + 3x = 2y$

$3x = 2y - xy = y(2 - x)$

$\dfrac{3x}{2-x} = y$

$f^{-1}(x) = \dfrac{3x}{2-x}$

PS5. $g(x) = \sqrt{x-2}$

$x - 2 \geq 0$

$x \geq 2$

The domain is $\{x \,|\, x \geq 2\}$.

PS6. The domain is the set of all positive real numbers.

Section 7.2

1. $1 = \log 10 \Rightarrow 10^1 = 10$

3. $2 = \log_8 64 \Rightarrow 8^2 = 64$

5. $0 = \log_7 x \Rightarrow 7^0 = x$

7. $\ln x = 4 \Rightarrow e^4 = x$

9. $\ln 1 = 0 \Rightarrow e^0 = 1$

11. $2 = \log(3x+1) \Rightarrow 10^2 = 3x+1$

13. $3^2 = 9 \Rightarrow \log_3 9 = 2$

15. $4^{-2} = \dfrac{1}{16} \Rightarrow \log_4 \dfrac{1}{16} = -2$

17. $b^x = y \Rightarrow \log_b y = x$

19. $y = e^x \Rightarrow \ln y = x$

21. $100 = 10^2 \Rightarrow \log 100 = 2$

23. $e^2 = x+5 \Rightarrow 2 = \ln(x+5)$

25. $\log_4 16 = 2$ because $4^2 = 16$

27. $\log_3 \dfrac{1}{243} = -5$ because $3^{-5} = \left(\dfrac{1}{3}\right)^5 = \dfrac{1}{243}$

29. $\ln e^3 = 3$ because $e^3 = e^3$

31. $\log \dfrac{1}{100} = -2$ because $10^{-2} = \dfrac{1}{10^2} = \dfrac{1}{100}$

33. $\log_{0.5} 16 = \log_{1/2} 16 = -4$ because $\left(\dfrac{1}{2}\right)^{-4} = 2^4 = 16$

35. $4\log 1000 = 12 \Rightarrow \log 1000^4 = 12$

because $10^{12} = \left(10^3\right)^4 = (1000)^4$

37. $2\log_7 2401 = 8 \Rightarrow \log_7 2401^2 = 8$

because $7^8 = \left(7^4\right)^2 = (2401)^2$

39. $\log_3 \sqrt[5]{9} = \dfrac{2}{5} \Rightarrow \log_3 9^{1/5} = \dfrac{2}{5}$

because $3^{2/5} = \left(3^2\right)^{1/5} = (9)^{1/5}$

41. $5\log_{13} \sqrt[3]{169} = \dfrac{10}{3} \Rightarrow \log_{13} 169^{5/3} = \dfrac{10}{3}$

because $13^{10/3} = \left(13^2\right)^{5/3} = (169)^{5/3}$

43. $y = \log_4 x$

$x = 4^y$

45. $y = \log_{12} x$

$x = 12^y$

47. $y = \log_{1/2} x$

$x = (1/2)^y$

49. $y = \log_{5/2} x$

$x = (5/2)^y$

51. $f(x) = \log_5(x-3)$

$x - 3 > 0$

$\quad x > 3$

The domain is $(3, \infty)$.

53. $k(x) = \log_{2/3}(11-x)$

$11 - x > 0$

$\quad -x > -11$

$\quad x < 11$

The domain is $(-\infty, 11)$.

55. $P(x) = \ln(x^2 - 4)$

$x^2 - 4 > 0$

$(x + 2)(x - 2) > 0$

The critical values are −2 and 2.
The product is positive.
The domain is $(-\infty,\ -2) \cup (2,\ \infty)$.

57. $h(x) = \ln\left(\dfrac{x^2}{x - 4}\right)$

$\dfrac{x^2}{x - 4} > 0$

The critical values are 0 and 4.
The quotient is positive.
$x > 4$
The domain is $(4,\ \infty)$.

59. $x^3 - x > 0$

$x(x^2 - 1) > 0$

$x(x + 1)(x - 1) > 0$

Critical values are 0, −1 and 1.
Product is positive.
$-1 < x < 0$ or $x > 1$
$(-1, 0) \cup (1, \infty)$

61. $2x - 11 > 0$

$2x > 11$

$x > \dfrac{11}{2}$

The domain is $\left(\dfrac{11}{2},\ \infty\right)$.

63. $3x - 7 > 0$

$3x > 7$

$x > \dfrac{7}{3}$

The domain is $\left(\dfrac{7}{3},\ \infty\right)$.

65.   Shift 3 units to the right.

67.   Shift 2 units up.

69.   Shift 3 units up.

71.   Shift 4 units to the right and 1 unit up

73.   The graph of $f(x) = \log_5(x - 2)$ is the graph of $y = \log_5 x$ shifted 2 units to the right.

The graph of $g(x) = 2 + \log_5 x$ is the graph of $y = \log_5 x$ shifted 2 units up.

The graph of $h(x) = \log_5(-x)$ is the graph of $y = \log_5 x$ reflected across the y-axis.

The graph of $k(x) = -\log_5(x + 3)$ is the graph of $y = \log_5 x$ reflected across the x-axis and shifted left 3 units.

a. $k(x)$ **b.** $f(x)$ **c.** $g(x)$ **d.** $h(x)$

75.

77.

79.

81.

83.

85. **a.** $r(t) = 0.69607 + 0.60781 \ln t \Rightarrow r(9) = 0.69607 + 0.60781 \ln 9 \approx 2.0\%$

b.

45 months

87. $N(x) = 2750 + 180 \ln\left(\dfrac{x}{1000} + 1\right)$

a. $N(20,000) = 2750 + 180 \ln\left(\dfrac{20,000}{1000} + 1\right) = 2750 + 180 \ln(21) \approx 3298$ units

$N(40,000) = 2750 + 180 \ln\left(\dfrac{40,000}{1000} + 1\right) = 2750 + 180 \ln(41) \approx 3418$ units

$N(60,000) = 2750 + 180 \ln\left(\dfrac{60,000}{1000} + 1\right) = 2750 + 180 \ln(61) \approx 3490$ units

b. $N(0) = 2750 + 180 \ln\left(\dfrac{0}{1000} + 1\right) = 2750 + 180 \ln(1) = 2750 + 180(0) = 2750 + 0 = 2750$ units

89. $BSA = 0.0003207 \cdot H^{0.3} \cdot W^{(0.7285 - 0.0188 \log W)}$

$BSA = 0.0003207 \cdot (185.42)^{0.3} \cdot (81,646.6)^{(0.7285 - 0.0188 \log 81,646.6)} \approx 2.05$ square meters

91. $N = \text{int}(x \log b) + 1$

a. $N = \text{int}(10 \log 2) + 1 = 3 + 1 = 4$ digits

b. $N = \text{int}(200 \log 3) + 1 = 95 + 1 = 96$ digits

c. $N = \text{int}(4005 \log 7) + 1 = 3384 + 1 = 3385$ digits

d. $N = \text{int}(2,0996,001 \log 2) + 1 = 6,320,429 + 1 = 6,320,430$ digits

93. $f(x)$ and $g(x)$ are inverse functions

Connecting Concepts

95. The domain of the inverse is the range of the function.
Range of f: $\{y \mid -1 < y \le 1\}$.
The domain of the function is the range of the inverse.
Range of g: all real numbers.

Prepare for Section 7.3

PS1. $\log 3 + \log 2 \approx 0.77815$
$\log 6 \approx 0.77815$

PS2. $\ln 8 - \ln 3 \approx 0.98083$
$\ln\left(\dfrac{8}{3}\right) \approx 0.98083$

PS3. $3 \log 4 \approx 1.80618$
$\log(4^3) \approx 1.80618$

PS4. $2 \ln 5 \approx 3.21888$
$\ln(5^2) \approx 3.21888$

PS5. $\ln 5 \approx 1.60944$
$\dfrac{\log 5}{\log e} \approx 1.60944$

PS6. $\log 8 \approx 0.90309$
$\dfrac{\ln 8}{\ln 10} \approx 0.90309$

Section 7.3

1. $\log_b(xyz) = \log_b x + \log_b y + \log_b z$

3. $\ln \dfrac{x}{z^4} = \ln x - \ln z^4$
$$= \ln x - 4\ln z$$

5. $\log_2 \dfrac{\sqrt{x}}{y^3} = \log_2 \sqrt{x} - \log_2 y^3$
$$= \log_2 x^{1/2} - \log_2 y^3$$
$$= \tfrac{1}{2}\log_2 x - 3\log_2 y$$

7. $\log_7 \dfrac{\sqrt{xz}}{y^2} = \log_7 \dfrac{(xz)^{1/2}}{y^2} = \log_7 \dfrac{x^{1/2}z^{1/2}}{y^2}$
$$= \log_7 x^{1/2} + \log_7 z^{1/2} - \log_7 y^2$$
$$= \tfrac{1}{2}\log_7 x + \tfrac{1}{2}\log_7 z - 2\log_7 y$$

9. $\ln(e^2 z) = \ln e^2 + \ln z = 2\ln e + \ln z = 2 + \ln z$

11. $\log_4 \left(\dfrac{\sqrt[3]{z}}{16y^3} \right) = \log_4 z^{1/3} - \log_4 4^2 - \log_4 y^3 = \tfrac{1}{3}\log_4 z - 2\log_4 4 - 3\log_4 y = \tfrac{1}{3}\log_4 z - 2 - 3\log_4 y$

13. $\log \sqrt{x\sqrt{z}} = \log(xz^{1/2})^{1/2} = \log x^{1/2} z^{1/4} = \tfrac{1}{2}\log x + \tfrac{1}{4}\log z$

15. $\ln\left(\sqrt[3]{z\sqrt{e}}\right) = \ln(ze^{1/2})^{1/3} = \ln z^{1/3}e^{1/6} = \ln z^{1/3} + \ln e^{1/6} = \tfrac{1}{3}\ln z + \tfrac{1}{6}\ln e = \tfrac{1}{3}\ln z + \tfrac{1}{6}$

17. $\log(x+5) + 2\log x = \log(x+5) + \log x^2 = \log[x^2(x+5)]$

19. $\ln(x^2 - y^2) - \ln(x-y) = \ln \dfrac{x^2 - y^2}{x-y} = \ln \dfrac{(x+y)(x-y)}{x-y} = \ln(x+y)$

21. $3\log x + \tfrac{1}{3}\log y + \log(x+1) = \log x^3 + \log y^{1/3} + \log(x+1) = \log x^3 + \log \sqrt[3]{y} + \log(x+1) = \log\left[x^3 \cdot \sqrt[3]{y}(x+1)\right]$

23. $\log(xy^2) - \log z = \log\left(\dfrac{xy^2}{z}\right)$

25. $2\left(\log_6 x + \log_6 y^2\right) - \log_6(x+2) = \log_6 x^2 + \log_6 y^4 - \log_6(x+2) = \log_6\left(\dfrac{x^2 y^4}{x+2}\right)$

27. $2\ln(x+4) - \ln x - \ln(x^2 - 3) = \ln(x+4)^4 - \ln x - \ln(x^2 - 3) = \ln\left[\dfrac{(x+4)^4}{x(x^2-3)}\right]$

29. $\ln(2x+5) - \ln y - 2\ln z + \tfrac{1}{2}\ln w = \ln(2x+5) - \ln y - \ln z^2 + \ln w^{1/2} = \ln\left[\dfrac{(2x+5)\sqrt{w}}{yz^2}\right]$

31. $\ln(x^2 - 9) - 2\ln(x-3) + 3\ln y = \ln(x+3)(x-3) - \ln(x-3)^2 + \ln y^3 = \ln\left[\dfrac{(x+3)(x-3)y^3}{(x-3)^2}\right] = \ln\left[\dfrac{(x+3)y^3}{x-3}\right]$

33. $\log_7 20 = \dfrac{\log 20}{\log 7} \approx 1.5395$

35. $\log_{11} 8 = \dfrac{\log 8}{\log 11} \approx 0.8672$

37. $\log_6 \tfrac{1}{3} = \dfrac{\log \tfrac{1}{3}}{\log 6} \approx -0.6131$

39. $\log_9 \sqrt{17} = \dfrac{\log \sqrt{17}}{\log 9} \approx 0.6447$

41. $\log_{\sqrt{2}} 17 = \dfrac{\log 17}{\log \sqrt{2}} \approx 8.1749$

43. $\log_\pi e = \dfrac{\log e}{\log \pi} \approx 0.8735$

45. $f(x) = \log_4 x = \dfrac{\log x}{\log 4}$

47. $g(x) = \log_8(x-3) = \dfrac{\log(x-3)}{\log 8}$

49. $h(x) = \log_3(x-3)^2$

$= \dfrac{\log(x-3)^2}{\log 3}$

$= \dfrac{2\log(x-3)}{\log 3}$

51. $F(x) = -\log_5|x-2| = -\dfrac{\log|x-2|}{\log 5}$

53. False. $\log 10 + \log 10 = 1 + 1 = 2$
but $\log(10+10) = \log 20 \neq 2$.

55. True.

57. False. $\log 100 - \log 10 = 2 - 1 = 1$
but $\log(100 - 10) = \log 90 \neq 1$

59. False. $\dfrac{\log 100}{\log 10} = \dfrac{2}{1} = 2$
but $\log 100 - \log 10 = 2 - 1 = 1$

61. False. $(\log 10)^2 = 1^2 = 1$ but $2\log 10 = 2(1) = 2$

63. $\log_3 5 \cdot \log_5 7 \cdot \log_7 9 = \dfrac{\log 5}{\log 3} \cdot \dfrac{\log 7}{\log 5} \cdot \dfrac{\log 9}{\log 7} = \dfrac{\log 5}{\log 3} \cdot \dfrac{\log 7}{\log 5} \cdot \dfrac{\log 9}{\log 7} = \dfrac{\log 9}{\log 3} = \dfrac{\log 3^2}{\log 3} = \dfrac{2\log 3}{\log 3} = \dfrac{2\log 3}{\log 3} = 2$

65. $\ln 500^{501} = 501\ln 500 \approx 3113.52$
$\ln 506^{500} = 500\ln 506 \approx 3113.27$
$\ln 500^{501}$ is larger.

67. $S_n = S_0 \cdot 10^{\frac{n}{N}(\log S_f - \log S_0)}$

$S_1 = 1{,}000{,}000 \cdot 10^{\frac{1}{5}(\log 500{,}000 - \log 1{,}000{,}000)}$
$= 870{,}551$

$S_2 = 1{,}000{,}000 \cdot 10^{\frac{2}{5}(\log 500{,}000 - \log 1{,}000{,}000)}$
$= 757{,}858$

$S_3 = 1{,}000{,}000 \cdot 10^{\frac{3}{5}(\log 500{,}000 - \log 1{,}000{,}000)}$
$= 659{,}754$

$S_4 = 1{,}000{,}000 \cdot 10^{\frac{4}{5}(\log 500{,}000 - \log 1{,}000{,}000)}$
$= 574{,}349$

$S_5 = 1{,}000{,}000 \cdot 10^{\frac{5}{5}(\log 500{,}000 - \log 1{,}000{,}000)}$
$= 500{,}000$

The scales are 1:870,551; 1:757,858; 1:659,754; 1:574,349;
1:500,000.

69. $pH = -\log[H^+]$

$pH = -\log[3.97 \times 10^{-11}]$

$pH = 10.4$

$10.4 > 7$; milk of magnesia is a base

71. $pH = -\log[H^+]$

$9.5 = -\log[H^+]$

$-9.5 = \log[H^+]$

$10^{-9.5} = 10^{\log[H^+]}$

$[H^+] = 3.16 \times 10^{-10}$ mole per liter

73. $dB(I) = 10\log\left(\dfrac{I}{I_0}\right)$

a. $dB(1.58 \times 10^8 \cdot I_0) = 10\log\left(\dfrac{1.58 \times 10^8 \cdot I_0}{I_0}\right)$

$= 10\log(1.58 \times 10^8)$

≈ 82.0 decibels

b. $dB(10{,}800 \cdot I_0) = 10\log\left(\dfrac{10{,}800 \cdot I_0}{I_0}\right)$

$= 10\log(10{,}800)$

≈ 40.3 decibels

c. $dB(3.16 \times 10^{11} \cdot I_0) = 10\log\left(\dfrac{3.16 \times 10^{11} \cdot I_0}{I_0}\right)$

$= 10\log(3.16 \times 10^{11})$

≈ 115.0 decibels

d. $dB(1.58 \times 10^{15} \cdot I_0) = 10\log\left(\dfrac{1.58 \times 10^{15} \cdot I_0}{I_0}\right)$

$= 10\log(1.58 \times 10^{15})$

≈ 152.0 decibels

75. $dB(I) = 10\log\left(\dfrac{I}{I_0}\right)$

$120 = 10\log\left(\dfrac{I_{120}}{I_0}\right)$

$12 = \log\left(\dfrac{I_{120}}{I_0}\right)$

$10^{12} = \dfrac{I_{120}}{I_0}$

$10^{12} \cdot I_0 = I_{120}$

$110 = 10\log\left(\dfrac{I_{110}}{I_0}\right)$

$11 = \log\left(\dfrac{I_{110}}{I_0}\right)$

$10^{11} = \dfrac{I_{110}}{I_0}$

$10^{11} \cdot I_0 = I_{110}$

$\dfrac{I_{120}}{I_{110}} = \dfrac{10^{12} \cdot I_0}{10^{11} \cdot I_0}$

$= \dfrac{10^{12}}{10^{11}} = 10^{12-11} = 10^1$

$= 10$ times more intense

77. $M = \log\left(\dfrac{100{,}000 I_0}{I_0}\right) = \log 100{,}000 = \log 10^5 = 5$

79. $\log\left(\dfrac{I}{I_0}\right) = M$

$\log\left(\dfrac{I}{I_0}\right) = 6.5$

$\dfrac{I}{I_0} = 10^{6.5}$

$I = 10^{6.5} I_0$

$I \approx 3{,}162{,}277.7 I_0$

81. $M = \log\left(\dfrac{I}{I_0}\right)$

$M_5 = \log\left(\dfrac{I_5}{I_0}\right) \Rightarrow 5 = \log\left(\dfrac{I_5}{I_0}\right) \Rightarrow 10^5 = \dfrac{I_5}{I_0} \Rightarrow 10^5 I_0 = I_5$

$M_3 = \log\left(\dfrac{I_3}{I_0}\right) \Rightarrow 3 = \log\left(\dfrac{I_3}{I_0}\right) \Rightarrow 10^3 = \dfrac{I_3}{I_0} \Rightarrow 10^3 I_0 = I_3$

$\dfrac{I_5}{I_3} = \dfrac{10^5 I_0}{10^3 I_0} = \dfrac{10^5}{10^3} = 10^{5-3} = 10^2 = 100$ to 1   • short cut: begin with this line

83. $\dfrac{10^{8.9}}{10^{7.1}} = \dfrac{10^{8.9-7.1}}{1} = 10^{1.8}$ to $1 \approx 63$ to 1

85. $M = \log A + 3\log 8t - 2.92$
$$= \log 18 + 3\log[8(31)] - 2.92 \approx 5.5$$

87. Let $r = \log_b M$ and $s = \log_b N$.

Then $M = b^r$ and $N = b^s$.
Consider the quotient of M and N

$$\frac{M}{N} = \frac{b^r}{b^s}$$

$$\frac{M}{N} = b^{r-s}$$

$$\log_b \frac{M}{N} = r - s$$

$$\log_b \frac{M}{N} = \log_b M - \log_b N$$

89. **a.** $M \approx 6$ **b.** $M \approx 4$

      c. When $t = 40$, $M = \log A + 3\log 8t - 2.92 = \log 50 + 3\log[8(40)] - 2.92 \approx 6.3$

           When $t = 30$, $M = \log A + 3\log 8t - 2.92 = \log 1 + 3\log[8(30)] - 2.92 \approx 4.2$

           The results from parts **a.** and **b.** are close to the magnitudes of 6.3 and 4.2 produced by the amplitude time difference formula.

●●●●●●●●●●●●●●●●●●●●●●●●●●●●●●●●●●●●●● **Prepare for Section 7.4**

PS1. $3^6 = 729 \Rightarrow \log_3 729 = 6$

PS2. $\log_5 625 = 4 \Rightarrow 5^4 = 625$

PS3. $a^{x+2} = b \Rightarrow \log_a b = x + 2$

PS4.
$$4a = 7bx + 2cx$$
$$7bx + 2cx = 4a$$
$$x(7b + 2c) = 4a$$
$$x = \frac{4a}{7b + 2c}$$

PS5.
$$165 = \frac{300}{1 + 12x}$$
$$165(1 + 12x) = 300$$
$$165 + 1980x = 300$$
$$1980x = 135$$
$$x = \frac{135}{1980} = \frac{3}{44}$$

PS6.
$$A = \frac{100 + x}{100 - x}$$
$$A(100 - x) = 100 + x$$
$$100A - Ax = 100 + x$$
$$100A - 100 = Ax + x$$
$$100(A - 1) = x(A + 1)$$
$$x = \frac{100(A - 1)}{A + 1}$$

Section 7.4

1. $2^x = 64$
$$2^x = 2^6$$
$$x = 6$$

3. $49^x = \dfrac{1}{343}$
$$7^{2x} = 7^{-3}$$
$$2x = -3$$
$$x = -\frac{3}{2}$$

5. $2^{5x+3} = \dfrac{1}{8}$
$$2^{5x+3} = 2^{-3}$$
$$5x + 3 = -3$$
$$5x = -6$$
$$x = -\frac{6}{5}$$

7. $\left(\dfrac{2}{5}\right)^x = \dfrac{8}{125}$
$$\left(\frac{2}{5}\right)^x = \left(\frac{2}{5}\right)^3$$
$$x = 3$$

9.
$$5^x = 70$$
$$\log(5^x) = \log 70$$
$$x \log 5 = \log 70$$
$$x = \frac{\log 70}{\log 5}$$

11.
$$3^{-x} = 120$$
$$\log(3^{-x}) = \log 120$$
$$-x \log 3 = \log 120$$
$$-x = \frac{\log 120}{\log 3}$$
$$x = -\frac{\log 120}{\log 3}$$

13.
$$10^{2x+3} = 315$$
$$\log 10^{2x+3} = \log 315$$
$$(2x+3)\log 10 = \log 315$$
$$2x + 3 = \log 315$$
$$x = \frac{\log 315 - 3}{2}$$

15.
$$e^x = 10$$
$$\ln e^x = \ln 10$$
$$x = \ln 10$$

17.
$$2^{1-x} = 3^{x+1}$$
$$\log 2^{1-x} = \log 3^{x+1}$$
$$(1-x)\log 2 = (x+1)\log 3$$
$$\log 2 - x \log 2 = x \log 3 + \log 3$$
$$\log 2 - x \log 2 - x \log 3 = \log 3$$
$$-x \log 2 - x \log 3 = \log 3 - \log 2$$
$$-x(\log 2 + \log 3) = \log 3 - \log 2$$
$$x = -\frac{(\log 3 - \log 2)}{(\log 2 + \log 3)}$$
$$x = \frac{\log 2 - \log 3}{\log 2 + \log 3} \text{ or } \frac{\log 2 - \log 3}{\log 6}$$

19.
$$2^{2x-3} = 5^{-x-1}$$
$$\log 2^{2x-3} = \log 5^{-x-1}$$
$$(2x-3)\log 2 = (-x-1)\log 5$$
$$2x \log 2 - 3 \log 2 = -x \log 5 - \log 5$$
$$2x \log 2 + x \log 5 - 3 \log 2 = -\log 5$$
$$2x \log 2 + x \log 5 = 3 \log 2 - \log 5$$
$$x(2 \log 2 + \log 5) = 3 \log 2 - \log 5$$
$$x = \frac{3 \log 2 - \log 5}{2 \log 2 + \log 5}$$

21.
$$\log(4x - 18) = 1$$
$$4x - 18 = 10^1$$
$$4x - 18 = 10$$
$$4x = 28$$
$$x = 7$$

23.
$$\ln(x^2 - 12) = \ln x$$
$$x^2 - 12 = x$$
$$x^2 - x - 12 = 0$$
$$(x-4)(x+3) = 0$$
$$x = 4 \quad \text{or} \quad x = -3 \text{ (No; not in domain.)}$$
$$x = 4$$

25.
$$\log_2 + \log_2(x - 4) = 2$$
$$\log_2 x(x - 4) = 2$$
$$\log_2(x^2 - 4x) = 2$$
$$2^2 = x^2 - 4x$$
$$0 = x^2 - 4x - 4$$
$$x = \frac{4 \pm \sqrt{16 - 4(1)(-4)}}{2}$$
$$x = \frac{4 \pm 4\sqrt{2}^2}{2}$$
$$x = 2 \pm 2\sqrt{2}$$

$2 - 2\sqrt{2}$ is not a solution because the logarithm of a negative number is not defined. The solution is $x = 2 + 2\sqrt{2}$.

27.
$$\log(5x-1) = 2 + \log(x-2)$$
$$\log(5x-1) - \log(x-2) = 2$$
$$\log\frac{(5x-1)}{(x-2)} = 2$$
$$10^2 = \frac{(5x-1)}{(x-2)}$$
$$100(x-2) = 5x-1$$
$$100x - 200 = 5x - 1$$
$$95x = 199$$
$$x = \frac{199}{95}$$

29.
$$\ln(1-x) + \ln(3-x) = \ln 8$$
$$\ln[(1-x)(3-x)] = \ln 8$$
$$(1-x)(3-x) = 8$$
$$3 - 4x + x^2 = 8$$
$$x^2 - 4x - 5 = 0$$
$$(x+1)(x-5) = 0$$
$$x = -1 \text{ or } x = 5 \text{ (No; not in domain.)}$$
The solution is $x = -1$.

31.
$$\log\sqrt{x^3-17} = \frac{1}{2}$$
$$\frac{1}{2}\log(x^3-17) = \frac{1}{2}$$
$$10^1 = x^3 - 17$$
$$27 = x^3$$
$$\sqrt[3]{27} = \sqrt[3]{x^3}$$
$$3 = x$$
The solution is $x = 3$.

33.
$$\log(\log x) = 1$$
$$10^1 = \log x$$
$$10^{10} = x$$

35.
$$\ln(e^{3x}) = 6$$
$$3x\ln e = 6$$
$$3x(1) = 6$$
$$3x = 6$$
$$x = 2$$

37.
$$\log_7(5x) - \log_7 3 = \log_7(2x+1)$$
$$\log_7\left(\frac{5x}{3}\right) = \log_7(2x+1)$$
$$\frac{5x}{3} = 2x+1$$
$$5x = 6x+3$$
$$-3 = x$$
$x = -3$ is not a solution because $\log_7(-15)$ is undefined.
No solution.

39.
$$e^{\ln(x-1)} = 4$$
$$\ln e^{\ln(x-1)} = \ln 4$$
$$\ln(x-1)\ln e = \ln 4$$
$$\ln(x-1)(1) = \ln 4$$
$$(x-1) = 4$$
$$x = 5$$

41.
$$\frac{10^x - 10^{-x}}{2} = 20$$
$$10^x(10^x - 10^{-x}) = 40(10^x)$$
$$10^{2x} - 1 = 40(10^x)$$
$$10^{2x} - 40(10)^x - 1 = 0$$
Let $u = 10^x$.
$$u^2 - 40u - 1 = 0$$
$$u = \frac{40 \pm \sqrt{40^2 - 4(1)(-1)}}{2}$$
$$= \frac{40 \pm \sqrt{1600 + 4}}{2}$$
$$= \frac{40 \pm \sqrt{1604}}{2}$$
$$= \frac{40 \pm 2\sqrt{401}}{2}$$
$$= 20 \pm \sqrt{401}$$
$$10^x = 20 + \sqrt{401}$$
$$\log 10^x = \log\left(20 + \sqrt{401}\right)$$
$$x = \log\left(20 + \sqrt{401}\right)$$

43.
$$\frac{10^x + 10^{-x}}{10^x - 10^{-x}} = 5$$
$$10^x + 10^{-x} = 5(10^x - 10^{-x})$$
$$10^x(10^x + 10^{-x}) = 5(10^x - 10^{-x})10^x$$
$$10^{2x} + 1 = 5(10^{2x} - 1)$$
$$4(10^{2x}) = 6$$
$$2(10^{2x}) = 3$$
$$(10^x)^2 = \frac{3}{2}$$
$$10^x = \sqrt{\frac{3}{2}}$$
$$x \log 10 = \log\sqrt{\frac{3}{2}}$$
$$x = \log\sqrt{\frac{3}{2}}$$
$$x = \frac{1}{2}\log\left(\frac{3}{2}\right)$$

45.
$$\frac{e^x + e^{-x}}{2} = 15$$
$$e^x(e^x + e^{-x}) = (30)e^x$$
$$e^{2x} + 1 = e^x(30)$$
$$e^{2x} - 30e^x + 1 = 0$$
Let $u = e^x$.
$$u^2 - 30u + 1 = 0$$
$$u = \frac{30 \pm \sqrt{900 - 4}}{2}$$
$$u = \frac{30 \pm \sqrt{896}}{2}$$
$$u = \frac{30 \pm 8\sqrt{14}}{2}$$
$$u = 15 \pm 4\sqrt{14}$$
$$e^x = 15 \pm 4\sqrt{14}$$
$$x \ln e = \ln(15 \pm 4\sqrt{14})$$
$$x = \ln(15 \pm 4\sqrt{14})$$

47.
$$\frac{1}{e^x - e^{-x}} = 4$$
$$1 = 4(e^x - e^{-x})$$
$$1(e^x) = 4(e^x)(e^x - e^{-x})$$
$$e^x = 4(e^{2x} - 1)$$
$$e^x = 4e^{2x} - 4$$
$$0 = 4e^{2x} - e^x - 4$$
Let $u = e^x$.
$$0 = 4u^2 - u - 4$$
$$u = \frac{1 \pm \sqrt{1 - 4(4)(-4)}}{8}$$
$$u = \frac{1 \pm \sqrt{65}}{8}$$
$$e^x = \frac{1 + \sqrt{65}}{8}$$
$$x \ln e = \ln\left(\frac{1 + \sqrt{65}}{8}\right)$$
$$x = \ln(1 + \sqrt{65}) - \ln 8$$

49. $2^{-x+3} = x + 1$

Graph $f = 2^{-x+3} - (x+1)$.

Its x-intercept is the solution.

$x \approx 1.61$

Xmin = −4, Xmax = 4, Xscl = 1,

Ymin = −4, Ymax = 4, Yscl = 1

51. $e^{3-2x} - 2x = 1$

Graph $f = e^{3-2x} - 2x - 1$.

Its x-intercept is the solution.

$x \approx 0.96$

Xmin = −4, Xmax = 4, Xscl = 1,

Ymin = −4, Ymax = 4, Yscl = 1

53. $3 \log_2(x-1) = -x + 3$

Graph $f = \dfrac{3 \log(x-1)}{\log 2} + x - 3$.

Its x-intercept is the solution.

$x \approx 2.20$

Xmin = −4, Xmax = 4, Xscl = 1,

Ymin = −4, Ymax = 4, Yscl = 1

55. $\ln(2x+4) + \dfrac{1}{2}x = -3$

Graph $f = \ln(2x+4) + \dfrac{1}{2}x + 3$.

Its x-intercept is the solution.

$x \approx -1.93$

Xmin = −4, Xmax = 4, Xscl = 1,

Ymin = −4, Ymax = 4, Yscl = 1

57. $2^{x+1} = x^2 - 1$

Graph $f = 2^{x+1} - x^2 + 1$.

Its x-intercept is the solution.

$x \approx -1.34$

Xmin = −4, Xmax = 4, Xscl = 1,

Ymin = −4, Ymax = 4, Yscl = 1

59. a. $P(0) = 8500(1.1)^0 = 8500(1) = 8500$

$P(2) = 8500(1.1)^2 = 10,285$

b.
$$15,000 = 8500(1.1)^t$$
$$\ln 15,000 = 8500(1.1)^t$$
$$\ln 51,000 = \ln 8500 + t \ln(1.1)$$
$$\frac{\ln 15,000 - \ln 8500}{\ln(1.1)} = t$$
$$6 \approx t$$

The population will reach 15,000 in 6 years.

61. a. $T(10) = 36 + 43e^{-0.058(10)} = 36 + 43e^{-0.58}$

$T \approx 60°F$

b.
$$45 = 36 + 43e^{-0.058t}$$
$$\ln(45 - 36) = \ln 43 - 0.058t \ln e$$
$$\frac{\ln(45 - 36) - \ln 43}{-0.058} = t$$
$$t \approx 27 \text{ minutes}$$

63.
$$114 = 198 - (198 - 0.9)e^{-0.23x}$$
$$-84 = -197.1e^{-0.23x}$$
$$\frac{84}{197.1} = e^{-0.23x}$$
$$\ln\left(\frac{84}{197.1}\right) = -0.23x$$
$$x = \frac{\frac{84}{197.1}}{-0.23} \approx 3.7 \text{ years}$$

65. $5 + 29\ln(t+1) = 65$

Graph $f = 29\ln(x+1) - 60$.

Its x-intercept is the solution.

$x \approx 6.9$ months

Xmin $= -4$, Xmax $= 10$, Xscl $= 1$,

Ymin $= -4$, Ymax $= 4$, Yscl $= 1$

67. Consider the first function for time less than 10 seconds.

$$275 = -2.25x^2 + 56.26x - 0.28$$

$$0 = -2.25x^2 + 56.26x - 275.28$$

$$x = \frac{-56.26 \pm \sqrt{(56.26)^2 - 4(-2.25)(-275.28)}}{2(-2.25)}$$

$x = 6.67$ or 18.33

18.33 s > 10 s, so it is not a solution. The solution is 6.67 s.

Consider the second function for time greater than 10 seconds.

$$275 = 8320(0.73)^x$$

$$\frac{275}{8320} = (0.73)^x$$

$$\ln\left(\frac{275}{8320}\right) = \ln 0.73^x$$

$$\ln\left(\frac{275}{8320}\right) = x\ln 0.73$$

$$x = \frac{\ln\left(\dfrac{275}{8320}\right)}{\ln 0.73} \approx 10.83$$

The solutions are 6.67 s and 10.83 s.

69. **a.**

b. 48 hours

c. $P = 100$

d. As the number of hours of training increases, the test scores approach 100%.

73. **a.**

c. When $T = 100$, $r \approx 0.019$, or 1.9%

71. **a.**

b. In 27 years or 2026

c. $B = 1000$

d. As the number of years increases, the bison population approaches but never exceeds 1000.

b. When $r = 3\%$, or 0.03, $T \approx 78$ years

75. a.

$$v = 100\left(\frac{e^{0.64t}-1}{e^{0.64t}+1}\right)$$

$$50 = 100\left(\frac{e^{0.64t}-1}{e^{0.64t}+1}\right)$$

$$\frac{50}{100} = \frac{e^{0.64t}-1}{e^{0.64t}+1}$$

$$0.5 = \frac{e^{0.64t}-1}{e^{0.64t}+1}$$

$$0.5(e^{0.64t}+1) = e^{0.64t}-1$$

$$0.5e^{1.64t}+0.5 = e^{0.64t}-1$$

$$0.5e^{0.64t}-e^{0.64t} = -1.5$$

$$-0.5e^{0.64t} = -1.5$$

$$e^{0.64t} = 3$$

$$0.64t = \ln 3$$

$$t = \frac{\ln 3}{0.64}$$

$$t \approx 1.72$$

In approximately 1.72 seconds, the velocity will be 50 feet per second.

b. The horizontal asymptote is the value of

$$100\left[\frac{e^{0.64t}-1}{e^{0.64t}+1}\right] \quad \text{as } t \to \infty. \text{ Therefore, the}$$

horizontal asymptote is $v = 100$ feet per second.

c. The object cannot fall faster than 100 feet per second.

77. Graph $V = 400{,}000 - 150{,}000(1.005)^x$

and $V = 100{,}000$.

They intersect when $x \approx 138.97$.

After 138 withdrawals, the account has \$101,456.39.

After 139 withdrawals, the account has \$99,963.67.

The designer can make at most 138 withdrawals and still have \$100,000.

Xmin = 0, Xmax = 200, Xscl = 25

Ymin = −50000, Ymax = 350000, Yscl = 50000

Connecting Concepts

79. The second step because log 0.5 < 0. Thus the inequality sign must be reversed.

81.

$$\log(x+y) = \log x + \log y$$
$$\log(x+y) = \log xy$$

Therefore $x + y = xy$

$$x - xy = -y$$
$$x(1-y) = -y$$
$$x = \frac{-y}{1-y}$$
$$x = \frac{y}{y-1}$$

83. Since $e^{0.336} \approx 1.4$,

$$F(x) = (1.4)^x \approx (e^{0.336})^x = e^{0.336x} = G(x)$$

Prepare for Section 7.5

PS1. $A = 1000\left(1+\dfrac{0.1}{12}\right)^{12(2)} = 1220.39$

PS2. $A = 600\left(1+\dfrac{0.04}{4}\right)^{4(8)} = 824.96$

PS3.
$$0.5 = e^{14k}$$
$$\ln 0.5 = \ln e^{14k}$$
$$\ln 0.5 = 14k$$
$$\frac{\ln 0.5}{14} = k$$
$$-0.0495 \approx k$$

PS4.
$$0.85 = 0.5^{t/5730}$$
$$\ln 0.85 = \ln 0.5^{t/5730}$$
$$\ln 0.85 = \frac{t}{5730}\ln 0.5$$
$$\frac{5730\ln 0.85}{\ln 0.5} = t$$
$$1340 \approx t$$

PS5.
$$6 = \frac{70}{5 + 9e^{-12k}}$$
$$6(5 + 9e^{-12k}) = 70$$
$$30 + 54e^{-12k} = 70$$
$$54e^{-12k} = 40$$
$$e^{-12k} = \frac{20}{27}$$
$$\ln e^{-12k} = \ln\frac{20}{27}$$
$$-12k = \ln\frac{20}{27}$$
$$k = -\frac{1}{12}\ln\frac{20}{27}$$
$$k \approx 0.025$$

PS6.
$$2,000,000 = \frac{3^{n+1} - 3}{2}$$
$$4,000,000 = 3^{n+1} - 3$$
$$3,999,997 = 3^{n+1}$$
$$\ln 3,999,997 = \ln 3^{n+1}$$
$$\ln 3,999,997 = (n+1)\ln 3$$
$$\frac{\ln 3,999,997}{\ln 3} = n+1$$
$$\frac{\ln 3,999,997}{\ln 3} - 1 = n$$
$$12.8 \approx n$$

Section 7.5

1. a. $t = 0$ hours, $N(0) = 2200(2)^0 = 2200$ bacteria

3. a.
$$N(t) = N_0 e^{kt} \text{ where } N_0 = 24600$$
$$N(5) = 22,600e^{k(5)}$$
$$24,200 = 22,600e^{5k}$$
$$\frac{24,200}{22,600} = e^{5k}$$
$$\ln\left(\frac{24,200}{22,600}\right) = \ln\left(e^{5k}\right)$$
$$\ln\left(\frac{24,200}{22,600}\right) = 5k$$
$$\frac{1}{5}\left[\ln\frac{24,200}{22,600}\right] = k$$
$$0.01368 \approx k$$
$$N(t) = 22,600e^{0.01368t}$$

b. $t = 3$ hours, $N(3) = 2200(2)^3 = 17,600$ bacteria

b.
$$t = 15$$
$$N(15) = 22,600e^{0.01368(15)}$$
$$= 22,600e^{0.2052}$$
$$\approx 27,700$$

5.

$$N(t) = N_0 e^{kt} \text{ where } N_0 = 362,300$$

$$N(4) = 362,300 e^{k(4)}$$

$$379,700 = 362,300 e^{4k}$$

$$\frac{379,700}{362,300} = e^{4k}$$

$$\ln\left(\frac{379,700}{362,300}\right) = \ln\left(e^{4k}\right)$$

$$\ln\left(\frac{379,700}{362,300}\right) = 4k$$

$$\frac{1}{4}\left[\ln\frac{379,700}{362,300}\right] = k$$

$$0.011727 \approx k$$

$$N(t) = 362,300 e^{0.011727\, t}$$

$$t = 9$$

$$N(9) = 362,300 e^{0.011727(9)}$$

$$= 362,300 e^{0.105543}$$

$$\approx 402,600$$

7.   **a.**

Microgoms of Na — Time, in hours

b. $A(5) = 4e^{-0.23} \approx 3.18$ micrograms

c. Since $A = 4$ micrograms are present when $t = 0$, find the time t at which half remains—that is when $A = 2$.

$$2 = 4e^{-0.046t}$$

$$\frac{1}{2} = e^{-0.046t}$$

$$\ln\left(\frac{1}{2}\right) = -0.046t$$

$$\frac{\ln\left(\frac{1}{2}\right)}{-0.046} = t$$

$$15.07 \approx t$$

The half-life of sodium-24 is about 15.07 hours.

d.

$$1 = 4e^{-0.046t}$$

$$\frac{1}{4} = e^{-0.046t}$$

$$\ln\left(\frac{1}{4}\right) = -0.046t$$

$$\frac{\ln\left(\frac{1}{4}\right)}{-0.046} = t$$

$$30.14 \approx t$$

The amount of sodium-24 will be 1 microgram after 30.14 hours.

9.

$$N(t) = N_0 (0.5)^{t/5730}$$

$$N(t) = 0.45 N_0$$

$$0.45 N_0 = N_0 (0.5)^{t/5730}$$

$$\ln(0.45) = \frac{t}{5730}\ln 0.5$$

$$5730\frac{\ln 0.45}{\ln 0.5} = t$$

$$6601 \approx t$$

The bone is about 6601 years old.

11.

$$N(t) = N_0 (0.5)^{t/5730}$$

$$N(t) = 0.75 N_0$$

$$0.75 N_0 = N_0 (0.5)^{t/5730}$$

$$\ln 0.75 = \frac{t}{5730}\ln 0.5$$

$$5730\frac{\ln 0.75}{\ln 0.5} = t$$

$$2378 \approx t$$

The Rhind papyrus is about 2378 years old.

13.   **a.** $P = 8000,\ r = 0.05,\ t = 4,\ n = 1$

$$B = 8000\left(1 + \frac{0.05}{1}\right)^4 \approx \$9724.05$$

b. $t = 7,\ B = 8000\left(1 + \frac{0.05}{1}\right)^7 \approx \$11,256.80$

15.   **a.** $P = 38,000,\ r = 0.065,\ t = 4,\ n = 1$

$$B = 38,000\left(1 + \frac{0.065}{1}\right)^4 \approx \$48,885.72$$

b. $n = 365,\ B = 38,000\left(1 + \frac{0.065}{365}\right)^{4(365)} \approx \$49,282.20$

c. $n = 8760,\ B = 38,000\left(1 + \frac{0.065}{8760}\right)^{4(8760)} \approx \$49,283.30$

266

Chapter 7: Exponential and Logarithmic Functions

17. $P = 15{,}000, \quad r = 0.1, t = 5$

$B = 15{,}000e^{5(0.1)} \approx \$24{,}730.82$

19. $t = \dfrac{\ln 2}{r} \qquad r = 0.0784$

$t = \dfrac{\ln 2}{0.0784}$

$t \approx 8.8$ years

21. $B = Pe^{rt} \qquad$ Let $B = 3P$

$3P = Pe^{rt}$

$3 = e^{rt}$

$\ln 3 = rt \ln e$

$t = \dfrac{\ln 3}{r}$

23. $t = \dfrac{\ln 3}{r} \qquad r = 0.076$

$t = \dfrac{\ln 3}{0.076}$

$t \approx 14$ years

25.
a. 1900
b. 0.16
c. $P(0) = \dfrac{1900}{1 + 8.5e^{-0.16(0)}} = 200$

27.
a. 157,500
b. 0.04
c. $P(0) = \dfrac{157{,}500}{1 + 2.5e^{-0.04(0)}} = 45{,}000$

29.
a. 2400
b. 0.12
c. $P(0) = \dfrac{2400}{1 + 7e^{-0.12(0)}} = 300$

31. $a = \dfrac{c - P_0}{P_0} = \dfrac{5500 - 400}{400} = 12.75$

$P(t) = \dfrac{c}{1 + ae^{-bt}}$

$P(2) = \dfrac{5500}{1 + 12.75e^{-b(2)}}$

$780 = \dfrac{5500}{1 + 12.75e^{-2b}}$

$780(1 + 12.75e^{-2b}) = 5500$

$780 + 9945e^{-2b} = 5500$

$9945e^{-2b} = 4720$

$e^{-2b} = \dfrac{4720}{9945}$

$\ln e^{-2b} = \ln \dfrac{4720}{9945}$

$-2b = \ln \dfrac{4720}{9945}$

$b = -\dfrac{1}{2}\ln \dfrac{4720}{9945}$

$b \approx 0.37263$

$P(t) = \dfrac{5500}{1 + 12.75e^{-0.37263t}}$

33. $a = \dfrac{c - P_0}{P_0} = \dfrac{100 - 18}{18} = 4.55556$

$P(t) = \dfrac{c}{1 + ae^{-bt}}$

$P(3) = \dfrac{100}{1 + 4.55556e^{-b(3)}}$

$30 = \dfrac{100}{1 + 4.55556e^{-3b}}$

$30(1 + 4.55556e^{-3b}) = 100$

$30 + 136.67e^{-3b} = 100$

$136.67e^{-3b} = 70$

$e^{-3b} = \dfrac{70}{136.67}$

$\ln e^{-3b} = \ln \dfrac{70}{136.67}$

$-3b = \ln \dfrac{70}{136.67}$

$b = -\dfrac{1}{3}\ln \dfrac{70}{136.67}$

$b \approx 0.22302$

$P(t) = \dfrac{100}{1 + 4.55556e^{-0.22302t}}$

Copyright © Houghton Mifflin Company. All rights reserved.

35. a. $R(t) = \dfrac{625,000}{1 + 3.1e^{-0.045t}}$

$R(1) = \dfrac{625,000}{1 + 3.1e^{-0.045(1)}} \approx \$158,000$

$R(2) = \dfrac{625,000}{1 + 3.1e^{-0.045(2)}} \approx \$163,000$

37. a. $a = \dfrac{c - P_0}{P_0} = \dfrac{1600 - 312}{312} = 4.12821$

$P(t) = \dfrac{c}{1 + ae^{-bt}}$

$P(6) = \dfrac{1600}{1 + 4.12821e^{-b(6)}}$

$416 = \dfrac{1600}{1 + 4.12821e^{-6b}}$

$416(1 + 4.12821e^{-6b}) = 1600$

$416 + 1717.34e^{-6b} = 1600$

$1717.34e^{-6b} = 1184$

$e^{-6b} = \dfrac{1184}{1717.34}$

$\ln e^{-6b} = \ln \dfrac{1184}{1717.34}$

$-6b = \ln \dfrac{1184}{1717.34}$

$b = -\dfrac{1}{6} \ln \dfrac{1184}{1717.34}$

$b \approx 0.06198$

$P(t) = \dfrac{1600}{1 + 4.12821e^{-0.06198t}}$

b. $R(t) = \dfrac{625,000}{1 + 3.1e^{-0.045t}}$, as $t \to \infty$, $R(t) \to \$625,000$

b. $P(10) = \dfrac{1600}{1 + 4.12821e^{-0.06198(10)}} \approx 497$ wolves

39. a. $a = \dfrac{c - P_0}{P_0} = \dfrac{8500 - 1500}{1500} = 4.66667$

$P(t) = \dfrac{c}{1 + ae^{-bt}}$

$P(2) = \dfrac{8500}{1 + 4.66667e^{-b(2)}}$

$1900 = \dfrac{8500}{1 + 4.66667e^{-2b}}$

$1900(1 + 4.66667e^{-2b}) = 8500$

$1900 + 8866.673e^{-2b} = 8500$

$8866.673e^{-2b} = 6600$

$e^{-2b} = \dfrac{6600}{8866.673}$

$\ln e^{-2b} = \ln \dfrac{6600}{8866.673}$

$-2b = \ln \dfrac{6600}{8866.673}$

$b = -\dfrac{1}{2} \ln \dfrac{6600}{8866.673}$

$b \approx 0.14761$

$P(t) = \dfrac{8500}{1 + 4.66667e^{-0.14761t}}$

b. $4000 = \dfrac{8500}{1 + 4.66667e^{-0.14761t}}$

$4000(1 + 4.66667e^{-0.14761t}) = 8500$

$1 + 4.66667e^{-0.14761t} = 2.125$

$4.66667e^{-0.14761t} = 1.125$

$e^{-0.14761t} = \dfrac{1.125}{4.66667}$

$\ln e^{-0.14761t} = \ln \dfrac{1.125}{4.66667}$

$-0.14761t = \ln \dfrac{1.125}{4.66667}$

$t = -\dfrac{1}{0.14761} \ln \dfrac{1.125}{4.66667}$

$t \approx 9.6$

The population will exceed 4000 in $2007 + 9 = 2016$.

41. **a.** $A = 34°\text{F}, T_0 = 75°\text{F}, T_t = 65°\text{F}, t = 5.$ Find k.

$$65 = 34 + (75 - 34)\,e^{-5k}$$

$$31 = 41e^{-5k}$$

$$\frac{31}{41} = e^{-5k}$$

$$\ln\left(\frac{31}{41}\right) = -5k$$

$$k = -\frac{1}{5}\ln\left(\frac{31}{41}\right)$$

$$k \approx 0.056$$

  b. $A = 34°\text{F}, k = 0.056,\ T_0 = 75°\text{F}, t = 30$

$$T_t = 34 + (75 - 34)\,e^{-30(0.056)}$$

$$T_t = 34 + (41)e^{-1.68}$$

$$T_t \approx 42°\text{F}$$

  c. $T_t = 36°\text{F}, k = 0.056,\ T_t = 75°\text{F}, A = 34°\text{F}$

$$36 = 34 + (75 - 34)e^{-0.056t}$$

$$2 = 41e^{-0.056t}$$

$$t \approx 54 \text{ minutes}$$

45. $V(t) = V_0(1 - r)^t$

$$0.5V_0 = V_0(1 - 0.20)^t$$

$$0.5 = (1 - 0.20)^t$$

$$0.5 = 0.8^t$$

$$\ln 0.5 = \ln 0.8^t$$

$$\ln 0.5 = t\ln 0.8$$

$$\frac{\ln 0.5}{\ln 0.8} = t$$

$$3.1 \text{ years} \approx t$$

43. **a.**   10% of 80,000 is 8000.

$$8000 = 80,000\left(1 - e^{-0.0005t}\right)$$

$$0.1 = 1 - e^{-0.0005t}$$

$$-0.9 = -e^{-0.0005t}$$

$$0.9 = e^{-0.0005t}$$

$$\ln 0.9 = -0.0005t \ln e$$

$$\ln 0.9 = -0.0005t$$

$$\frac{\ln 0.9}{-0.0005} = t$$

$$211\,\text{h} \approx t$$

  b.   50% of 80,000 is 40,000.

$$40,000 = 80,000\left(1 - e^{-0.0005t}\right)$$

$$0.5 = 1 - e^{-0.0005t}$$

$$-0.5 = -e^{-0.0005t}$$

$$0.5 = e^{-0.0005t}$$

$$\ln(0.5) = \ln\left(e^{-0.0005t}\right)$$

$$\ln(0.5) = -0.0005t$$

$$\frac{\ln(0.5)}{-0.0005} = t$$

$$1386\,\text{h} \approx t$$

47. **a.**

  b. $20 = 64(1 - e^{-t/2})$

$$0.625 = 1 - e^{-t/2}$$

$$e^{-t/2} = 0.375$$

$$-t/2 = \ln 0.375$$

$$t \approx 0.98 \text{ seconds}$$

  c.   The horizontal asymptote is $v = 32$.

  d.   As time increases, the object's velocity approaches but never exceeds 32 ft/sec.

49. **a.**

  b.   The graphs of $s = 32t + 32(e^{-t} - 1)$ and $s = 50$ intersect when $t \approx 2.5$ seconds.

  c.   The slope m of the secant line containing $(1, s(1))$ and $(2, s(2))$ is $m = \dfrac{s(2) - s(1)}{2 - 1} \approx 24.56$ ft/sec

  d.   The average speed of the object was 24.56 feet per second between $t = 1$ and $t = 2$.

51.

Xmin $= 0$, Xmax $= 80$, Xscl $= 10$,

Ymin $= -10$, Ymax $= 110$, Yscl $= 15$

When $P = 75\%$, $t \approx 45$ hours.

53. a. $A(1) = 0.5^{1/2}$

≈ 0.71 gram

b. $A(4) = 0.5^{4/2} + 0.5^{(4-3)/2}$

$= 0.5^2 + 0.5^{1/2}$

≈ 0.96 gram

c. $A(9) = 0.5^{9/2} + 0.5^{(9-3)/2} + 0.5^{(9-6)/2}$

$= 0.5^{4.5} + 0.5^3 + 0.5^{1.5}$

≈ 0.52 gram

55. $N(t) = 22,755e^{0.0287\,t}$

$= 22,755\left(e^{0.0287}\right)^t$

$\approx 22,755(1.0291)^t$

The annual growth rate is 2.91%.

PS1. Decreasing

PS3. $P(0) = \dfrac{108}{1+2e^{-0.1(0)}} = \dfrac{108}{1+2} = 36$

PS5.

$$10 = \frac{20}{1+2.2e^{-0.05t}}$$

$$10(1+2.2e^{-0.05t}) = 20$$

$$10+22e^{-0.05t} = 20$$

$$e^{-0.05t} = \frac{10}{22}$$

$$\ln e^{-0.05t} = \ln\frac{10}{22}$$

$$-0.05t = \ln\frac{10}{22}$$

$$t = -20\ln\frac{10}{22}$$

$$t \approx 15.8$$

PS2. Decreasing

PS4. $N(0) = 840e^{1.05(0)} = 840$

PS6. $P(t) = \dfrac{55}{1+3e^{-0.08t}}$

There is a horizontal asymptote at $P = 55$.

Section 7.6

1.

increasing exponential function

3.

decreasing exponential function;
decreasing logarithmic function

5.

decreasing logarithmic function

7.

$y \approx 0.99628(1.20052)^x$; $r \approx 0.85705$

9.

$y \approx 1.81505(0.51979)^x$; $r \approx -0.99978$

11.

$y \approx 4.89060 - 1.35073\ln x$; $r \approx -0.99921$

13.

$y \approx 14.05858 + 1.76393\ln x$; $r \approx 0.99983$

15.

$$y \approx \frac{235.58598}{1 + 1.90188e^{-0.05101x}}$$

17.

$$y \approx \frac{2098.68307}{1 + 1.19794e^{-0.06004x}}$$

19.

a. Linear: $p \approx 0.22129t + 3.99190$; $r \approx 0.99288$. Exponential: $p \approx 4.05326(1.04460)^t$; $r \approx 0.99412$

b. The exponential model's r is closer to 1.

c. $p = 4.05326(1.04460)^{16} \approx \$8.15$

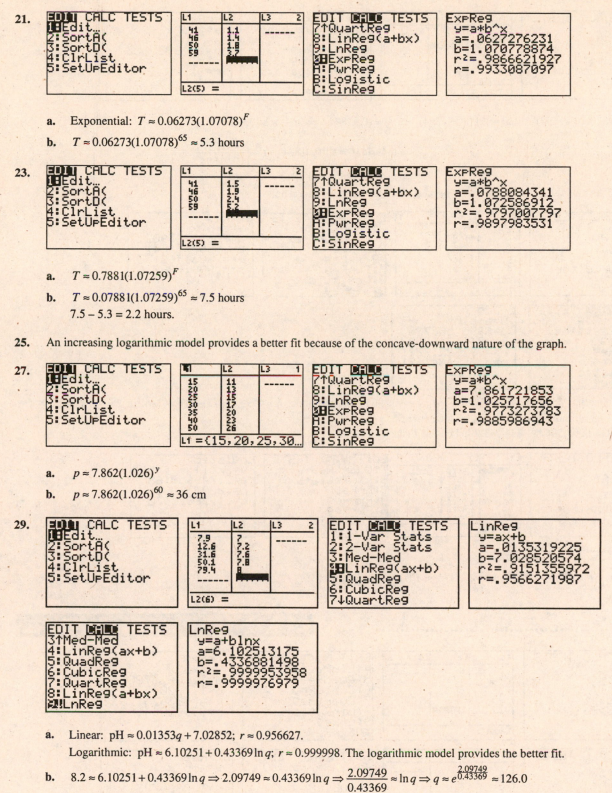

21.

a. Exponential: $T \approx 0.06273(1.07078)^F$

b. $T \approx 0.06273(1.07078)^{65} \approx 5.3$ hours

23.

a. $T \approx 0.7881(1.07259)^F$

b. $T \approx 0.07881(1.07259)^{65} \approx 7.5$ hours

$7.5 - 5.3 = 2.2$ hours.

25. An increasing logarithmic model provides a better fit because of the concave-downward nature of the graph.

27.

a. $p \approx 7.862(1.026)^y$

b. $p \approx 7.862(1.026)^{60} \approx 36$ cm

29.

a. Linear: $pH \approx 0.01353q + 7.02852$; $r \approx 0.956627$.

Logarithmic: $pH \approx 6.10251 + 0.43369 \ln q$; $r \approx 0.999998$. The logarithmic model provides the better fit.

b. $8.2 \approx 6.10251 + 0.43369 \ln q \Rightarrow 2.09749 \approx 0.43369 \ln q \Rightarrow \dfrac{2.09749}{0.43369} \approx \ln q \Rightarrow q \approx e^{\frac{2.09749}{0.43369}} \approx 126.0$

31.

a. $p \approx 3200(0.91894)^t$; $200 \approx 3200(0.91894)^t \Rightarrow \frac{1}{16} \approx (0.91894)^t \Rightarrow \ln\frac{1}{16} \approx \ln(0.91894)^t$

$$\Rightarrow \ln 1 - \ln 16 \approx t \ln 0.91894 \Rightarrow t \approx \frac{-\ln 16}{\ln 0.91894}$$

$$\Rightarrow t \approx 32.8 \text{ years after } 1980 \Rightarrow \text{ in } 2012$$

b. No. The model fits the data perfectly because there are only two data points.

33.

a. Exponential: $S \approx 7062.46390(0.956776)^t$, $r \approx -0.89618$; Logarithmic: $S \approx 6995.50673 - 841.74326\ln t$, $r \approx -0.96240$

b. The logarithmic regression provides the better fit.

c. $S(11) \approx 6995.50673 - 841.74326\ln 11 \approx 4977$ sites

35.

a. $P(t) \approx \dfrac{11.26828}{1 + 2.74965e^{-0.02924t}}$

b. As $t \to \infty$, $P(t) \to 11$ billion people $t \to \infty$, $P(t) \to 11$ billion people

Connecting Concepts

37.

A and B have different exponential regression functions.

39.

a. Exponential: $y \approx 1.81120(1.61740)^x$; $r \approx 0.96793$. Power: $y \approx 2.09385(x^{1.40246})$; $r \approx 0.99999$.

b. The power regression provides the better fit.

....

Exploring Concepts with Technology

Using a Semilog Graph to Model Exponential Decay

1. **a.**

Xmin $= -1$, Xmax $= 31$, Xscl $= 4$,
Ymin $= -1$, Ymax $= 7$, Yscl $= 1$

b. $m = \dfrac{4.3 - 3.3}{4 - 15} \approx -0.0909$

c. $\ln A - 4.3 = -0.0909(t - 4)$
$\ln A = -0.0909t + 4.664$

d. $e^{\ln A} = e^{-0.0909\,t + 4.664}$
$A = e^{-0.0909\,t} e^{4.664}$
$A = e^{-0.0909\,t} e^{4.664}$

e.

Xmin $= -5$, Xmax $= 35$, Xscl $= 5$,
Ymin $= -10$, Ymax $= 110$, Yscl $= 15$

f. $A(t) = 106e^{-0.0909t}$

At $t = 0$ there is $A = 106$ mg present

We must find t where $A = \dfrac{1}{2}(106) = 53$ mg.

$53 = 106e^{-0.0909\,t}$

$\dfrac{53}{106} = e^{-0.0909\,t}$

$\dfrac{1}{2} = e^{-0.0909\,t}$

$\ln 0.5 = -0.0909t$

$t = \dfrac{\ln 0.5}{-0.0909} \approx 7.6$ days

....

Assessing Concepts

1. False; $f(x) = x^2$ does not have an inverse function.

2. True

3. True

4. False; $h(x)$ is not an increasing function for $0 < b < 1$.

5. False; $j(x)$ is not an increasing function for $0 < b < 1$.   **6.** c

7. b   **8.** f   **9.** a   **10.** e

11. d   **12.** g

●●●   **Chapter Review**

1. $\log_5 25 = x$ [7.2]
$$5^x = 25$$
$$5^x = 5^2$$
$$x = 2$$

2. $\log_3 81 = x$ [7.2]
$$3^x = 81$$
$$3^x = 3^4$$
$$x = 4$$

3. $\ln e^3 = x$ [7.2]
$$e^x = e^3$$
$$x = 3$$

4. $\ln e^\pi = x$ [7.2]
$$e^x = e^\pi$$
$$x = \pi$$

5. $3^{2x+7} = 27$ [7.4]
$$3^{2x+7} = 3^3$$
$$2x + 7 = 3$$
$$2x = -4$$
$$x = -2$$

6. $5^{x-4} = 625$ [7.4]
$$5^{x-4} = 5^4$$
$$x - 4 = 4$$
$$x = 8$$

7. $2^x = \dfrac{1}{8}$ [7.4]
$$2^x = 2^{-3}$$
$$x = -3$$

8. $27(3^x) = 3^{-1}$ [7.4]
$$27(3^x) = \dfrac{1}{3}$$
$$3^x = \dfrac{1}{81}$$
$$3^x = 3^{-4}$$
$$x = -4$$

9. $\log x^2 = 6$ [7.4]
$$10^6 = x^2$$
$$1{,}000{,}000 = x^2$$
$$\pm\sqrt{1{,}000{,}000} = x$$
$$\pm 1000 = x$$

10. $\dfrac{1}{2}\log|x| = 5$ [7.4]
$$\log|x| = 10$$
$$10^{10} = |x|$$
$$x = \pm 10^{10}$$

11. $10^{\log 2x} = 14$ [7.4]
$$2x = 14$$
$$x = 7$$

12. $e^{\ln x^2} = 64$ [7.4]
$$x^2 = 64$$
$$x = \pm 8$$

13.

14.

15.

16.

17.

18.

19.

20.

21.

22.

23.

24.

25. $\log_4 64 = 3$ [7.2]

$4^3 = 64$

26. $\log_{1/2} 8 = -3$ [7.2]

$\left(\dfrac{1}{2}\right)^{-3} = 8$

27. $\log_{\sqrt{2}} 4 = 4$ [7.2]

$\left(\sqrt{2}\right)^4 = 4$

28. $\ln 1 = 0$ [7.2]

$e^0 = 1$

29. $5^3 = 125$ [7.2]

$\log_5 125 = 3$

30. $2^{10} = 1024$ [7.2]

$\log_2 1024 = 10$

31. $10^0 = 1$ [7.2]

$\log_{10} 1 = 0$

32. $8^{1/2} = 2\sqrt{2}$ [7.2]

$\log_8 2\sqrt{2} = \dfrac{1}{2}$

33. $\log_b \dfrac{x^2 y^3}{z} = 2\log_b x + 3\log_b y - \log_b z$ [7.3]

34. $\log_b \dfrac{\sqrt{x}}{y^2 z} = \dfrac{1}{2}\log_b x - \left(2\log_b y + \log_b z\right)$ [7.3]

$= \dfrac{1}{2}\log_b x - 2\log_b y - \log_b z$

35. $\ln xy^3 = \ln x + 3\ln y$ [7.3]

36. $\ln \dfrac{\sqrt{xy}}{z^4} = \dfrac{1}{2}\left(\ln x + \ln y\right) - 4\ln z$ [7.3]

$= \dfrac{1}{2}\ln x + \dfrac{1}{2}\ln y - 4\ln z$

37. $2\log x + \dfrac{1}{3}\log(x+1) = \log\left(x^2 \sqrt[3]{x+1}\right)$ [7.3]

38. $5\log x - 2\log(x+5) = \log\dfrac{x^5}{(x+5)^2}$ [7.3]

39. $\dfrac{1}{2}\ln 2xy - 3\ln z = \ln\dfrac{\sqrt{2xy}}{z^3}$ [7.3]

40. $\ln x - (\ln y - \ln z) = \ln\dfrac{x}{y/z} = \ln\dfrac{xz}{y}$ [7.3]

41. $\log_5 101 = \dfrac{\log 101}{\log 5} \approx 2.86754$ [7.3]

42. $\log_3 40 = \dfrac{\log 40}{\log 3} \approx 3.35776$ [7.3]

43. $\log_4 0.85 = \dfrac{\log 0.85}{\log 4} \approx -0.117233$ [7.3]

44. $\log_8 0.3 = \dfrac{\log 0.3}{\log 8} \approx -0.578989$ [7.3]

45. $4^x = 30$ [7.4]

$\log 4^x = \log 30$

$x\log 4 = \log 30$

$x = \dfrac{\log 30}{\log 4}$

46. $5^{x+1} = 41$ [7.4]

$(x+1)\log 5 = \log 41$

$x+1 = \dfrac{\log 41}{\log 5}$

$x = \dfrac{\log 41}{\log 5} - 1$

47. $\ln(3x) - \ln(x-1) = \ln 4$ [7.4]

$\ln\dfrac{3x}{x-1} = \ln 4$

$\dfrac{3x}{x-1} = 4$

$3x = 4(x-1)$

$3x = 4x - 4$

$4 = x$

48. $\ln(3x) + \ln 2 = \ln 1$ [7.4]

$\ln(3x \cdot 2) = 1$

$\ln(6x) = 1$

$e^1 = 6x$

$\dfrac{e}{6} = x$

49. $e^{\ln(x+2)} = 6$ [7.4]

$(x+2) = 6$

$x + 2 = 6$

$x = 4$

50. $10^{\log(2x+1)} = 31$ [7.4]

$2x + 1 = 31$

$2x = 30$

$x = 15$

51.
$$\frac{4^x + 4^{-x}}{4^x - 4^{-x}} = 2$$
$$4^x(4^x + 4^{-x}) = 2(4^x - 4^{-x})4^x$$
$$4^{2x} + 1 = 2(4^{2x} - 1)$$
$$4^{2x} + 1 = 2(4^{2x} - 1)$$
$$4^{2x} - 2 \cdot 4^{2x} + 3 = 0$$
$$4^{2x} = 3$$
$$2^x \ln 4 = \ln 3$$
$$x = \frac{\ln 3}{2 \ln 4} \qquad [7.4]$$

52.
$$\frac{5^x + 5^{-x}}{2} = 8$$
$$5^x(5^x + 5^{-x}) = 16(5^x)$$
$$5^{2x} + 1 = 16(5^x)$$
$$5^{2x} - 16(5^x) + 1 = 0$$
Let $5^x = u$
$$u^2 - 16u + 1 = 0$$
$$u = \frac{16 \pm \sqrt{16^2 - 4(1)(1)}}{2}$$
$$u = \frac{16 \pm \sqrt{252}}{2}$$
$$u = \frac{16 \pm 6\sqrt{7}}{2}$$
$$u = 8 \pm 3\sqrt{7}$$
$$5^x = 8 \pm 3\sqrt{7}$$
$$x = \frac{\ln(8 \pm 3\sqrt{7})}{\ln 5} \quad [7.4]$$

53. $\log(\log x) = 3$ [7.4]
$$10^3 = \log x$$
$$10^{(10^3)} = x$$
$$10^{1000} = x$$

54. $\ln(\ln x) = 2$ [7.4]
$$e^2 = \ln x$$
$$e^{(e^2)} = x$$

55. $\log\sqrt{x-5} = 3$ [7.4]
$$10^3 = \sqrt{x-5}$$
$$10^6 = x - 5$$
$$10^6 + 5 = x$$
$$x = 1,000,005$$

56. $\log x + \log(x - 15) = 1$
$$\log x(x - 15) = 1$$
$$10 = x^2 - 15x$$
$$0 = x^2 - 15x - 10$$
$$x = \frac{15 \pm \sqrt{15^2 - 4(1)(-10)}}{2}$$
$$x = \frac{15 \pm \sqrt{265}}{2}$$
$$x = \frac{15 + \sqrt{265}}{2} \qquad [7.4]$$

57. $\log_4(\log_3 x) = 1$ [7.4]
$$4 = \log_3 x$$
$$3^4 = x$$
$$81 = x \qquad [7.4]$$

58. $\log_7(\log_5 x^2) = 0$
$$7^0 = \log_5 x^2$$
$$1 = \log_5 x^2$$
$$5 = x^2$$
$$\pm\sqrt{5} = x \qquad [7.4]$$

59. $\log_5 x^3 = \log_5 16x$ [7.4]
$$x^3 = 16x$$
$$x^2 = 16$$
$$x = 4$$

60. $25 = 16^{\log_4 x}$ [7.4]
$$25 = 4^{2\log_4 x}$$
$$25 = 4^{\log_4 x^2}$$
$$25 = x^2$$
$$\pm 5 = x$$
$$5 = x$$

61. $m = \log\left(\dfrac{I}{I_0}\right)$ [7.3]
$$= \log\left(\frac{51,782,000 I_0}{I_0}\right)$$
$$= \log 51,782,000$$
$$\approx 7.7$$

62. $M = \log A + 3\log 8t - 2.92$ [7.3]

$\quad\quad = \log 18 + 3\log 8(21) - 2.92$

$\quad\quad = \log 18 + 3\log 168 - 2.92$

$\quad\quad \approx 5.0$

63. $\log\left(\dfrac{I_1}{I_0}\right) = 7.2$ and $\log\left(\dfrac{I_2}{I_0}\right) = 3.7$ [7.3]

$\quad\quad \dfrac{I_1}{I_0} = 10^{7.2} \quad\quad\quad\quad\quad \dfrac{I_2}{I_0} = 10^{3.7}$

$\quad\quad I_1 = 10^{7.2} I_0 \quad\quad\quad\quad\quad I_2 = 10^{3.7} I_0$

$\quad\quad \dfrac{I_1}{I_2} = \dfrac{10^{7.2} I_0}{10^{3.7} I_0} = \dfrac{10^{3.5}}{1} \approx \dfrac{3162}{1}$

3162 to 1

64. $\dfrac{I_1}{I_2} = 600 = 10^x$ [7.3]

$\quad \log 600 = \log 10^x$

$\quad \log 600 = x$

$\quad\quad 2.8 \approx x$

65. $pH = -\log\left[H_3O^+\right]$ [7.3]

$\quad\quad = -\log\left[6.28 \times 10^{-5}\right]$

$\quad\quad \approx 4.2$

66. $5.4 = -\log\left[H_3O^+\right]$ [7.3]

$\quad -5.4 = \log\left[H_3O^+\right]$

$\quad 10^{-5.4} = H_3O^+$

$\quad 0.00000398 \approx H_3O^+$

$\quad H_3O^+ \approx 3.98 \times 10^{-6}$

67. $P = 16,000, r = 0.08, t = 3$ [7.5]

a. $\quad B = 16,000\left(1 + \dfrac{0.08}{12}\right)^{36} \approx \$20,323.79$

b. $\quad B = 16,000e^{0.08(3)}$

$\quad\quad B = 16,000e^{0.24} \approx \$20,339.99$

68. $P = 19,000, r = 0.06, t = 5$ [7.5]

a. $\quad B = 19,000\left(1 + \dfrac{0.06}{365}\right)^{1825} \approx \$25,646.69$

b. $\quad B = 19,000e^{0.3} \approx \$25,647.32$

69. $S(n) = P(1 - r)^n, P = 12,400, r = 0.29, t = 3$ [7.5]

$\quad S(n) = 12,400(1 - 0.29)^3 \approx \$4438.10$

70. **a.** $\quad\quad N(t) = N_0 e^{-0.12t}$ [7.5]

$\quad\quad N(10) = N_0 e^{-0.12(10)}$

$\quad\quad \dfrac{N(10)}{N_0} = e^{-1.2}$

$\quad\quad\quad\quad\quad = .301$

$\quad\quad \dfrac{N(10)}{N_0} = 30.1\%$ healed

$\quad 100\% - 30.1\% = 69.9\%$ healed

b. $\quad \dfrac{N(t)}{N_0} = 0.5$

$\quad\quad 0.5 = e^{-0.12t}$

$\quad\quad \ln 0.5 = -0.12t$

$\quad\quad \dfrac{\ln 0.5}{-0.12} = t$

$\quad\quad\quad t \approx 6$ days

c. $\quad \dfrac{N(t)}{N_0} = 0.1$

$\quad\quad 0.1 = e^{-0.12t}$

$\quad\quad \ln 0.1 = -0.12t$

$\quad\quad \dfrac{\ln 0.1}{-0.12} = t$

$\quad\quad\quad t \approx 19$ days

71.

$N(0) = 1$ $N(2) = 5$

$1 = N_0 e^{k(0)}$ $5 = e^{2k}$

$1 = N_0$ $\ln 5 = 2k$

$$k = \frac{\ln 5}{2} \approx 0.8047$$

Thus $N(t) = e^{0.8047t}$ [7.5]

72. $N(0) = N_0 = 2$ and $N(3) = N_0 e^{3k} = 2e^{3k} = 11$

$$e^{3k} = \frac{11}{2}$$

$$e^{3k} = \frac{11}{2}$$

$$3k = \ln\left(\frac{11}{2}\right)$$

$$k = \frac{1}{3}\ln\left(\frac{11}{2}\right)$$

$$\approx 0.5682$$

Thus $N(t) = 2e^{0.5682t}$ [7.5]

73. $4 = N(1) = N_0 e^k$ and thus $\dfrac{4}{N_0} = e^k$. Now, we also

have $N(5) = 5 = N_0 e^{5k} = N_0 \left(\dfrac{4}{N_0}\right)^5 = \dfrac{1024}{N_0^4}$.

$$N_0 = \sqrt[4]{\frac{1024}{5}} \approx 3.783$$

Thus $4 = 3.783 e^k$.

$$\ln\left(\frac{4}{3.783}\right) = k$$

$$k \approx 0.0558$$

Thus $N_0 = 3.783 e^{0.0558t}$. [7.5]

74. $1 = N(0) = N_0$ and $2 = N(-1) = N_0 e^{-k}$.

Since $N_0 = 1$, we have $2 = 1 \cdot e^{-k}$.

$\ln 2 = -k$

$k \approx -0.6931$

Thus $N(t) = e^{-0.6931\,t}$. [7.5]

75. **a.** $N(1) = 25{,}200 e^{k(1)} = 26{,}800$ [7.5]

$$e^k = \frac{26{,}800}{25{,}200}$$

$$\ln e^k = \ln\left(\frac{26{,}800}{25{,}200}\right)$$

$$k \approx 0.061557893$$

$$N(t) = 25{,}200 e^{0.061557893\,t}$$

b. $N(7) = 25{,}200 e^{0.061557893(7)}$

$$= 25{,}200 e^{0.430905251}$$

$$\approx 38{,}800$$

76. $P(t) = 0.5^{\,t/5730} = 0.96$ [7.5]

$$\log\left(0.5^{\,t/5730}\right) = \log 0.96$$

$$\frac{t}{5730}\log 0.5 = \log 0.96$$

$$\frac{t}{5730} = \frac{\log 0.96}{\log 0.5}$$

$$t = 5730\left(\frac{\log 0.96}{\log 0.5}\right) \approx 340 \text{ years}$$

77. Answers will vary.

78. **a.**

exponential: $R \approx 179.949\left(0.968094^t\right)$, $r \approx -0.99277$

logarithmic: $R \approx 171.19665 - 35.71341\ln t$, $r \approx -0.98574$

b.   The exponential equation provides a better fit to the data.

c. $R \approx 179.949\left(0.968094^{108}\right) \approx 5.4$ per 1000 live births   [7.6]

79. a.

$$P(t) = \frac{mP_0}{P_0 + (m - P_0)e^{-kt}}$$

$$P(3) = 360 = \frac{1400(210)}{210 + (1400 - 210)e^{-k(3)}}$$

$$360 = \frac{294000}{210 + 1190e^{-3k}}$$

$$360\left(210 + 1190e^{-3k}\right) = 294000$$

$$210 + 1190e^{-3k} = \frac{294000}{360}$$

$$1190e^{-3k} = \frac{29400}{36} - 210$$

$$e^{-3k} = \frac{29400/36 - 210}{1190}$$

$$\ln e^{-3k} = \ln\left(\frac{29400/36 - 210}{1190}\right)$$

$$-3k = \ln\left(\frac{29400/36 - 210}{1190}\right)$$

$$k = -\frac{1}{3}\ln\left(\frac{29400/36 - 210}{1190}\right)$$

$$k \approx 0.2245763649$$

$$P(t) = \frac{294000}{210 + 1190e^{-0.22458t}} = \frac{1400}{1 + \frac{17}{3}e^{-0.22458t}}$$

b.

$$P(13) = \frac{294000}{210 + 1190e^{-0.22458(13)}}$$

$$= \frac{294000}{210 + 1190e^{-2.919492744}}$$

$$\approx 1070 \text{ coyotes} \qquad [7.5]$$

80. a.

$$P(0) = \frac{128}{1 + 5e^{-0.27(0)}} = \frac{128}{1 + 5e^0} = \frac{128}{1 + 5} = \frac{128}{6} = 21\frac{1}{3}$$

b. As $t \to \infty$, $e^{-0.27t} \to 0$.

$$P(t) \to \frac{128}{1 + 5(0)} = \frac{128}{1} = 128 \quad [7.5]$$

$$\mathbf{Q}\text{uantitative } \mathbf{R}\text{easoning}$$

QR1.

Exponential: $S \approx 5.94860(1.72192)^t$ Logistic: $S \approx \dfrac{244.56468}{1 + 39.38651e^{-0.57829t}}$

QR3.

$S \approx 58.73554 + 16.75801\ln t$

QR5.

```
EDIT CALC TESTS
1:Edit…
2:SortA(
3:SortD(
4:ClrList
5:SetUpEditor
```

```
L1    L2    L3    2
1     59.3  -------
2     69.2
3     77.2
4     82.5
      -------
L2(5) =
```

```
EDIT CALC TESTS
7↑QuartReg
8:LinReg(a+bx)
9:LnReg
0:ExpReg
A:PwrReg
B:Logistic
C:SinReg
```

```
Logistic
y=c/(1+ae^(-bx))
a=.9340870164
b=.521706492
c=92.11991877
```

$$S \approx \frac{85.24460}{1+14.0040e^{-0.70591t}}$$

•••

Chapter Test

1.　a.　$\log_b(5x-3)=c$　[7.2]

　　　　$b^c=5x-3$

b.　$3^{x/2}=y$

　　　$\log_3 y = \dfrac{x}{2}$

2.　$\log_b \dfrac{z^2}{y^3\sqrt{x}} = \log_b z^2 - \log_b y^3 - \log_b x^{1/2}$　[7.3]

　　　　　　　　　$= 2\log_b z - 3\log_b y - \dfrac{1}{2}\log_b x$

3.　$\log_{10}(2x+3) - 3\log_{10}(x-2) = \log_{10}(2x+3) - \log_{10}(x-2)^3$

　　　　　　　　　　　　　　　$= \log_{10}\dfrac{2x+3}{(x-2)^3}$　[7.3]

4.　$\log_4 12 = \dfrac{\log 12}{\log 4}$　[7.3]

　　　　≈ 1.7925

5.

6.

7.　$5^x = 22$　[7.4]

　　　$x\log 5 = \log 22$

　　　$x = \dfrac{\log 22}{\log 5}$

　　　$x \approx 1.9206$

8.　$4^{5-x} = 7^x$　[7.4]

　　　$\ln 4^{5-x} = \ln 7^x$

　　$(5-x)\ln 4 = x\ln 7$

　　$5\ln 4 - x\ln 4 = x\ln 7$

　　　$5\ln 4 = x\ln 7 + x\ln 4$

　　　$5\ln 4 = x(\ln 7 + \ln 4)$

　　　$\dfrac{5\ln 4}{\ln 28} = x$

9.　$\log(x+99) - \log(3x-2) = 2$　[7.4]

　　$\log\dfrac{x+99}{3x-2} = 2$

　　$\dfrac{x+99}{3x-2} = 10^2$

　　$x+99 = 100(3x-2)$

　　$x+99 = 300x - 200$

　　$-299x = -299$

　　　　$x = 1$

10.　$\ln(2-x) + \ln(5-x) = \ln(37-x)$

　　　$\ln(2-x)(5-x) = \ln(37-x)$

　　　$(2-x)(5-x) = (37-x)$

　　　$10 - 7x + x^2 = 37 - x$

　　　$x^2 - 6x - 27 = 0$

　　　$(x-9)(x+3) = 0$

　　$x = 9$ (not in domain)　or　$x = -3$

　　　$x = -3$　[7.4]

11. a. $A = P\left(1 + \dfrac{r}{n}\right)^{nt}$ [7.5]

$\quad = 20{,}000\left(1 + \dfrac{0.078}{12}\right)^{12(5)}$

$\quad = 20{,}000(1.0065)^{60}$

$\quad = \$29{,}502.36$

b. $A = Pe^{rt}$

$\quad = 20{,}000e^{0.078(5)}$

$\quad = 20{,}000e^{0.39}$

$\quad = \$29{,}539.62$

12. $A = P\left(1 + \dfrac{r}{n}\right)^{nt}$ [7.5]

$2P = P\left(1 + \dfrac{0.04}{12}\right)^{12t}$

$2 = \left(1 + \dfrac{0.04}{12}\right)^{12t}$

$\ln 2 = \ln\left(1 + \dfrac{0.04}{12}\right)^{12t}$

$\ln 2 = 12t\ln\left(1 + \dfrac{0.04}{12}\right)$

$12t = \dfrac{\ln 2}{\ln\left(1 + \dfrac{0.04}{12}\right)}$

$t = \dfrac{1}{12}\cdot\dfrac{\ln 2}{\ln\left(1 + \dfrac{0.04}{12}\right)}$

$t \approx 17.36 \text{ years}$

13. a. $M = \log\left(\dfrac{I}{I_0}\right)$ [7.3]

$\quad = \log\left(\dfrac{42{,}304{,}000 I_0}{I_0}\right)$

$\quad = \log 42{,}304{,}000$

$\quad \approx 7.6$

b. $\log\left(\dfrac{I_1}{I_0}\right) = 6.3 \qquad \text{and} \qquad \log\left(\dfrac{I_2}{I_0}\right) = 4.5$

$\dfrac{I_1}{I_0} = 10^{6.3} \qquad\qquad\qquad \dfrac{I_2}{I_0} = 10^{4.5}$

$I_1 = 10^{6.3} I_0 \qquad\qquad\qquad I_2 = 10^{4.5} I_0$

$\dfrac{I_1}{I_2} = \dfrac{10^{6.3} I_0}{10^{4.5} I_0} = \dfrac{10^{1.8}}{1} \approx \dfrac{63}{1}$

Therefore the ratio is 63 to 1.

14. a. $N(3) = 34600e^{k(3)} = 39800$

$34600e^{3k} = 39800$

$e^{3k} = \dfrac{39800}{34600}$

$\ln e^{3k} = \ln\left(\dfrac{398}{346}\right)$

$3k = \ln\left(\dfrac{398}{346}\right)$

$k = \dfrac{1}{3}\ln\left(\dfrac{398}{346}\right)$

$k \approx 0.0466710767$

$N(t) = 34600e^{0.0466710767\,t}$ [7.5]

b. $N(10) = 34600e^{0.0466710767(10)}$

$\quad = 34600e^{0.466710767}$

$\quad \approx 55{,}000$

15. $P(t) = 0.5^{\,t/5730} = 0.92$ [7.5]

$\log 0.5^{\,t/5730} = \log 0.92$

$\dfrac{t}{5730}\log 0.5 = \log 0.92$

$\dfrac{t}{5730} = \dfrac{\log 0.92}{\log 0.5}$

$t = 5730\left(\dfrac{\log 0.92}{\log 0.5}\right)$

$t \approx 690 \text{ years}$

16.

L1	L2	L3	1
2.5	16	------	
3.7	48		
5	155		
6.5	571		
6.9	896		
------	------		
L1(1)=2.5			

```
ExpReg
y=a*b^x
a=1.671991998
b=2.471878247
r²=.9996384751
r=.9998192212
```

a. $y = 1.671991998(2.471878247)^x$

b. $y = 1.671991998(2.471878247)^{7.8}$ [7.6]

$\quad \approx 1945$

17.

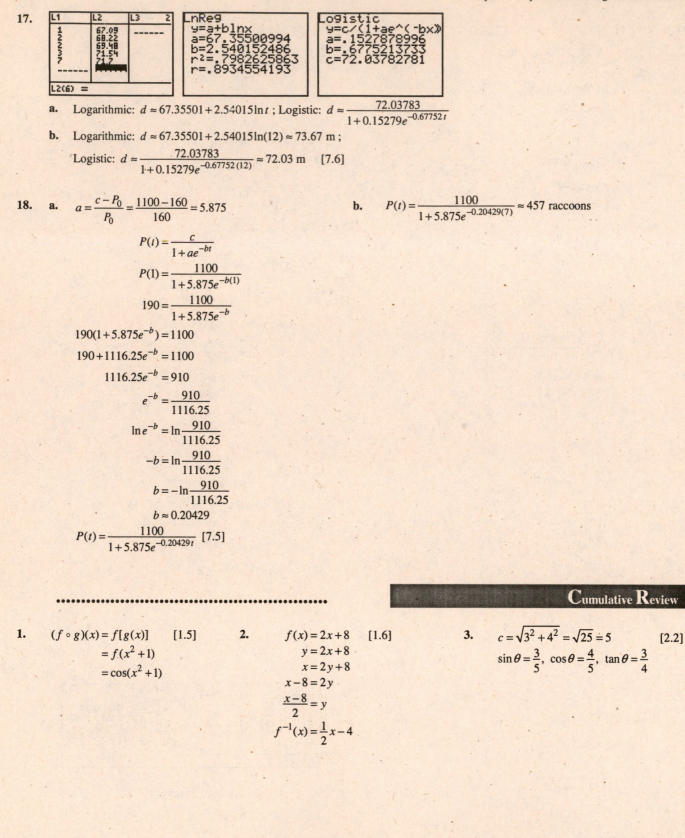

a. Logarithmic: $d \approx 67.35501 + 2.54015\ln t$; Logistic: $d \approx \dfrac{72.03783}{1 + 0.15279e^{-0.67752t}}$

b. Logarithmic: $d \approx 67.35501 + 2.54015\ln(12) \approx 73.67$ m ;

Logistic: $d \approx \dfrac{72.03783}{1 + 0.15279e^{-0.67752(12)}} \approx 72.03$ m   [7.6]

18. **a.** $\quad a = \dfrac{c - P_0}{P_0} = \dfrac{1100 - 160}{160} = 5.875$

$$P(t) = \dfrac{c}{1 + ae^{-bt}}$$

$$P(1) = \dfrac{1100}{1 + 5.875e^{-b(1)}}$$

$$190 = \dfrac{1100}{1 + 5.875e^{-b}}$$

$$190(1 + 5.875e^{-b}) = 1100$$

$$190 + 1116.25e^{-b} = 1100$$

$$1116.25e^{-b} = 910$$

$$e^{-b} = \dfrac{910}{1116.25}$$

$$\ln e^{-b} = \ln \dfrac{910}{1116.25}$$

$$-b = \ln \dfrac{910}{1116.25}$$

$$b = -\ln \dfrac{910}{1116.25}$$

$$b \approx 0.20429$$

$$P(t) = \dfrac{1100}{1 + 5.875e^{-0.20429t}} \quad [7.5]$$

b. $\quad P(t) = \dfrac{1100}{1 + 5.875e^{-0.20429(7)}} \approx 457$ raccoons

··   **Cumulative Review**

1. $(f \circ g)(x) = f[g(x)]$ [1.5]

$\qquad\qquad = f(x^2 + 1)$

$\qquad\qquad = \cos(x^2 + 1)$

2. $f(x) = 2x + 8$ [1.6]

$\qquad y = 2x + 8$

$\qquad x = 2y + 8$

$\qquad x - 8 = 2y$

$\qquad \dfrac{x - 8}{2} = y$

$\qquad f^{-1}(x) = \dfrac{1}{2}x - 4$

3. $c = \sqrt{3^2 + 4^2} = \sqrt{25} = 5$ [2.2]

$\sin\theta = \dfrac{3}{5}, \ \cos\theta = \dfrac{4}{5}, \ \tan\theta = \dfrac{3}{4}$

4. $\cos 26° = \dfrac{15}{a}$ [2.2]

$a = \dfrac{15}{\cos 26°} \approx 16.7$ cm

5. $y = 3\sin\left(\dfrac{1}{3}x - \dfrac{\pi}{2}\right)$ [2.5]

$0 \le 2x - \dfrac{\pi}{2} \le 2\pi$

$\dfrac{\pi}{2} \le 2x \le \dfrac{5\pi}{2}$

$\dfrac{\pi}{4} \le x \le \dfrac{5\pi}{4}$

amplitude = 4, period = π,

phase shift = $\dfrac{\pi}{4}$

6. $y = \sin x + \cos x$ [3.4]

Amplitude: $\sqrt{2}$, period: 2π

7. $f(-x) = \sin(-x) = -\sin x = -f(x)$
odd function [2.4]

8. $\dfrac{1}{\sin x} - \sin x = \dfrac{1 - \sin^2 x}{\sin x}$ [3.1]

$= \dfrac{\cos^2 x}{\sin x}$

$= \cos x \dfrac{\cos x}{\sin x}$

$= \cos x \cot x$

9. $\tan\left(\sin^{-1}\left(\dfrac{12}{13}\right)\right)$ [3.5]

Let $\alpha = \sin^{-1}\dfrac{12}{13}$, $\sin\alpha = \dfrac{12}{13} = \dfrac{\text{opp}}{\text{hyp}}$

$\tan\alpha = \dfrac{\text{opp}}{\text{adj}} = \dfrac{12}{\sqrt{13^2 - 12^2}} = \dfrac{12}{5}$

$\tan\left[\sin^{-1}\left(\dfrac{12}{13}\right)\right] = \dfrac{12}{5}$

10. $2\cos^2 x + \sin x - 1 = 0$ [3.6]

$2(1 - \sin^2 x) + \sin x - 1 = 0$

$2\sin^2 x - \sin x - 1 = 0$

$(2\sin x + 1)(\sin x - 1) = 0$

$2\sin x + 1 = 0 \qquad \sin x - 1 = 0$

$\sin x = -\dfrac{1}{2} \qquad \sin x = 1$

$x = \dfrac{7\pi}{6}, \dfrac{11\pi}{6} \qquad x = \dfrac{\pi}{2}$

11. $\|\mathbf{v}\| = \sqrt{(-3)^2 + 4^2}$

$\|\mathbf{v}\| = \sqrt{9 + 16}$

$\|\mathbf{v}\| = 5$

$\alpha = \tan^{-1}\left|\dfrac{4}{-3}\right| = \tan^{-1}\dfrac{4}{3}$ [4.3]

$\alpha \approx 53.1°$

$\theta = 180° - \alpha$

$\theta \approx 180° - 53.1°$

$\theta \approx 126.9°$

12. $\cos\theta = \dfrac{\mathbf{v} \cdot \mathbf{w}}{\|\mathbf{v}\|\,\|\mathbf{w}\|}$ [4.3]

$\cos\theta = \dfrac{(2\mathbf{i} - 3\mathbf{j}) \cdot (-3\mathbf{i} + 4\mathbf{j})}{\sqrt{2^2 + (-3)^2}\sqrt{(-3)^2 + 4^2}}$

$\cos\theta = \dfrac{2(-3) + (-3)(4)}{\sqrt{13}\sqrt{25}}$

$\cos\theta = \dfrac{-18}{\sqrt{325}} = -0.9985$

$\theta = 176.8°$

13 $\mathbf{AB} = 400(\cos 42\mathbf{i} + \sin 42\mathbf{j}) \approx 297.3\mathbf{i} + 267.7\mathbf{j}$ [4.3]

$\mathbf{AD} = 55[\cos(-25°)\mathbf{i} + \sin(-25°)\mathbf{j}] \approx 49.8\mathbf{i} - 23.2\mathbf{j}$

$\mathbf{AC} = \mathbf{AB} + \mathbf{AD}$

$\mathbf{AC} = 297.3\mathbf{i} + 267.7\mathbf{j} + 49.8\mathbf{i} - 23.2\mathbf{j}$

$\mathbf{AC} \approx 347.1\mathbf{i} + 244.5\mathbf{j}$

$\|\mathbf{AC}\| = \sqrt{347.1^2 + (244.5)^2}$

$\|\mathbf{AC}\| \approx 425$ mph

$\alpha = 90° - \theta = 90° - \tan^{-1}\left(\dfrac{244.5}{347.1}\right) \approx 55°$

14. $\dfrac{a}{\sin A} = \dfrac{b}{\sin B}$ [4.1]

$\sin A = \dfrac{a\sin B}{b} = \dfrac{42\sin 32°}{50} \approx 0.445$

$A \approx 26°$

15. $z = 1 - i$ [5.3]

$r = \sqrt{1^2 + (-1)^2}$

$r = \sqrt{2}$

$\alpha = \tan^{-1}\left|\dfrac{-1}{1}\right| = 45°$

$\theta = 315°$

$z = \sqrt{2}(\cos 315° + i\sin 315°)$

$(1 - i)^8 = [\sqrt{2}(\cos 315° + i\sin 315°)]^8$

$= 16[\cos(8 \cdot 315°) + i\sin(8 \cdot 315°)]$

$= 16(\cos 2520° + i\sin 2520°)$

$= 16(\cos 0° + i\sin 0°)$

$= 16 + 0i = 16$

16. $i = 1(\cos 90° + i\sin 90°)$ [5.3]

$$w_k = 1^{1/2}\left(\cos\frac{90° + 360°k}{2} + i\sin\frac{90° + 360°k}{2}\right) \quad k = 0, 1$$

$w_0 = \cos\frac{90°}{2} + i\sin\frac{90°}{2}$ $\qquad\qquad w_1 = \cos\frac{90°+360°}{2} + i\sin\frac{90°+360°}{2}$

$w_0 = \cos 45° + i\sin 45°$ $\qquad\qquad\qquad w_1 = \cos 225° + i\sin 225°$

$w_0 = \frac{\sqrt{2}}{2} + \frac{\sqrt{2}}{2}i$ $\qquad\qquad\qquad\quad w_1 = -\frac{\sqrt{2}}{2} - \frac{\sqrt{2}}{2}i$

17. $r = \sqrt{x^2 + y^2}$ $\qquad\qquad\qquad \theta = \tan^{-1}\frac{y}{x}$ [6.5]

$\quad = \sqrt{1^2 + 1^2}$ $\qquad\qquad\qquad\quad = \tan^{-1}\left(\frac{1}{1}\right)$

$\quad = \sqrt{2}$ $\qquad\qquad\qquad\qquad\qquad = 45°$

The polar coordinates of the point are $(\sqrt{2}, 45°)$.

18. $r = 2 - 2\cos\theta$ [6.5]

19. $5^x = 10$ [7.4]

$\log 5^x = \log 10$

$x\log 5 = 1$

$\quad x = \frac{1}{\log 5} \approx 1.43$

20. $N(t) = N_0 e^{kt}$

$N(138) = N_0 e^{138k}$

$0.5N_0 = N_0 e^{138k}$

$0.5 = e^{138k}$

$\ln 0.5 = \ln e^{138k}$

$\ln 0.5 = 138k\ln e$

$\ln 0.5 = 138k$

$\frac{\ln 0.5}{138} = k$

$-0.005023 \approx k$

$N(t) = N0(0.5)^{t/138} \approx N_0 e^{-0.005023t}$

$N(100) = 3(0.5)^{100/138} \approx 1.8 \text{ mg}$ [7.5]

Chapter Tests

Functions and Graphs

1. Solve $4(3x-2)-3(2x-5)=25$

2. Solve by factoring and apply the zero product property.
$12x^2 - 17x + 6 = 0$

3. Use interval notation to express the solution set.
$|x-3| \le 2$

4. Find the midpoint and the length of the line segment with endpoints $(-3, 4)$ and $(-7, -5)$.

5. Determine the x- and y-intercepts, and then graph the equation $x + y^2 = 6$.

6. Graph the equation $y = -2|x+3| + 4$.

7. Find the center and radius of the circle that has the general form $x^2 + 6x + y^2 - 8y - 11 = 0$.

8. Determine the domain of the function $f(x) = \sqrt{25 - x^2}$.

9. Use the formula $d = \dfrac{|mx_1 + b - y_1|}{\sqrt{1 + m^2}}$ to find the distance from the point $(-2, 5)$ to the line given by the equation $y = 4x - 3$.

10. Graph $f(x) = \begin{cases} x^2 & \text{if } x \le -1 \\ |x| & \text{if } -1 < x < 1 \\ 1 & \text{if } x \ge 1 \end{cases}$.

    Identify the intervals over which the function is
    a. increasing
    b. constant
    c. decreasing

11. Graph the function $f(x) = (x-2)^2 - 1$. From the graph, find the domain and range of the function.

12. Use the graph of $f(x) = x^2$ to graph $y = 3f(x+1) + 2$.

13. Classify each of the following as either an even function, an odd function, or neither an even nor an odd function.
    a. $f(x) = x^2 + 3$
    b. $f(x) = x^2 + 2x$
    c. $f(x) = 4x$

14. Let $f(x) = 4x^2 - 3$ and $g(x) = x + 6$. Find $(f - g)$ and (fg).

15. Find the difference quotient of the function
$f(x) = x^2 - x - 4$.

16. Evaluate $(f \circ g)$, where $f(x) = \sqrt{2x}$ and $g(x) = \frac{1}{2}x^2$.

17. Find the inverse of $f(x) = \frac{2}{3}x - 4$. Graph f and f^{-1} on the same coordinate axes.

18. Find the inverse of $f(x) = \dfrac{x-3}{3x+6}$. State the domain and the range of f^{-1}.

19. The distance traveled by a ball rolling down a ramp is given by $s(t) = 4t^2$, where t is the time in seconds after the ball is released and $s(t)$ is measured in feet. Evaluate the average velocity of the ball for each of the following time intervals.
    a. [1, 2]   b. [1, 1.5]   c. [1, 1.01]

20. a. Determine the linear regression equation for the given set.

x	2	3	4	6	10
y	3	4	6	10	16

    b. Using the linear model from part **a**, find the expected y-value when the x-value is 8.

Trigonometric Functions

In Problems 1 and 2, convert the measure of the given angle from degrees to radians.

1. $-135°$

2. $240°$

In Problems 3 and 4, convert the measure of the given angle from radians to degrees.

3. $\dfrac{9\pi}{2}$

4. $\dfrac{16\pi}{3}$

5. Determine the measure in radians and in degrees of the central angle that cuts an arc of length 12π centimeters in a circle with radius 8 centimeters.

In Problems 6-9, determine the given trigonometric function value.

6. $\sin\dfrac{5\pi}{4}$

7. $\cos\left(-\dfrac{4\pi}{3}\right)$

8. $\sec\dfrac{19\pi}{2}$

9. $\tan\left(-\dfrac{17\pi}{3}\right)$

In Problems 10 and 11, the given point is located on the terminal side of angle θ. Find $\sin\theta$ and $\tan\theta$.

10. $(-10, 24)$

11. $(-12, -9)$

In Problems 12 and 13, evaluate each quantity.

12. $\cos(-270°)$

13. $\sin 180°$

In Problems 14-17, sketch the graph on the interval $[0, 2\pi]$.

14. $y = -2 + 3\cos(-x)$

15. $y = -1 + \sin(2x - \pi)$

16. $y = \sec\dfrac{x}{2}$

17. $y = \cot(2x + \pi)$

18. Determine the amplitude and period and then sketch the graph of the function $y = 3\cos\dfrac{1}{2}x$.

19. Determine the amplitude, period, and phase shift and then sketch the graph of the function $y = 14\sin(2x + \pi)$.

20. Sketch the graph of the function $y = -3\sin\dfrac{1}{2}x + x$.

21. What is the angle of elevation of the sun at the time that a flagpole 68 feet tall casts a shadow 25 feet long?

22. A ladder 25 feet long is leaning against a building. If the foot of the ladder is 6 feet from the base of the building (on ground level), what acute angle does the ladder make with the ground?

23. The equation $x = 4\sin\left(\dfrac{t}{2} - 2\pi\right)$ represents simple harmonic motion. Determine the amplitude, period, and frequency. Sketch the graph.

Trigonometric Identities and Equations

In Problems 1 and 2, verify the identity.

1. $\dfrac{1-\tan x}{\sec x}+\sin x=\cos x$

2. $\dfrac{\tan^2 x+1}{\tan^2 x+2}=\dfrac{1}{1+\cos^2 x}$

3. Evaluate $\cot\dfrac{5\pi}{12}$ without using a calculator.

In Problems 4 and 5, verify the identity.

4. $\dfrac{\sin(a+b)}{\cos a\cos b}=\tan a+\tan b$

5. $\dfrac{\tan a-\tan b}{\tan a+\tan b}=\dfrac{\sin(a-b)}{\sin(a+b)}$

6. Given that $\alpha=315°$, find $\sin 2\alpha$, $\cos 2\alpha$, and $\tan 2\alpha$, using the double-angle identities.

In Problems 7 and 8, from the given information, determine $\cos\dfrac{\alpha}{2}$.

7. $\sin\alpha=-\dfrac{2}{3}$, α in Quadrant III

8. $\tan\alpha=-\dfrac{3}{2}$, α in Quadrant II

In Problems 9 and 10, write the given expression as a single trigonometric function.

9. $\dfrac{\cot\varphi-\tan\varphi}{\cot\varphi+\tan\varphi}$

10. $\dfrac{\sin 2\alpha}{1-\cos 2\alpha}$

In Problems 11 and 12, verify the identity.

11. $\dfrac{1}{\sin x}-\sin x=\dfrac{\sin x}{\tan^2 x}$

12. $\sin\left(\dfrac{\pi}{4}-\alpha\right)=\cos\left(\alpha+\dfrac{\pi}{4}\right)$

In Problems 13 and 14, solve for x. Express the answer in radians.

13. $2\cos^2 x-\cos x=1$

14. $\sin^2 x+3\sin x=4$

In Problems 15 and 16, given that $\sin 72°=0.95$ and $\cos 72°=0.31$, find the function value. Round your answer to two decimal places.

15. $\sec 18°$

16. $\cot 18°$

In Problems 17 to 20, find the exact value of each expression.

17. $\sin^{-1}\left(-\dfrac{\sqrt{2}}{2}\right)$

18. $\cos^{-1}\left(-\dfrac{1}{2}\right)$

19. $\cos^{-1}\left(\cos\left(-\dfrac{\pi}{4}\right)\right)$

20. $\sin^{-1}\left(\cos\dfrac{\pi}{6}\right)$

Applications of Trigonometry

In Problems 1 - 9, solve triangle *ABC*.

1. $a = 17$, $B = 46°$, $C = 81°$

2. $a = 11$, $b = 16$, $A = 40°$

3. $a = 14$, $A = 38°$, $B = 73°$

4. $a = 18$, $b = 7.0$, $B = 23°$

5. $A = 41°$, $B = 83°$, $c = 44.6$

6. $A = 120°$, $b = 20$, $c = 16$

7. $a = 78$, $B = 128°$, $c = 125$

8. $a = 14$, $b = 19$, $c = 13$

9. $a = 10$, $b = 13$, $C = 100°$

In Problems 10 - 12, find the area of triangle *ABC*.

10. $B = 30°$, $C = 135°$, $b = 4\sqrt{2}$

11. $a = 7.0$, $b = 9.0$, $C = 26°$

12. $a = 12$, $b = 15$, $c = 23$

13. The vector **u** has initial point $(0, 0)$ and terminal point $(-3, 3)$. The vector **v** has initial point $(0, 0)$ and terminal point $(1, 1)$.
     a. Determine the magnitude and direction of vector **u**.
     b. Draw $\mathbf{u} - 2\mathbf{v}$.

In Problems 14 and 15, determine the dot product of the given vectors. Find the angle between the vectors to the nearest degree.

14. $\mathbf{u} = \langle 3, -1 \rangle$, $\mathbf{v} = \langle -2, -2 \rangle$

15. $\mathbf{u} = \langle 4, 6 \rangle$, $\mathbf{v} = \langle -3, 2 \rangle$

In Problems 16 and 17, use the vectors $\mathbf{u} = \left\langle -\dfrac{1}{2}, \dfrac{7}{2} \right\rangle$ and $\mathbf{v} = \langle 6, -1 \rangle$ to determine the following.

16. $2\mathbf{u} - \dfrac{1}{2}\mathbf{v}$

17. $\mathbf{u} - 4\mathbf{v}$

18. A plane's air speed is 500 mph and its bearing is N 75° E. An 6-mph wind is blowing in the direction of N 79° W. Determine the ground speed and true bearing of the plane.

19. Find the force required to hold a 120-pound crate stationary on a ramp inclined at 25°.

Complex Numbers

In Problems 1 – 14, perform the indicated operation; then write the answer in standard form.

1. i^{51}

2. $(-i)^{35}$

3. $(3i)^4 \cdot i^{44}$

4. $(1-3i)+(-2+i)$

5. $(4-3i)-(6-2i)$

6. $5i(4-2i)$

7. $2i-i(8+3i)$

8. $(1-4i)(3+2i)$

9. $(3+5i)(7-4i)$

10. $(2-9i)(2+9i)$

11. $(1+i\sqrt{5})(1-i\sqrt{5})$

12. $\dfrac{4}{3i}$

13. $\dfrac{1}{3+11i}$

14. $\dfrac{6-2i}{2+i}$

In Problems 15 and 16, express the complex number in trigonometric form.

15. $3-i\sqrt{3}$

16. $-4-4i$

17. Write $z = 4 \text{ cis } 45°$ in standard form.

18. Multiply: $(-2 \text{ cis } 225°)(\sqrt{3} \text{ cis } 75°)$. Write the answer in standard form.

19. Divide: $\dfrac{6\sqrt{3}+6i}{1-i\sqrt{3}}$. Write the answer in trigonometric form.

20. Find $\left(\dfrac{-\sqrt{3}}{2}+\dfrac{1}{2}i\right)^5$. Write the answer in standard form.

21. Find the fourth roots of $-1+i\sqrt{3}$. Write the answer in trigonometric form.

Topics in Analytic Geometry

In Problems 1 – 4, write the equation of each conic in standard form and state the type of conic, the foci, and the vertices of each.

1. $x^2 + 4y^2 - 36 = 0$
2. $y^2 + 2y - x + 3 = 0$
3. $x^2 - 4y^2 - 36 = 0$
4. $y^2 - 6y + x = 0$

In Problems 5 – 8, find the foci and vertices of each conic. If the conic is a hyperbola, find the asymptotes. Graph each equation.

5. $y^2 - 6y - x + 5 = 0$

6. $\dfrac{(x-5)^2}{25} + \dfrac{y^2}{4} = 1$

7. $\dfrac{(y+2)^2}{24} - \dfrac{(x-1)^2}{4} = 1$

8. $4x^2 - 9y^2 - 8x + 18y = 149$

In Problems 9 and 10, the rectangular coordinates of a point are given. Determine two polar representations (r, θ) for the point with $0 \le \theta < 2\pi$.

9. $(-1, 1)$

10. $(-\sqrt{3}, -1)$

In Problems 11 – 14, convert the given polar equation into rectangular form.

11. $r = 3\sin\theta$
12. $r = 2\cos\theta$
13. $\theta = \pi/6$
14. $r^2 + \sin 2\theta = 0$

In Problems 15 and 16, graph each polar equation.

15. $r = 2 + \sin\theta$

16. $r = 3 - 2\cos\theta$

In Problems 17 and 18, graph the conic given by each polar equation.

17. $r = \dfrac{5}{1 + \sin\theta}$

18. $r = \dfrac{18}{2 + 3\cos\theta}$

19. Write the equation $13x^2 + 6\sqrt{3}xy + 7y^2 - 32 = 0$ without an xy term. Name the graph of the equation.

20. Eliminate the parameter and graph the curve given by the parametric equations $x = 2t - 1$, $y = 2t^2 + t$.

Exponential and Logarithmic Functions

1. Graph : $f(x) = 2^{-x}$

2. Graph : $f(x) = e^{2x}$

3. Write $\log_4 (3x + 2) = d$ in exponential form.

4. Write $4^{x+3} = y$ in logarithmic form.

5. Write $\log_b \dfrac{\sqrt{y}}{x^3 z^2}$ in terms of logarithms x, y, and z.

6. Write $\log_b (x-1) + \frac{1}{2}\log_b (y+2) - 4\log_b (z+3)$ as a single logarithm with a coefficient of 1.

7. Use the change-of-base formula and a calculator to approximate $\log_5 15$. Round your result to the nearest 0.0001.

8. Graph: $f(x) = \ln(x + 2)$

9. Solve: $6^x = 60$. Round your solution to the nearest 0.0001.

10. Find the *exact* solution(s) of
    $$\frac{10^x - 10^{-x}}{10^x + 10^{-x}} = \frac{1}{4}$$

11. Solve: $\log(2x+1) - 2 = \log(3x-1)$

12. Solve: $\ln(x+1) + \ln(x-2) = \ln(4x-8)$

13. Find the balance on $12,000 invested at an annual interest rate of 6.7% for 3 years:
    a. compounded quarterly.
    b. compounded continuously

14. a. What, to the nearest 0.1, will an earthquake measure on the Richter scale if it has an intensity of $I = 35,792,000I_0$?

    b. Compare the intensity of an earthquake that measures 6.4 on the Richter scale to the intensity of an earthquake that measures 5.1 on the Richter scale by finding the ratio of the larger intensity to the smaller intensity. Round to the nearest whole number.

15. a. Find the exponential growth function for a city whose population was 51,300 in 1980 and 240,500 in 2000. Use $t = 0$ to represent the year 1980.
    b. Use the growth function to predict the population of the city in 2010. Round to the nearest 100.

16. Determine, to the nearest 10 years, the age of a bone if it now contains 89% of its original amount of carbon-14. The half-life of carbon-14 is 5730 years.

17. a. Use a graphing utility to find the exponential regression function for the following data:
        {(1.4, 3), (2.6, 7), (3.9, 15), (4.3, 20), (5.2, 26)}
    b. Use the function to predict, to the nearest whole number, the y-value associated with $x = 8.2$.

18. An altimeter is used to determine the height of an airplane above sea level. The following table shows the values for the pressure p and altitude h of an altimeter.

Pressure p in pounds per square inch	13.1	12.7	12.3	11.4	10.9
Altitude h in feet	3000	3900	4800	6800	8000

a. Find the logarithmic regression function for the data and use the function to estimate, to the nearest 100 feet, the height of the airplane when the pressure is 12.0 pounds per square inch.
b. According to your function, what will the pressure be, to the nearest 0.1 per square inch, when the airplane is 6200 feet above sea level?

Solutions to
Chapter Tests

Functions and Graphs

1. $4(3x-2)-3(2x-5)=25$
$12x-8-6x+15=25$
$6x+7=25$
$6x=18$
$x=3$

2. $12x^2-17x+6=0$
$(3x-2)(4x-3)=0$
$3x-2=0 \quad \text{or} \quad 4x-3=0$
$3x=2 \qquad\qquad 4x=3$
$x=\frac{2}{3} \qquad\qquad x=\frac{3}{4}$

3. $|x-3|\le 2$
$-2\le x-3\le 2$
$1\le \ x \ \le 5$

$[1, 5]$

4. midpoint $=\left(\dfrac{x_1+x_2}{2}, \ \dfrac{y_1+y_2}{2}\right)=\left(\dfrac{-3+(-7)}{2}, \ \dfrac{4+(-5)}{2}\right)=\left(\dfrac{-3-7}{2}, \ \dfrac{4-5}{2}\right)=\left(\dfrac{-10}{2}, \ \dfrac{-1}{2}\right)=\left(-5, \ -\dfrac{1}{2}\right)$

length $= d = \sqrt{(x_1-x_2)^2+(y_1-y_2)^2} = \sqrt{[-3-(-7)]^2+[4-(-5)]^2} = \sqrt{(-3+7)^2+(4+5)^2} = \sqrt{(4)^2+(9)^2} = \sqrt{16+81} = \sqrt{97}$

5. $x+y^2=6$

$y=0 \Rightarrow x+(0)^2=6 \Rightarrow x=6$
Thus the x-intercept is $(6, 0)$.
$x=0 \Rightarrow 0+y^2=6$
$y^2=6$
$y=\pm\sqrt{6}$
Thus the y-intercepts are $\left(0, \ -\sqrt{6}\right)$ and $\left(0, \ \sqrt{6}\right)$.

6. $y=-2|x+3|+4$

7. $x^2+6x+y^2-8y-11=0.$
$(x^2+6x)+(y^2-8y)=11$
$(x^2+6x+9)+(y^2-8y+16)=11+9+16$
$(x+3)^2+(y-4)^2=36$
$(x+3)^2+(y-4)^2=6^2$
center $(-3, 4)$, radius 6

8. $25-x^2\ge 0$
$(5-x)(5+x)\ge 0$
The product is positive or zero.
The critical values are 5 and -5.

The domain is $\{x|-5\le x\le 5\}$.

9. $y=4x-3, \ (x_1, y_1)=(-2, 5)$

$d=\dfrac{|mx_1+b-y_1|}{\sqrt{1+m^2}}=\dfrac{|4(-2)+(-3)-5|}{\sqrt{1+4^2}}=\dfrac{|-8-3-5|}{\sqrt{1+16}}=\dfrac{|-16|}{\sqrt{17}}=\dfrac{16}{\sqrt{17}}=\dfrac{16\sqrt{17}}{17}$

10.
$f(x)=\begin{cases} x^2 & \text{if } x\le -1 \\ |x| & \text{if } -1<x<1 \\ 1 & \text{if } x\ge 1 \end{cases}$

a. increasing on $[0, 1]$
b. constant on $[1, \infty)$
c. decreasing on $(-\infty, 0]$

11. $f(x) = (x-2)^2 - 1$

domain: all real numbers
range: $\{y \mid y \geq -1\}$

12. The graph is shifted 1 unit to the left and 2 units up, and is stretched vertically by multiplying each y-coordinate by 3.

13. a. $f(x) = x^2 + 3$

$f(-x) = (-x)^2 + 3 = x^2 + 3 = f(x)$

$f(x)$ is an even function.

b. $f(x) = x^2 + 2x$

$f(-x) = (-x)^2 + 2(-x) = x^2 - 2x$

$f(-x) = x^2 - 2x \neq f(x)$ not an even function

$f(-x) = x^2 - 2x \neq -f(x)$ not an odd function

$f(x)$ is neither an even function nor an odd function.

c. $f(x) = 4x$

$f(-x) = 4(-x) = -4x = -f(x)$

$f(x)$ is an odd function.

$f(3) = -(3)^2 + 6(3) + 7$

$\quad = -9 + 18 + 7$

$\quad = 16$

The maximum value of the function is 16.

14. $(f - g)(x) = f(x) - g(x)$

$\qquad = (4x^2 - 3) - (x + 6)$

$\qquad = 4x^2 - 3 - x - 6$

$\qquad = 4x^2 - x - 9$

$(fg) = f(x) \cdot g(x)$

$\qquad = (4x^2 - 3)(x + 6)$

$\qquad = 4x^3 + 24x^2 - 3x - 18$

15. $f(x) = x^2 - x - 4$

$\dfrac{f(x+h) - f(x)}{h} = \dfrac{(x+h)^2 - (x+h) - 4 - (x^2 - x - 4)}{h}$

$\qquad = \dfrac{x^2 + 2xh + h^2 - x - h - 4 - x^2 + x + 4}{h}$

$\qquad = \dfrac{2xh + h^2 - h}{h}$

$\qquad = \dfrac{h(2x + h - 1)}{h}$

$\qquad = 2x + h - 1$

16. $f(x) = \sqrt{2x}$ and $g(x) = \frac{1}{2}x^2$

$(f \circ g)(x) = f[g(x)] = f\left(\frac{1}{2}x^2\right)$

$\qquad = \left(\sqrt{2\left(\frac{1}{2}x^2\right)}\right)$

$\qquad = \sqrt{x^2}$

$\qquad = x$

17. $y = \frac{2}{3}x - 4$

Interchange x and y, then solve for y.

$x = \frac{2}{3}y - 4$

$3(x) = 3\left(\frac{2}{3}y - 4\right)$

$3x = 2y - 12$

$3x + 12 = 2y$

$\frac{1}{2}(3x + 12) = \frac{1}{2}(2y)$

$\frac{3}{2}x + 6 = y$

$f^{-1}(x) = \frac{3}{2}x + 6$

18. $y = \dfrac{x-3}{3x+6}$

Interchange x and y, then solve for y.

$$x = \frac{y-3}{3y+6}$$

$$x(3y+6) = y-3$$

$$3xy + 6x = y - 3$$

$$3xy - y = -6x - 3$$

$$y(3x-1) = -6x - 3$$

$$y = \frac{-6x-3}{3x-1}$$

$$f^{-1}(x) = \frac{-6x-3}{3x-1}$$

Domain $f^{-1}(x)$: all real numbers except $\frac{1}{3}$

Range $f^{-1}(x)$: all real numbers except -2

19. $s(t) = 4t^2$

a. Average velocity $= \dfrac{4(2)^2 - 4(1)^2}{2-1}$

$$= \frac{4(4) - 4(1)}{1}$$

$$= 16 - 4$$

$$= 12 \text{ ft/sec}$$

b. Average velocity $= \dfrac{4(1.5)^2 - 4(1)^2}{1.5-1}$

$$= \frac{4(2.25) - 4(1)}{1.5-1}$$

$$= \frac{9-4}{0.5}$$

$$= \frac{5}{0.5}$$

$$= 10 \text{ ft/sec}$$

c. Average velocity $= \dfrac{4(1.01)^2 - 4(1)^2}{1.01-1}$

$$= \frac{4(1.0201) - 4(1)}{0.01}$$

$$= \frac{4.0804 - 4}{0.01}$$

$$= \frac{0.0804}{0.01}$$

$$= 8.04 \text{ ft/sec}$$

20. a. Enter the data on your calculator. The technique for a TI-83 calculator is illustrated here.

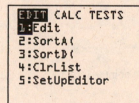

$$y = 1.675x - 0.575$$

b. $y = 1.675(8) - 0.575$

$$= 12.825$$

Trigonometric Functions

1. $\alpha = \dfrac{d}{180°} \cdot \pi = -\dfrac{135°}{180°} \cdot \pi = -\dfrac{3\pi}{4}$

2. $\alpha = \dfrac{d}{180°} \cdot \pi = \dfrac{240°}{180°} \cdot \pi = \dfrac{4\pi}{3}$

3. $d = \dfrac{\alpha}{\pi} \cdot 180° = \dfrac{\frac{9\pi}{2}}{\pi} \cdot 180° = 810°$

4. $d = \dfrac{\alpha}{\pi} \cdot 180° = \dfrac{\frac{16\pi}{3}}{\pi} \cdot 180° = 960°$

5. $\theta = \dfrac{s}{r} = \dfrac{12\pi}{8} = \dfrac{3\pi}{2}$

   The angle measures $\dfrac{3\pi}{2}$ radians, or $270°$.

6. $\sin \dfrac{5\pi}{4} = -\sin\left(\dfrac{5\pi}{4} - \pi\right) = -\sin \dfrac{\pi}{4} = -\dfrac{\sqrt{2}}{2}$

7. $\cos\left(-\dfrac{4\pi}{3}\right) = \cos\left(2\pi - \dfrac{4\pi}{3}\right) = \cos\left(\dfrac{2\pi}{3}\right) = -\cos\left(\pi - \dfrac{2\pi}{3}\right) = -\cos \dfrac{\pi}{3} = -\dfrac{1}{2}$

8. $\sec\left(\dfrac{19\pi}{2}\right) = \sec\left(5 \cdot 2\pi - \dfrac{19}{2}\pi\right) = \sec \dfrac{\pi}{2}$: undefined

9. $\tan\left(-\dfrac{17\pi}{6}\right) = \tan\left(2 \cdot 2\pi - \dfrac{17\pi}{6}\right) = \tan \dfrac{7\pi}{6} = \tan\left(\dfrac{7\pi}{6} - \pi\right) = \tan \dfrac{\pi}{6} = \dfrac{\sqrt{3}}{3}$

10. $r = \sqrt{x^2 + y^2} = \sqrt{(-10)^2 + 24^2}$
    $= \sqrt{676} = 26$

    $\sin \theta = \dfrac{y}{r} = \dfrac{24}{26} = \dfrac{12}{13}$

    $\tan \theta = \dfrac{y}{x} = \dfrac{24}{-10} = -\dfrac{12}{5}$

11. $r = \sqrt{x^2 + y^2} = \sqrt{(-12)^2 + (-9)^2}$
    $= \sqrt{225} = 15$

    $\sin \theta = \dfrac{y}{r} = \dfrac{-9}{15} = -\dfrac{3}{5}$

    $\tan \theta = \dfrac{y}{x} = \dfrac{-9}{-12} = \dfrac{3}{4}$

12. For the angle $-270°$, use the point $(0, 1)$.

    $\cos(-270°) = \dfrac{x}{r} = \dfrac{0}{1} = 0$

13. For the angle $180°$, use the point $(-1, 0)$.

    $\sin 180° = \dfrac{y}{r} = \dfrac{0}{1} = 0$

14.

15.

16.

17.

18. Amplitude 3, period 4π

19. Amplitude 4, period π, phase shift $-\dfrac{\pi}{2}$

20.

21.

$\tan\theta = \dfrac{68}{25}$

$\tan\theta = 2.72$

$\theta \approx 70°$

The angle of elevation is approximately $70°$.

22.

$\cos\theta = \dfrac{6}{25}$

$\cos\theta = 0.24$

$\theta \approx 76°$

The ladder makes an angle of approximately $76°$ with the ground.

23. Amplitude 4, period 4π, frequency $\dfrac{1}{4\pi}$

Trigonometric Identities and Equations

1. $\dfrac{1-\tan x}{\sec x} + \sin x = \dfrac{1-\dfrac{\sin x}{\cos x}}{\dfrac{1}{\cos x}} + \sin x = \cos x - \sin x + \sin x = \cos x$

2. $\dfrac{\tan^2 x + 1}{\tan^2 x + 2} = \dfrac{\sec^2 x}{\left(\tan^2 x + 1\right) + 1} = \dfrac{\sec^2 x}{\sec^2 x + 1} = \dfrac{\dfrac{1}{\cos^2 x}}{\dfrac{1}{\cos^2 x} + 1} \cdot \dfrac{\cos^2 x}{\cos^2 x} = \dfrac{1}{1+\cos^2 x}$

3. $\cot \dfrac{5\pi}{12} = \cot\left(\dfrac{\pi}{4} + \dfrac{\pi}{6}\right) = \dfrac{1}{\tan\left(\dfrac{\pi}{4}+\dfrac{\pi}{6}\right)} = \dfrac{1}{\dfrac{\tan\frac{\pi}{4}+\tan\frac{\pi}{6}}{1-\tan\frac{\pi}{4}\tan\frac{\pi}{6}}} = \dfrac{1-\tan\frac{\pi}{4}\tan\frac{\pi}{6}}{\tan\frac{\pi}{4}+\tan\frac{\pi}{6}} = \dfrac{1-(1)\left(\frac{\sqrt{3}}{3}\right)}{1+\frac{\sqrt{3}}{3}} = \dfrac{3-\sqrt{3}}{3+\sqrt{3}}$

4. $\dfrac{\sin(a+b)}{\cos a \cos b} = \dfrac{\sin a \cos b + \cos a \sin b}{\cos a \cos b} = \dfrac{\sin a \cos b}{\cos a \cos b} + \dfrac{\cos a \sin b}{\cos a \cos b} = \dfrac{\sin a}{\cos a} + \dfrac{\sin b}{\cos b} = \tan a + \tan b$

5. $\dfrac{\tan a - \tan b}{\tan a + \tan b} = \dfrac{\dfrac{\sin a}{\cos a} - \dfrac{\sin b}{\cos b}}{\dfrac{\sin a}{\cos a} + \dfrac{\sin b}{\cos b}} \cdot \dfrac{\cos a \cos b}{\cos a \cos b} = \dfrac{\sin a \cos b - \sin b \cos a}{\sin a \cos b + \sin b \cos a} = \dfrac{\sin(a-b)}{\sin(a+b)}$

6. $\sin(2 \cdot 315°) = 2\sin 315° \cos 315° = 2\left(-\dfrac{\sqrt{2}}{2}\right)\left(\dfrac{\sqrt{2}}{2}\right) = -1$

   $\cos(2 \cdot 315°) = 2\cos^2 315° - 1 = 2\left(\dfrac{\sqrt{2}}{2}\right)^2 - 1 = 0$

   $\tan(2 \cdot 315°) = \dfrac{2\tan 315°}{1-\tan^2 315°} = \dfrac{2(-1)}{1-1^2} = -\dfrac{2}{0}$ Undefined

7. $\sin \alpha = -\dfrac{2}{3}$

   $\dfrac{y}{r} = \dfrac{-2}{3}$

   $x^2 = r^2 - y^2 = 3^2 - (-2)^2 = 5$

   $x = -\sqrt{5}$

   $\cos \alpha = \dfrac{x}{r} = -\dfrac{\sqrt{5}}{3}$

   $\cos \dfrac{\alpha}{2} = -\sqrt{\dfrac{1+\cos \alpha}{2}} = -\sqrt{\dfrac{1-\frac{\sqrt{5}}{3}}{2}}$

   $= -\sqrt{\dfrac{\frac{3-\sqrt{5}}{3}}{2}} = -\sqrt{\dfrac{3-\sqrt{5}}{6}}$

8. $\tan \alpha = -\dfrac{3}{2}$

   $\dfrac{y}{x} = \dfrac{3}{-2}$

   $r = \sqrt{x^2 + y^2} = \sqrt{(-2)^2 + 3^2} = \sqrt{13}$

   $\cos \alpha = \dfrac{x}{r} = \dfrac{-2}{\sqrt{13}} = -\dfrac{2\sqrt{13}}{13}$

   $\cos \dfrac{\alpha}{2} = \sqrt{\dfrac{1+\cos \alpha}{2}} = -\sqrt{\dfrac{1-\frac{2\sqrt{13}}{13}}{2}}$

   $= \sqrt{\dfrac{\frac{13-2\sqrt{13}}{13}}{2}} = \sqrt{\dfrac{13-2\sqrt{13}}{26}}$

9. $\dfrac{\cot \varphi - \tan \varphi}{\cot \varphi + \tan \varphi} = \dfrac{\dfrac{\cos \varphi}{\sin \varphi} - \dfrac{\sin \varphi}{\cos \varphi}}{\dfrac{\cos \varphi}{\sin \varphi} + \dfrac{\sin \varphi}{\cos \varphi}} \cdot \dfrac{\sin \varphi \cos \varphi}{\sin \varphi \cos \varphi} = \dfrac{\cos^2 \varphi - \sin^2 \varphi}{\cos^2 \varphi + \sin^2 \varphi} = \dfrac{\cos^2 \varphi}{1} = \cos 2\varphi$

10. $\dfrac{\sin 2\alpha}{1-\cos 2\alpha} = \dfrac{2\sin\alpha\cos\alpha}{1-(1-2\sin^2\alpha)} = \dfrac{2\sin\alpha\cos\alpha}{2\sin^2\alpha} = \dfrac{\cos\alpha}{\sin\alpha} = \cot\alpha$

11. $\dfrac{1}{\sin x} - \sin x = \dfrac{1-\sin^2 x}{\sin x} = \dfrac{\cos^2 x}{\sin x} = \dfrac{\cos^2 x}{\sin x} \cdot \dfrac{\dfrac{\sin x}{\cos^2 x}}{\dfrac{\sin x}{\cos^2 x}} = \dfrac{\sin x}{\dfrac{\sin^2 x}{\cos^2 x}} = \dfrac{\sin x}{\tan^2 x}$

12. $\sin\left(\dfrac{\pi}{4}-\alpha\right) = \sin\dfrac{\pi}{4}\cos\alpha - \cos\dfrac{\pi}{4}\sin\alpha = \dfrac{\sqrt{2}}{2}\cos\alpha - \dfrac{\sqrt{2}}{2}\sin\alpha = \cos\dfrac{\pi}{4}\cos\alpha - \sin\dfrac{\pi}{4}\sin\alpha = \cos\left(\dfrac{\pi}{4}+\alpha\right)$

13.
$$2\cos^2 x - \cos x = 1$$
$$2\cos^2 x - \cos x - 1 = 0$$
$$(2\cos x + 1)(\cos x - 1) = 0$$

$2\cos x + 1 = 0$	$\cos x - 1 = 0$
$\cos x = -\dfrac{1}{2}$	$\cos x = 1$
$x = \dfrac{2\pi}{3}$ or $\dfrac{4\pi}{3}$	$x = 0$

The solutions are $\dfrac{2\pi}{3}+2n\pi$, $\dfrac{4\pi}{3}+2n\pi$, and $0+2n\pi$.

14.
$$\sin^2 x + 3\sin x = 4$$
$$\sin^2 x + 3\sin x - 4 = 0$$
$$(\sin x - 1)(\sin x + 4) = 0$$

$\sin x - 1 = 0$	$\sin x + 4 = 0$
$\sin x = 1$	$\sin x = -4$
$x = \dfrac{\pi}{2}$	no solutions

The solution is $\dfrac{\pi}{2}+2n\pi$.

15. $\sec 18° = \dfrac{1}{\cos 18°} = \dfrac{1}{\sin(90°-18°)} = \dfrac{1}{\sin 72°} = \dfrac{1}{0.95} \approx 1.05$

16. $\cot 18° = \tan(90°-18°) = \tan 72° = \dfrac{\sin 72°}{\cos 72°} = \dfrac{0.95}{0.31} \approx 3.06$

17. $\sin^{-1}\left(-\dfrac{\sqrt{2}}{2}\right) = -\dfrac{\pi}{4}$

18. $\cos^{-1}\left(-\dfrac{1}{2}\right) = \dfrac{2\pi}{3}$

19. $\cos^{-1}\left(\cos\left(-\dfrac{\pi}{4}\right)\right) = \cos^{-1}\left(\dfrac{\sqrt{2}}{2}\right) = \dfrac{\pi}{4}$

20. $\sin^{-1}\left(\cos\dfrac{\pi}{6}\right) = \sin^{-1}\left(\dfrac{\sqrt{3}}{2}\right) = \dfrac{\pi}{3}$

Applications of Trigonometry

1. $A = 180° - (46° + 81°) = 53°$

$$\frac{17}{\sin 53°} = \frac{b}{\sin 46°}$$

$$\frac{17}{0.7986} = \frac{b}{0.7193}$$

$$b \approx 15$$

$$\frac{17}{\sin 53°} = \frac{c}{\sin 81°}$$

$$\frac{17}{0.7986} = \frac{c}{0.9877}$$

$$c \approx 21$$

2. $$\frac{11}{\sin 40°} = \frac{16}{\sin B}$$

$$\frac{11}{0.6428} = \frac{16}{\sin B}$$

$$\sin B \approx 0.9350$$

$$B \approx 69° \text{ or } 111°$$

Case 1:

$B \approx 69°$

$C = 180° - (40° + 69°) = 71°$

$$\frac{11}{\sin 40°} = \frac{c}{\sin 71°}$$

$$\frac{11}{0.6428} = \frac{c}{0.9455}$$

$$c \approx 16$$

Case 2:

$B \approx 111°$

$C = 180° - (40° + 111°) = 29°$

$$\frac{11}{\sin 40°} = \frac{c}{\sin 29°}$$

$$\frac{11}{0.6428} = \frac{c}{0.4848}$$

$$c \approx 8.3$$

Thus, there are two possible triangles: one with $B \approx 69°, C \approx 71°$, and $c \approx 16$, and one with $B \approx 111°, C \approx 29°$, and $c \approx 8.3$.

3. $C = 180° - (38° + 73°) = 69°$

$$\frac{14}{\sin 38°} = \frac{b}{\sin 73°}$$

$$\frac{14}{0.6157} = \frac{b}{0.9563}$$

$$b \approx 22$$

$$\frac{14}{\sin 38°} = \frac{c}{\sin 69°}$$

$$\frac{14}{0.6157} = \frac{c}{0.9336}$$

$$c \approx 21$$

4. $$\frac{18}{\sin A} = \frac{7.0}{\sin 23°}$$

$$\frac{18}{\sin A} = \frac{7.0}{0.3907}$$

$$\sin A \approx 1.0047$$

Since there is no value A such that $\sin A > 1$, no triangle is possible.

5. $C = 180° - (41° + 83°) = 56°$

$$\frac{a}{\sin 41°} = \frac{44.6}{\sin 56°}$$

$$\frac{a}{0.6561} = \frac{44.6}{0.8290}$$

$$a \approx 35$$

$$\frac{b}{\sin 83°} = \frac{44.6}{\sin 56°}$$

$$\frac{b}{0.9925} = \frac{44.6}{0.8290}$$

$$b \approx 53$$

6. $a^2 = b^2 + c^2 - 2bc \cos A$

$a^2 = 20^2 + 16^2 - 2(20)(16)\cos 120°$

$a^2 = 656 - 640(-0.5)$

$a^2 \approx 976$

$a \approx 31$

$b^2 = a^2 + c^2 - 2ac \cos B$

$20^2 = 31^2 + 16^2 - 2(31)(16)\cos B$

$\cos B \approx 0.8236$

$B \approx 35°$

$C \approx 180° - (120° + 35°) = 25°$

7.
$$b^2 = a^2 + c^2 - 2ac\cos B$$
$$b^2 = 78^2 + 125^2 - 2(78)(125)\cos 128°$$
$$b^2 \approx 21,709 - 19,500(-0.6157)$$
$$b^2 \approx 33,715.15$$
$$b \approx 184$$

$$a^2 = b^2 + c^2 - 2bc\cos A$$
$$78^2 = 184^2 + 125^2 - 2(184)(125)\cos A$$
$$\cos A \approx 0.9434$$
$$A \approx 19°$$

$$C \approx 180° - (19° + 128°) = 33°$$

8.
$$a^2 = b^2 + c^2 - 2bc\cos A$$
$$14^2 = 19^2 + 13^2 - 2(19)(13)\cos A$$
$$\cos A \approx 0.6791$$
$$A \approx 47°$$

$$b^2 = a^2 + c^2 - 2ac\cos B$$
$$19^2 = 14^2 + 13^2 - 2(14)(13)\cos B$$
$$\cos B \approx 0.0110$$
$$B \approx 89°$$

$$C \approx 180° - (47° + 89°) = 44°$$

9.
$$c^2 = a^2 + b^2 - 2ab\cos C$$
$$c^2 = 10^2 + 13^2 - 2(10)(13)\cos 100°$$
$$c^2 \approx 269 - 260(-0.1736)$$
$$c^2 \approx 314.1$$
$$c \approx 18$$

$$a^2 = b^2 + c^2 - 2bc\cos A$$
$$10^2 = 13^2 + 18^2 - 2(13)(18)\cos A$$
$$\cos A \approx 0.8397$$
$$A \approx 33°$$

$$B \approx 180° - (100° + 33°) = 47°$$

10.
$$\frac{4\sqrt{2}}{\sin 30°} = \frac{c}{\sin 135°}$$
$$\frac{4\sqrt{2}}{\frac{1}{2}} = \frac{c}{\frac{\sqrt{2}}{2}}$$
$$c = 8.0$$

$$A = 180° - (30° + 135°) = 15°$$

$$K = \frac{1}{2}bc\sin A$$
$$= \frac{1}{2}(4\sqrt{2})(8.0)\sin 15°$$
$$\approx 6 \text{ square units}$$

11.
$$K = \frac{1}{2}ab\sin C$$
$$= \frac{1}{2}(7.0)(9.0)\sin 26°$$
$$\approx 14 \text{ square units}$$

12.
$$s = \frac{1}{2}(a+b+c)$$
$$= \frac{1}{2}(12+15+23)$$
$$= 25$$

$$K = \sqrt{s(s-a)(s-b)(s-c)}$$
$$= \sqrt{25(25-12)(25-15)(25-23)}$$
$$= \sqrt{25(13)(10)(2)}$$
$$= \sqrt{6500}$$
$$\approx 81 \text{ square units}$$

13. a.
$$d = \sqrt{(-3-0)^2 + (3-0)^2} = \sqrt{18} = 3\sqrt{2}$$
$$\tan\theta = \frac{y}{x}$$
$$\tan\theta = \frac{3}{-3}$$
$$\tan\theta = -1$$
$$\theta = 135°$$

b.

14. $\langle 3,-1\rangle \cdot \langle -2,-2\rangle = 3(-2)+(-1)(-2)$
$$= -4$$

$$\cos\theta = \frac{\mathbf{u}\cdot\mathbf{v}}{\|\mathbf{u}\|\cdot\|\mathbf{v}\|}$$

$$\cos\theta = \frac{-\sqrt{4}}{(\sqrt{3^2+(-1)^2})(\sqrt{(-2)^2+(-2)^2})}$$

$$\cos\theta \approx -0.4472$$

$$\theta \approx 117°$$

15. $\langle 4,6\rangle \cdot \langle -3,2\rangle = 4(-3)+6(2)$
$$= 0$$

$$\cos\theta = \frac{\mathbf{u}\cdot\mathbf{v}}{\|\mathbf{u}\|\cdot\|\mathbf{v}\|}$$

$$\cos\theta = \frac{0}{(\sqrt{4^2+6^2})(\sqrt{(-3)^2+2^2})}$$

$$\cos\theta = 0$$

$$\theta = 90°$$

16. $2\mathbf{u}-\dfrac{1}{2}\mathbf{v} = 2\left\langle -\dfrac{1}{2},\dfrac{7}{2}\right\rangle -\dfrac{1}{2}\langle 6,-1\rangle$

$$= \langle -1,7\rangle + \left\langle -3,\dfrac{1}{2}\right\rangle$$

$$= \left\langle -4,\dfrac{15}{2}\right\rangle$$

17. $\mathbf{u}-4\mathbf{v} = \langle 6,-1\rangle -4\left\langle -\dfrac{1}{2},\dfrac{7}{2}\right\rangle$

$$= \langle 6,-1\rangle + \langle 2,-14\rangle$$

$$= \langle 8,-15\rangle$$

18. $\|\mathbf{v}\|^2 = 500^2 + 8^2 - 2(500)(8)\cos 35°$

$$\|\mathbf{v}\|^2 = 250{,}064 - 8000(0.8192)$$

$$\|\mathbf{v}\|^2 = 243{,}510.4$$

$$\|\mathbf{v}\| = 493$$

$$\frac{\sin\theta}{8} = \frac{\sin 35°}{493}$$

$$\frac{\sin\theta}{8} = \frac{0.5736}{493}$$

$$\sin\theta \approx 0.0093$$

$$\theta \approx 0°32'$$

$$75° - \theta \approx 75° - 0°32' = 74°28'$$

The jet's ground speed is approximately 493 mph, and its true bearing is approximately N 74°28' E.

19. $\|\mathbf{F}\| = (\sin 25°)(120) = (0.4226)(120) \approx 51$

The required force is approximately 51 pounds.

Complex Numbers

1. $i^{51} = i^{48}i^3 = (1)(-i) = -i$

2. $(-i)^{35} = (-1)^{35}(i)^{35} = (-1)i^{32}i^3 = (-1)(1)(-i) = -i$

3. $(3i)^4 \cdot i^{44} = 3^4 i^4 i^{44} = 81(1)(1) = 81$

4. $(1-3i) + (-2+i) = (1-2) + (-3+1)i = -1-2i$

5. $(4-3i) - (6-2i) = (4-6) + (-3+2)i = -2-i$

6. $5i(4-2i) = 20i - 10i^2 = 20i - 10(-1) = 10 + 20i$

7. $2i - i(8+3i) = 2i - 8i - 3i^2$
$$= -6i - 3(-1)$$
$$= 3 - 6i$$

8. $(1-4i)(3+2i) = 3 + 2i - 12i - 8i^2$
$$= 3 - 10i - 8(-1)$$
$$= 3 - 10i + 8$$
$$= 11 - 10i$$

9. $(3+5i)(7-4i) = 21 - 12i + 35i - 20i$
$$= 21 + 23i - 20(-1)$$
$$= 21 + 23i + 20$$
$$= 41 + 23i$$

10. $(2-9i)(2+9i) = 4 - 81i^2$
$$= 4 - 81(-1)$$
$$= 4 + 81$$
$$= 85$$

11. $(1+i\sqrt{5})(1-i\sqrt{5}) = 1 - 5i^2 = 1 - 5(-1)$
$$= 1 + 5$$
$$= 6$$

12. $\dfrac{4}{3i} = \dfrac{4}{3i} \cdot \dfrac{i}{i} = \dfrac{4i}{3i^2} = \dfrac{4i}{3(-1)} = \dfrac{4i}{-3}$
$$= -\dfrac{4}{3}i$$

13. $\dfrac{1}{3+11i} = \dfrac{1}{3+11i} \cdot \dfrac{3-11i}{3-11i} = \dfrac{3-11i}{9-121i^2} = \dfrac{3-11i}{9-121(-1)}$
$$= \dfrac{3-11i}{9+121} = \dfrac{3-11i}{130}$$
$$= \dfrac{3}{130} - \dfrac{11}{130}i$$

14. $\dfrac{6-2i}{2+i} = \dfrac{6-2i}{2+i} \cdot \dfrac{2-i}{2-i} = \dfrac{12-6i-4i+2i^2}{4-i^2}$
$$= \dfrac{12-10i+2(-1)}{4-(-1)} = \dfrac{12-10i-2}{4+1}$$
$$= \dfrac{10-10i}{5} = \dfrac{10}{5} - \dfrac{10}{5}i$$
$$= 2 - 2i$$

In Problems 15 and 16, express the complex number in trigonometric form.

15. $3 - i\sqrt{3} \Rightarrow a = 3, \ b = -\sqrt{3}$
$$r = \sqrt{3^2 + (-\sqrt{3})^2} = \sqrt{12} = 2\sqrt{3}$$
$$\tan\theta = -\dfrac{\sqrt{3}}{3}$$
$$\theta = \dfrac{11\pi}{6}$$
$$3 - i\sqrt{3} = 2\sqrt{3}\left(\cos\dfrac{11\pi}{6} + i\sin\dfrac{11\pi}{6}\right)$$
$$= 2\sqrt{3} \operatorname{cis}\dfrac{11\pi}{6}, \quad \text{or} \quad 2\sqrt{3} \operatorname{cis} 330°$$

16. $-4 - 4i \Rightarrow a = -4, \ b = -4$
$$r = \sqrt{(-4)^2 + (-4)^2} = \sqrt{32} = 4\sqrt{2}$$
$$\tan\theta = \dfrac{-4}{-4}$$
$$\tan\theta = 1$$
$$\theta = \dfrac{5\pi}{4}$$
$$-4 - 4i = 4\sqrt{2}\left(\cos\dfrac{5\pi}{4} + i\sin\dfrac{5\pi}{4}\right)$$
$$= 4\sqrt{2} \operatorname{cis}\dfrac{5\pi}{4}, \quad \text{or} \quad 4\sqrt{2} \operatorname{cis} 225°$$

17. $z = 4 \text{ cis } 45°$

$z = 4(\cos 45° + i \sin 45°)$

$z = 4\left(\dfrac{\sqrt{2}}{2} + \dfrac{i\sqrt{2}}{2}\right)$

$z = 2\sqrt{2} + 2i\sqrt{2}$

18. $(-2 \text{ cis } 225°)(\sqrt{3} \text{ cis } 75°) = -2\sqrt{3} \text{ cis } (225° + 75°)$

$= -2\sqrt{3} \text{ cis } (300°)$

$= -2\sqrt{3}(\cos 300° + i \sin 300°)$

$= -2\sqrt{3}\left(\dfrac{\sqrt{3}}{2} - \dfrac{i}{2}\right)$

$= -3 + i\sqrt{3}$

19.

$\tan\theta = \dfrac{6}{6\sqrt{3}} = \dfrac{1}{\sqrt{3}} = \dfrac{\sqrt{3}}{3}$

$\theta = 30°$

$r = \sqrt{\left(6\sqrt{3}\right)^2 + (6)^2}$

$= \sqrt{36(3) + 36}$

$= \sqrt{144}$

$= 12$

$\tan\alpha = \dfrac{-\sqrt{3}}{1} = -\sqrt{3}$

$\alpha = -60°$

$\beta = 360° - 60°$

$\beta = 300°$

$r = \sqrt{1^2 + \left(-\sqrt{3}\right)^2}$

$= \sqrt{1 + 3}$

$= \sqrt{4}$

$= 2$

$6\sqrt{3} + 6i = 12 \text{ cis } 30°$

$1 - i\sqrt{3} = 2 \text{ cis } 300°$

$\dfrac{6\sqrt{3} + 6i}{1 - i\sqrt{3}} = \dfrac{12 \text{ cis } 30°}{2 \text{ cis } 300°}$

$= \dfrac{12}{2} \text{ cis } (30° - 300°)$

$= 6 \text{ cis } (-270°)$

$= 6 \text{ cis } 90°$

20. $\left(\dfrac{-\sqrt{3}}{2} + \dfrac{1}{2}i\right)^5 = (\text{cis } 150°)^5$

$= \text{cis } 5(150°)$

$= \text{cis } 750°$

$= (\cos 750° + i \sin 750°)$

$= (\cos 30° + i \sin 30°)$

$= \dfrac{\sqrt{3}}{2} + \dfrac{1}{2}i$

21. $-1 + i\sqrt{3} = 2 \text{ cis } 120°$

$w_k = 2^{1/4} \text{ cis } \dfrac{120° + 360°k}{4} \qquad k = 0, 1, 2, 3$

$w_0 \approx 1.19 \text{ cis } \dfrac{120°}{4}$

$w_0 \approx 1.19 \text{ cis } 30°$

$w_1 \approx 1.19 \text{ cis } \dfrac{120° + 360°}{4}$

$w_1 \approx 1.19 \text{ cis } 120°$

$w_2 \approx 1.19 \text{ cis } \dfrac{120° + 720°}{4}$

$w_2 \approx 1.19 \text{ cis } 210°$

$w_3 \approx 1.19 \text{ cis } \dfrac{120° + 1080°}{4}$

$w_3 \approx 1.19 \text{ cis } 300°$

Topics in Analytic Geometry

1. $x^2 + 4y^2 - 36 = 0$

$x^2 + 4y^2 = 36$

$\dfrac{x^2}{36} + \dfrac{4y^2}{36} = \dfrac{36}{36}$

$\dfrac{x^2}{36} + \dfrac{y^2}{9} = 1$

The graph is an ellipse.

vertices: (−6, 0) and (6, 0)

foci: $(-3\sqrt{3}, 0)$ and $(3\sqrt{3}, 0)$

2. $y^2 + 2y - x + 3 = 0$

$y^2 + 2y = x - 3$

$y^2 + 2y + 1 = x - 3 + 1$

$(y + 1)^2 = x - 2$

The graph is a parabola.

$4p = 1$

$p = \dfrac{1}{4}$

vertex: (2, −1)

focus: $(2 + \frac{1}{4}, -1) = \left(\frac{9}{4}, -1\right)$

3. $x^2 - 4y^2 - 36 = 0$

$x^2 - 4y^2 = 36$

$\dfrac{x^2}{36} - \dfrac{4y^2}{36} = \dfrac{36}{36}$

$\dfrac{x^2}{36} - \dfrac{y^2}{9} = 1$

The graph is a hyperbola.

vertices: (−6, 0) and (6, 0)

$c^2 = a^2 + b^2$

$c^2 = 36 + 9$

$c^2 = 45$

$c = \sqrt{45} = 3\sqrt{5}$

foci: $(-3\sqrt{5}, 0)$ and $(3\sqrt{5}, 0)$

4. $y^2 - 6y + x = 0$

$y^2 - 6y = -x$

$y^2 - 6y + 9 = -x + 9$

$(y - 3)^2 = -(x - 9)$

The graph is a parabola

vertex: (9, 3)

$4p = -1$

$p = -\dfrac{1}{4}$

focus: $\left(9 - \frac{1}{4}, 3\right) = \left(\frac{35}{4}, 3\right)$

5. $y^2 - 6y - x + 5 = 0$

$y^2 - 6y = x - 5$

$y^2 - 6y + 9 = x - 5 + 9$

$(y - 3)^2 = x + 4$

$4p = 1$

$p = \dfrac{1}{4}$

vertex: (−4, 3)

focus: $\left(-4 + \frac{1}{4}, 3\right) = \left(-\frac{15}{4}, 3\right)$

6. $\dfrac{(x - 5)^2}{25} + \dfrac{y^2}{4} = 1$

ellipse with center at (5, 0)

$c^2 = a^2 - b^2$

$c^2 = 25 - 4$

$c^2 = 21$

$c = \sqrt{21}$

vertices: (0, 0) and (10, 0)

foci: $(5 - \sqrt{21}, 0)$ and $(5 + \sqrt{21}, 0)$

7. $\dfrac{(y+2)^2}{24} - \dfrac{(x-1)^2}{4} = 1$

hyperbola with vertical transverse axis, center at $(1, -2)$

$a^2 = 24$ $\qquad$ $b^2 = 4$ $\qquad$ $c^2 = a^2 + b^2$ $\qquad$ $\dfrac{a}{b} = \dfrac{2\sqrt{6}}{2}$

$a = 2\sqrt{6}$ $\qquad$ $b = 2$ $\qquad$ $c^2 = 28$ $\qquad\qquad$ $= \sqrt{6}$

$\qquad\qquad\qquad\qquad\qquad\qquad c = 2\sqrt{7}$

vertices: $(1, -2 - 2\sqrt{6})$ and $(1, -2 + 2\sqrt{6})$

foci: $(1, -2 - 2\sqrt{7})$ and $(1, -2 + 2\sqrt{7})$

asymptotes: $y + 2 = \pm\sqrt{6}(x - 1)$

8. $4x^2 - 9y^2 - 8x + 18y = 149$

$4(x^2 - 2x + 1) - 9(y^2 - 2y + 1) = 149 + 4 - 9$

$4(x - 1)^2 - 9(y - 1)^2 = 144$

$\dfrac{4(x - 1)^2}{144} - \dfrac{9(y - 1)^2}{144} = \dfrac{144}{144}$

$\dfrac{(x - 1)^2}{36} - \dfrac{(y - 1)^2}{16} = 1$

hyperbola with horizontal transverse axis, center at $(1, 1)$ a

$a^2 = 36$ $\qquad$ $b^2 = 16$ $\qquad$ $c^2 = a^2 + b^2$ $\qquad$ $\dfrac{b}{a} = \dfrac{4}{6}$

$a = 6$ $\qquad$ $b = 4$ $\qquad$ $c^2 = 36 + 16$ $\qquad\qquad = \dfrac{2}{3}$

$\qquad\qquad\qquad\qquad\qquad c^2 = 52$

$\qquad\qquad\qquad\qquad\qquad c = \sqrt{52}$

$\qquad\qquad\qquad\qquad\qquad\quad = 2\sqrt{13}$

vertices: $(-5, 1)$ and $(7, 1)$

foci: $(1 - 2\sqrt{13}, 1)$ and $(1 + 2\sqrt{13}, 1)$

asymptotes: $y - 1 = \pm\dfrac{2}{3}(x - 1)$

9. $r = \pm\sqrt{(-1)^2 + 1^2} = \pm\sqrt{2}$

$\tan\theta = \dfrac{y}{x} = \dfrac{1}{-1} = -1$

$\theta = \dfrac{3\pi}{4}$

Two polar representations are $\left(\sqrt{2}, \dfrac{3\pi}{4}\right)$ and $\left(-\sqrt{2}, \dfrac{7\pi}{4}\right)$.

10. $r = \pm\sqrt{(-\sqrt{3})^2 + (-1)^2} = \pm 2$

$\tan\theta = \dfrac{y}{x} = \dfrac{-1}{-\sqrt{3}} = \dfrac{\sqrt{3}}{3}$

$\theta = \dfrac{7\pi}{6}$

Two polar representations are $\left(2, \dfrac{7\pi}{6}\right)$ and $\left(-2, \dfrac{\pi}{6}\right)$.

11. $r = 3\sin\theta$

$r^2 = 3r\sin\theta$

$x^2 + y^2 = 3y$

12. $r = 2\cos\theta$

$r^2 = 2r\cos\theta$

$x^2 + y^2 = 2x$

13. $\theta = \pi/6$

$\tan\theta = \tan\pi/6$

$\dfrac{y}{x} = \dfrac{\sqrt{3}}{3}$

$y = \dfrac{\sqrt{3}}{3}x$

14. $r^2 + \sin 2\theta = 0$

$r^2 + 2\sin\theta\cos\theta = 0$

$(r^2)(r^2) + r^2(2\sin\theta\cos\theta) = 0$

$(r^2)^2 + 2(r\sin\theta)(r\cos\theta) = 0$

$x^2 + y^2 + 2xy = 0$

15.

16.

17.

18.

19. $13x^2 + 6\sqrt{3}xy + 7y^2 - 32 = 0$

$A = 13,\ B = 6\sqrt{3},\ C = 7,\ D = 0,\ E = 0,\ F = -32$

$$\cot 2\alpha = \frac{13-7}{6\sqrt{3}} = \frac{6}{6\sqrt{3}} = \frac{1}{\sqrt{3}}$$

$$\csc^2 2\alpha = 1 + \cot^2 2\alpha = 1 + \frac{1}{3} = \frac{4}{3}$$

$\csc 2\alpha = \dfrac{2}{\sqrt{3}}$. Therefore, $\sin 2\alpha = \dfrac{\sqrt{3}}{2}$ and $\cos 2\alpha = \dfrac{1}{2}$.

$$\cos\alpha = \sqrt{\frac{1+\frac{1}{2}}{2}} = \frac{\sqrt{3}}{2} \qquad \sin\alpha = \sqrt{\frac{1-\frac{1}{2}}{2}} = \frac{1}{2}$$

$$A' = 13\left(\frac{\sqrt{3}}{2}\right)^2 + 6\sqrt{3}\left(\frac{\sqrt{3}}{2}\right)\left(\frac{1}{2}\right) + 7\left(\frac{1}{2}\right)^2 = 16$$

$$B' = 0$$

$$C' = 13\left(\frac{1}{2}\right)^2 - 6\sqrt{3}\left(\frac{\sqrt{3}}{2}\right)\left(\frac{1}{2}\right) + 7\left(\frac{\sqrt{3}}{2}\right)^2 = 4$$

$$D' = 0$$

$$E' = 0$$

$$F' = -32$$

$$16x'^2 + 4y'^2 - 32 = 0 \qquad \text{or} \qquad \frac{x'^2}{2} + \frac{y'^2}{8} = 1$$

The graph is an ellipse.

20. $x = 2t - 1 \qquad\qquad y = 2t^2 + t$

$t = \dfrac{x+1}{2} \qquad\qquad y = 2\left(\dfrac{x+1}{2}\right)^2 + \dfrac{x+1}{2}$

$2y = x^2 + 3x + 2$

$y = \dfrac{1}{2}x^2 + \dfrac{3}{2}x + 1$

Exponential and Logarithmic Functions

1.

2.

3. $\quad 4^d = 3x + 2$

4. $\quad \log_4 y = x + 3$

5. $\quad \frac{1}{2}\log_b y - 3\log_b x - 2\log_b z$

6. $\quad \log_b \dfrac{(x-1)\sqrt{y+2}}{(z+3)^4}$

7. $\quad \log_5 15 = \dfrac{\log 15}{\log 5} = \dfrac{1.176091259}{0.6989700043} \approx 1.6826$

Note that you could also have used natural logarithms.

$$\log_5 15 = \dfrac{\ln 15}{\ln 5} = \dfrac{2.708050201}{1.609437912} \approx 1.6826$$

8.

9. $\quad 6^x = 60$

$\quad\quad x = \log_6 60$

$\quad\quad x = \dfrac{\log 60}{\log 6} = \dfrac{1.77815125}{0.7781512504} \approx 2.2851$

Note that you could also have used natural logarithms.

$\quad\quad x = \log_6 60 = \dfrac{\ln 60}{\ln 6} = \dfrac{4.094344562}{1.791759469} \approx 2.2851$

10. $\quad \left(\dfrac{10^x}{10^x}\right)\left(\dfrac{10^x - 10^{-x}}{10^x + 10^{-x}}\right) = \dfrac{1}{4}$

$$\dfrac{10^{2x} - 1}{10^{2x} + 1} = \dfrac{1}{4}$$

$$4(10^{2x} - 1) = 10^{2x} + 1$$

$$4(10^{2x}) - 4 = 10^{2x} + 1$$

$$4(10^{2x}) - 10^{2x} = 1 + 4$$

$$10^{2x}(4 - 1) = 1 + 4$$

$$10^{2x}(3) = 5$$

$$10^{2x} = \dfrac{5}{3}$$

$$\log\!\left(10^{2x}\right) = \log\!\left(\dfrac{5}{3}\right)$$

$$2x = \log 5 - \log 3$$

$$x = \tfrac{1}{2}\log 5 - \tfrac{1}{2}\log 3$$

11. $\log(2x+1) - \log(3x-1) = 2$

$$\log\frac{2x+1}{3x-1} = 2$$

$$10^2 = \frac{2x+1}{3x-1}$$

$$100(3x-1) = 2x+1$$

$$300x - 100 = 2x+1$$

$$298x = 101$$

$$x = \frac{101}{298}$$

12. $\ln[(x+1)(x-2)] = \ln(4x-8)$

$$(x+1)(x-2) = 4x-8$$

$$x^2 - x - 2 = 4x-8$$

$$x^2 - 5x + 6 = 0$$

$$(x-3)(x-2) = 0$$

$x-3 = 0 \Rightarrow x = 3$

$x-2 = 0 \Rightarrow x = 2$ [No; 2 is not in the domain of $\log(x-2)$ and not in the domain of $\log(4x-8)$,]

The solution is 3.

13. a. Use $A = P\left(1+\frac{r}{n}\right)^{nt}$ with $P = \$12,000$, $r = 0.067$, $n = 4$, $t = 3$.

$$A = 12,000\left(1+\frac{0.067}{4}\right)^{4(3)}$$

$$= 12,000(1+0.01675)^{12}$$

$$= 12,000(1.01675)^{12}$$

$$= 12,000(1.220591027)$$

$$\approx \$14,647.09$$

b. Use $A = Pe^{rt}$ with $P = \$12,000$, $r = 0.067$, $t = 3$.

$$A = 12,000e^{0.067(3)}$$

$$= 12,000e^{0.201}$$

$$= 12,000(1.222624772)$$

$$\approx \$14,671.50$$

14. a. $M = \log\left(\frac{I}{I_0}\right)$

$$= \log\left(\frac{35,792,000 I_0}{I_0}\right)$$

$$= \log 35,792,000$$

$$\approx 7.6 \text{ on the Richter scale}$$

b. $\log\left(\frac{I_1}{I_0}\right) = 6.4$ and $\log\left(\frac{I_2}{I_0}\right) = 5.1$

$$\frac{I_1}{I_0} = 10^{6.4} \qquad \frac{I_2}{I_0} = 10^{5.1}$$

$$I_1 = 10^{6.4} I_0 \qquad I_2 = 10^{5.1} I_0$$

To compare the intensities, compute the ratio $\frac{I_1}{I_2}$.

$$\frac{I_1}{I_2} = \frac{10^{6.4} I_0}{10^{5.1} I_0} = \frac{10^{6.4}}{10^{5.1}} = 10^{6.4-5.1} = 10^{1.3} \approx 20$$

An earthquake that measures 6.4 on the Richter scale is approximately 20 times as intense as an earthquake that measures 5.1 on the Richter scale.

15. a. $N(t) = N_0 e^{kt}$

$N_0 = N(0) = 51,300$

$N(20) = 240,500$

To determine k, substitute $t = 20$ and $N_0 = 51,300$ into $N(t) = N_0 e^{kt}$ to produce

$$N(20) = 51,300 e^{k(20)}$$

$$240,500 = 51,300 e^{20k}$$

$$\frac{240,500}{51,300} = e^{20k}$$

$$\ln\frac{240,500}{51,300} = \ln e^{20k}$$

$$\ln\frac{240,500}{51,300} = 20k$$

$$\frac{1}{20}\ln\frac{240,500}{51,300} = k$$

$$0.0773 \approx k$$

The exponential growth function is $N(t) = 51,300 e^{0.0773t}$.

b. Use $N(t) = 51,300 e^{0.0773t}$ with $t = 30$.

$$N(30) = 51,300 e^{0.0773(30)}$$

$$\approx 521,500$$

The exponential growth function yields 521,500 as the approximate population of the town in 2010.

16. The percent of carbon-14 present at time t is
$$P(t) = 0.5^{t/5730}.$$

$$0.89 = 0.5^{t/5730}$$

$$\ln 0.89 = \ln 0.5^{t/5730}$$

$$\ln 0.89 = \left(\frac{t}{5730}\right)\ln 0.5$$

$$5730\left(\frac{\ln 0.89}{\ln 0.5}\right) = t$$

$$960 \approx t$$

The bone is about 960 years old.

17. a.

```
EDIT CALC TESTS
1:Edit
2:SortA(
3:SortD(
4:ClrList
5:SetUpEditor
```

```
L1      L2      L3    2
1.4     3       ------
2.6     7
3.9     15
4.3     20
5.2     26
------
L2(6)=
```

```
EDIT CALC TEST
4↑LinReg(ax+b)
5:QuadReg
6:CubicReg
7:QuartReg
8:LinReg(a+bx)
9:LnReg
0↓ExpReg
```

```
ExpReg
y=a*b^x
a=1.447348651
b=1.79279858
r²=.9837633643
r=.9918484583
```

$$y = 1.447348651(1.79279858)^x$$

b. Use $y = 1.447348651(1.79279858)^x$ with $x = 8.2$.

$$y = 1.447348651(1.79279858)^{8.2}$$

$$\approx 174$$

18. a.

```
EDIT CALC TESTS
1:Edit
2:SortA(
3:SortD(
4:ClrList
5:SetUpEditor
```

```
L1      L2      L3    2
13.1    3000    ------
12.7    3900
12.3    4800
11.4    6800
10.9    8000
------
L2(6)=
```

```
EDIT CALC TEST
4↑LinReg(ax+b)
5:QuadReg
6:CubicReg
7:QuartReg
8:LinReg(a+bx)
9:LnReg
0↓ExpReg
```

```
LnReg
y=a+blnx
a=72610.5637
b=-27040.63079
r²=.9996955915
r=-.9998477842
```

The logarithmic regression function is $h = 72610.5637 - 27040.63079\ln p$.

When $p = 12.0$, $h = 72610.5637 - 27040.63079\ln 12.0 \approx 5400$ feet.

b. $6200 = 72610.5637 - 27040.63079\ln p \Rightarrow \dfrac{6200 - 72610.5637}{-27040.63079} = \ln p$

$\ln p = \dfrac{-66410.5637}{-27040.63079} = 2.455954679 \Rightarrow e^{2.455954679} = e^{\ln p} \Rightarrow e^{2.455954679} = p \approx 11.7$

When the altitude is 6200 feet, the pressure is about 11.7 pounds per square inch.